Chemical Modeling for Air Resources

Chemical Modeling for Air Resources
Fundamentals, Applications, and Corroborative Analysis

Jinyou Liang

ZHEJIANG UNIVERSITY PRESS
浙江大学出版社

ELSEVIER

AMSTERDAM • BOSTON • HEIDELBERG • LONDON • NEW YORK
OXFORD • PARIS • SAN DIEGO • SAN FRANCISCO • SINGAPORE
SYDNEY • TOKYO
Academic Press is an Imprint of Elsevier

Academic Press is an imprint of Elsevier
The Boulevard, Langford Lane, Kidlington, Oxford OX5 1GB, UK
225 Wyman Street, Waltham, MA 02451, USA

First edition 2013

Notice
No responsibility is assumed by the publisher for any injury and/or damage to persons or property as a matter of products liability, negligence or otherwise, or from any use or operation of any methods, products, instructions or ideas contained in the material herein. Because of rapid advances in the medical sciences, in particular, independent verification of diagnoses and drug dosages should be made

British Library Cataloguing-in-Publication Data
A catalogue record for this book is available from the British Library

Library of Congress Cataloging-in-Publication Data
A catalog record for this book is available from the Library of Congress

ISBN–13: 978-0-12-408135-2

For information on all Academic Press publications
visit our website at books.elsevier.com

Printed and bound in the US

13 14 15 16 17   10 9 8 7 6 5 4 3 2 1

Working together
to grow libraries in
developing countries

www.elsevier.com • www.bookaid.org

Contents

Part Two   Applications

Part Three   Analysis

Preface

Air is an invaluable resource for humans. It participates in maintaining human life chemically and shields humans from harmful radiation, but also contains toxic constituents to be cleaned. To understand a chemical phenomenon in the air, nearby or far away, or to assess implications of marketing a new chemical, or to evaluate the investment of implementing a thorough clean air policy, chemical modeling provides a powerful tool for integrated analyses.

Built on over 20 years of experience in developing, applying, and analyzing chemical models for air resource research and regulatory purposes at the Chinese Research Academy of Environmental Sciences, Harvard/Stanford/Zhejiang Universities, and California Environmental Protection Agency of the USA, I wrote this book during the summer of 2012. This book is written for graduate students and junior researchers in a manner similar to assembling puzzle blocks: many pieces have been arranged, while some remaining pieces are identified for interested readers to research.

To provide a concise tutorial on chemical modeling for air resources, this book is divided into three parts:

- The first part focuses on fundamentals required for air resource chemical modeling. The chemical composition of the air is described in Chapter 1, chemical reactions in the air are discussed in Chapter 2, radiation in the air is considered in Chapter 3, and modeling chemical changes in the air is described in Chapter 4.
- The second part focuses on application cases of air resource chemical modeling. The ozone hole is considered in Chapter 5, acid rain is discussed in Chapter 6, climate change is the subject of Chapter 7, surface oxidants are described in Chapter 8, particulate matter is discussed in Chapter 9, and other toxins in the air are considered in Chapter 10.
- The third part, Chapter 11, introduces methods to corroborate analyses of data from models and observations for serious simulations, such as in support of governmental regulations.

At the end of each chapter, a handful of exercises are provided.

Additional information is available from *Introduction to Atmospheric Chemistry* by Professor Daniel J. Jacob at Harvard University, *Fundamentals of Atmospheric Modeling* by Professor Mark Z. Jacobson at Stanford University, air resource documents from regional environmental protection agencies, and a number of professional journals, such as *Atmospheric Environment*, *JGR-Atmospheres*, *China Environmental*

Sciences, as well as other national journals. Readers who steward air resources from a chemical perspective at regional, national, or global level will hopefully find this book helpful.

Jinyou Liang
California, USA
April 2013
(E-mail address: jinyou.liang@gmail.com)

Part One

Fundamentals

1 Chemical composition of the atmosphere of the Earth

Humans can only survive for 1–2 minutes without taking oxygen (O_2) into the bloodstream and, in the explored universe to date, there is no direct reservoir for O_2 other than the atmosphere of the Earth. Thus, the Earth's atmosphere, which mainly consists of N_2 and O_2, is the most important resource for humans, though it is the least commercialized resource compared with food, land, and water.

Chemicals in the modern atmosphere are versatile, and their evolution from primitive Earth is still largely a mystery. In the modern atmosphere, O_2 serves as the only fuel to maintain the biochemical engines of humans and most animals, and H_2O is the most important gas to adjust air temperature in the lower atmosphere to comfort humans and animals living near the surface, while CO_2 and H_2O are necessary nutrients for plants and crops to grow. Numerous chemical species, besides N_2, O_2, H_2O, CO_2, the "noble gases" (He, Ne, Ar, Kr, Xe, Rn), and H_2, have been identified in the atmosphere since industrialization, when analytical instruments such as chromatographs and spectrometers were invented. Among them, ozone (O_3) is found to be necessary in the middle atmosphere to protect humans from harmful ultraviolet radiation during daytime. However, O_3 is harmful to humans when present near the Earth's surface. When a chemical in the air, such as O_3, has a so-called "dose–response relationship", mostly determined from animal experiments, it is called an air toxin.

Chemical Modeling for Air Resources. http://dx.doi.org/10.1016/B978-0-12-408135-2.00001-X

There are still many chemicals that are suspected to be present in the atmosphere but not detectable due to limitations in instrumentation or theoretical methods.

1.1   Atmospheric composition from observation and theory

In the solar system, the Earth is a unique blue planet looking from space, and the blue color is a result of scattering of sunlight by the atmospheric chemicals of the Earth. The Earth's atmosphere extends from the surface to an ambiguous outer bound, ~100 km above average sea level (ASL). If using the indicator "1 mole air $= 0.79$ $N_2 + 0.21$ O_2", then the Earth's atmosphere may be characterized by four to six vertical layers upward, namely the troposphere (0–10 km), stratosphere (10–50 km), mesosphere (50–85 km), ionosphere (80–90 km), thermosphere (85–100 km), and exosphere (100–500 km), as shown in Figure 1.1. The troposphere may be further divided into the planetary boundary layer (PBL; 0–1.5 km) at the bottom and the free troposphere (1.5–10 km) at the top. The thickness of the planetary boundary layer shows strong diurnal and spatial variations, and is a hot topic for meteorologists, atmospheric scientists, and environmental professionals. The vertical structure of atmospheric temperature is regulated by chemical composition and radiations of the Sun and Earth, as well as other physical processes, such as surface characteristics,

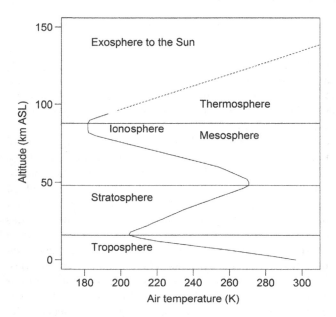

Figure 1.1 Vertical thermal structure of the atmosphere of the Earth. Air temperature near the surface is pertinent to subtropical land areas during spring and fall. Solid line: from measurements; dashed line: linear projection.

relative positions of planets and moons, and resulting dynamic patterns in atmospheric layers. Table 1.A1 at the end of this chapter lists vertical profiles of temperature as well as pressure and O_3 at 2-km intervals from the surface to 46 km ASL in the modern atmosphere, and more constituents of the atmosphere are discussed below.

1.1.1 Troposphere

The troposphere, where ~90% of air mass over the Earth resides, refers to the bottom ~10 km of the atmosphere (Figure 1.1). In the troposphere, atmospheric temperature descends upward with a slope of ~10 K km^{-1} for dry air and ~7 K km^{-1} for wet air. At night, air temperature at the surface may be lower than that up to ~100 m, due to the combination of long-wave radiation of Earth and the so-called greenhouse effect. In the troposphere, numerous field campaigns have been conducted to investigate air composition over developed areas, such as North America, Europe, East Asia, Australia and New Zealand, their downwind areas, such as the Atlantic Ocean and Pacific Ocean, and remote regions, such as the Arctic and Antarctic areas. While most observations have been made near the surface, significant efforts, such as the use of balloons, flights, rockets, and satellites, have also been made to observe the air composition above, especially in recent decades. In populated developing countries, such as China and India, field campaigns have also been conducted recently to survey the chemicals responsible for air pollution, such as O_3, acid rain, and particulate matter.

On a global, annual average basis, the modern tropospheric air composition excluding H_2O, CO_2, CH_4, and N_2O is listed in Table 1.1, which is termed "dry air".

It can be seen that N_2 is the most abundant chemical, followed by O_2, and in turn by noble gases and H_2. The chemical composition of the dry air, in terms of the mixing ratio, changes little in the open atmosphere of the Earth, or as defined, though the O_2 mixing ratio is perturbed by humans, animals, plants, and crops, and may be modulated by geochemical processes. There are a number of hypotheses with regard to how the chemical composition of the dry air has arrived at its current status. For example, in the very beginning, the dry air of the Earth could have been purely CO_2, similar to the current status of Mars; biogeochemical processes might have gradually

Table 1.1 Dry air composition

Dry air	Molar mixing ratio
N_2	7.81E-01
O_2	2.10E-01
"Noble gases"	9.32E-03
H_2	6.00E-07
Sum	1.00E+00

Note: 1E-01 denotes 1×10^{-1}, and molar mixing ratios of the noble gases He, Ne, Ar, Kr, Xe, and Rn are 5E-8, 1.5E-5, 0.93E-2, 1E-6, 5E-8, and 2E-19 respectively.

fixed carbon from the air to form fossil fuels underground and leaving O_2 in the air. The process involved is the photosynthesis in plants that converts CO_2 and H_2O into O_2, while other processes are the subject of Earth system modeling. Mixing ratios of N_2, H_2, and noble gases in the dry air are speculated to result from complex biogeochemical processes. At present levels, these gases, except Rn, have no reported adverse effects on human health, and humans and animals may have adapted to their levels in the air. As an industrial resource, N_2 is routinely used to make nitrogen fertilizers and is used as a liquid agent for small surgery, and He is used to fill balloons.

Besides the dry air, H_2O is an important component of the air in the troposphere. On one hand, it is the reservoir of precipitations that provide economic drinking water and water supplies for agricultural, industrial, and recreational purposes. On the other hand, it is a natural and the most important greenhouse gas in modern air that raises the temperature of surface air by over 30 K so that the Earth's surface is habitable for humans and animals. The mixing ratio of H_2O vapor in the troposphere ranges from <0.01 percent to a few percent, depending on elevation, latitude, longitude, surface temperature and other characteristics, such as closeness to bodies of water such as ponds, rivers, lakes, estuaries, seas, and oceans. The air may contain a small amount of liquid water as rain, cloud, fog, haze, or wet aerosol; when air is cold enough, such as in nontropical areas during winter or in the upper troposphere, it may also contain an even smaller amount of solid water as snow, hail, graupel, frost, cirrus cloud, contrails, or other icy particles suspended in the air. Table 1.2 lists typical seasonal saturated water vapor mixing ratio over the northern hemisphere, which ranges from 0.1% to 4%. Over global oceans, the relative humidity near the surface is close to 100%. Over the land, the relative humidity varies from below 5% over deserts to over 90% in coastal areas. Thus, water vapor is the third or fourth most abundant gas in surface air.

In general, the H_2O mixing ratio is higher over the tropics than over polar areas, higher in summer than in winter, higher over farmlands and forests than over deserts, and higher near the surface than further away from the surface; these phenomena reflect the facts that H_2O evaporates faster at higher temperatures and H_2O vapor is transported in the troposphere following air streams termed general circulations.

Table 1.2 Typical seasonal saturated water vapor mixing ratio

Latitude	DJF	MAM	JJA	SON
0	0.033	0.035	0.033	0.033
15	0.041	0.035	0.037	0.035
30	0.017	0.026	0.041	0.026
45	0.006	0.013	0.026	0.015
60	0.002	0.004	0.017	0.007
75	0.001	0.001	0.007	0.003

Note: Saturated water vapor pressure (pascals) was calculated as $610.94 \times \exp\{17.625 \times T\,(°C)/[T\,(°C) + 243.04]\}$. DJF, December, January, February; MAM, March, April, May; JJA, June, July, August; SON, September, October, November.

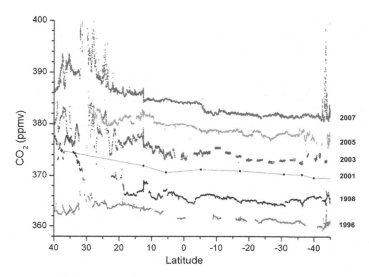

Figure 1.2 Observed atmospheric CO_2 mixing ratio.
Obtained from Longinelli et al. (2010).

CO_2, CH_4, and N_2O are the three most important greenhouse gases in the modern troposphere, as regional and global industrialization has accelerated their increasing trends, especially in recent decades. Anthropogenic activities involving combustion harness energy from fossil fuel and biomass and emit CO_2 into the atmosphere, mostly to the troposphere, except for aviation. Globally, anthropogenic emission of CO_2 has increased dramatically since the beginning of industrialization over a century ago, and amounted to ~40 billion tons per year recently. Freshly emitted CO_2 is partly fixed by plants over the land and in surface waters, and partly dissolved into water bodies. Atmospheric CO_2 may also transform some rocks on a geochemical time scale. The remainder stays in the atmosphere, mainly in the troposphere, and raises the mixing ratio of CO_2 there. Figure 1.2 shows the annual increase of CO_2 over world oceans in the years 1996–2007 (Longinelli et al., 2010). As the lifetime of CO_2 in the troposphere is an order of magnitude longer than the mixing time of tropospheric air, CO_2 is well mixed in the troposphere except at the surface with sinks or near emission sources. In fact, research has suggested that the CO_2 mixing ratio rose from ~280 ppmv in 1750 to ~310 ppmv in 1950, according to ice-core analyses, and to ~380 ppmv in 2010 based on measurements at a ground station of ~3 km ASL at the Mauna Loa Observatory in Hawaii (Intergovernmental Panel on Climate Change (IPCC), a Nobel Laureate, 2007). If anthropogenic CO_2 emission follows the current trend, the atmospheric CO_2 mixing ratio may reach 600 ppm before 2100; the exact response of atmospheric CO_2 to fossil fuel consumption depends on complex factors under active research. The increase of the atmospheric CO_2 mixing ratio has two opposite effects on humans: on one hand, a higher CO_2 mixing ratio may increase crop yields and warm up cold regions if other conditions are fixed; on the other hand, a higher CO_2 mixing ratio may have harmful consequences, such as the loss of coastal wetlands, more frequent storms or droughts, and more stagnant air near the surface.

The CH_4 mixing ratio in the troposphere is currently ~1.8 ppm, with a slightly higher mixing ratio in the northern hemisphere, where most sources are located, than in the southern hemisphere due to its relatively short lifetime (~10 years) compared with the timescale of interhemispheric air exchange (~1 year). CH_4 is the major component of natural gas, and is used widely as a clean fuel for residential, traffic, and industrial needs when available. For comparison, the CH_4 mixing ratio was estimated to be ~0.8 ppm in the middle of the eighteenth century. Tropospheric CH_4 may originate from leakages during the production, storage, transportation, and consumption of fossil fuels, and may also be emitted from rice paddies and swamps during certain periods, as well as from other sources. CH_4 is a potent greenhouse gas, e.g. with a 100-year global warming potential 21 times that of CO_2, according to the IPCC; it also contributes significantly to the photochemical production of O_3 in the troposphere on a global scale.

N_2O is rather stable in the troposphere and its current mixing ratio is ~0.32 ppm. In nature, it is a laughing gas, and is also emitted from farmlands. According to a recent survey in California, synthetic fertilizers and on-road vehicles have become dominant sources for N_2O emission there. It is estimated that tropospheric N_2O has increased by ~10% from preindustrial 1750. N_2O is a potent greenhouse gas, with a 100-year global warming potential 310 times that of CO_2, according to the IPCC.

1.1.2 Stratosphere

The stratosphere contains ~9.9% of air mass over the Earth, and ranges from ~10 to ~50 km ASL with ascending temperature up to ~270 K. Due to precipitation in the troposphere, H_2O can scarcely survive through vertical transport to reach the stratosphere. In the stratosphere, O_2 may be photolyzed by solar ultraviolet radiation to form ozone (O_3), which results in the so-called "O_3 layer". The O_3 layer itself absorbs solar ultraviolet radiation with a little longer wavelength, to close the Chapman cycle of O_3 formation in the natural stratosphere. As a result, solar UV radiation at the top of the stratosphere is much stronger than at the bottom of the stratosphere for wavelengths less than ~300 nm. Thus, humans and animals are effectively protected by the O_3 layer from harmful, solar UV radiation with wavelengths shorter than ~300 nm.

In the modern atmosphere, chemicals such as N_2O and chlorofluorocarbons (CFCs), which decompose slowly in the troposphere, may accumulate to a significant amount and enter the stratosphere via stratosphere–troposphere exchange events. Due to strong solar radiation in the stratosphere, these chemicals photolyze to form NO and halogen radicals, which then perturb the Chapman cycle to affect the thickness of the O_3 layer. The most important observation related to the O_3 layer in the stratosphere is the so-called "O_3 hole", initially observed over Antarctica during early spring in the early 1980s. Table 1.3 lists typical seasonal column O_3 over the Equator,

Table 1.3 Column O_3 in Dobson units over the Equator

Month	O_3	Month	O_3
Jan	251	July	265
Feb	252	Aug	269
Mar	257	Sept	271
Apr	260	Oct	265
May	260	Nov	258
June	262	Dec	253

and global distributions of column O_3 are presented in Chapter 5; ~90% of the column O_3 stays in the stratosphere.

Due to strong photochemical reactions in the stratosphere, mixing ratios of surface-originated compounds, such as CH_4, N_2O, and CFCs, are lower than in the troposphere. Meanwhile, their reaction products, such as NO, OH, and halogen radicals, may be more abundant in the stratosphere than part of the troposphere.

1.1.3 Mesosphere and above

The mesosphere contains ~0.1% of air mass over the Earth, ranging from ~50 to ~85 km ASL with descending temperature down to ~185 K. Besides dry air, H_2O has been observed with a mixing ratio of ~7 ppm there. The number of chemicals with significant mixing ratios becomes much smaller than in the troposphere, due to dynamic blocking of the stratosphere. Major photochemical species are O_2, O_3, H_2O, O, O^1D, H, OH, HO_2, CH_4, H_2, and H_2O_2, and 20 kinetic reactions were found to capture the major features of photochemistry there.

The atmosphere above the mesosphere contains ~ 0.001% of air mass over the Earth, which includes the ionosphere, thermosphere, and exosphere. Ions start to be effectively produced from air molecules by sunlight, and the shortest ultraviolet radiation from the Sun is absorbed by N_2 and O_2 from 90 to 100 km. Light gas species such as H may escape from the Earth's gravity to the Universe.

1.2 Trace chemicals observed in the troposphere

The troposphere directly interacts with humans, animals, and plants near the Earth's surface. In the troposphere, the number of trace chemicals detected has grown beyond elaboration here, due to a dramatic increase in anthropogenic emissions and advances in instrumental analyses.

1.2.1 Natural trace chemicals in the troposphere

In the natural troposphere, besides windblown dusts and sea sprays, wetlands emit CH_4, and plants emit ethene (C_2H_4), isoprene (C_5H_8), methylbutenol, terpenes

Table 1.4 Trace chemicals in "pristine" air

	Emission (Tg year^{-1})	Typical mixing ratio (ppb)
Ethene	25	0.2
Isoprene	500	0.1
Terpenes	130	0.1
Methylbutenol	50	0.1
$C_{15}H_{24}$	200	0.01
NH_3	4	1
CH_4	1900	1774
NO	11	0.1
DMS	34	0.1
N_2O	30	319
CO	400	100
SO_2	7	0.1

Note: Tg, teragram = 1 million tons; ppb, parts per billion.

($C_{10}H_{16}$), and sesquiterpenes ($C_{15}H_{24}$) in seasons; animals emit NH_3 and CH_4; lightning emits NO, and volcanoes emit sulfur compounds and ashes; oceanic biota emit dimethylsulfide (DMS), and farmlands emit N_2O. Table 1.4 lists the estimated global emission rates and typical mixing ratios of trace chemicals of natural origin in the troposphere.

1.2.2 Anthropogenic emission sources in the regional troposphere

Agricultural and industrial revolutions have greatly enhanced emissions near the Earth's surface, and the informational revolution has significantly increased anthropogenic emissions via flights and globalization. As a result, new chemicals, such as CFCs and their substitutes, have been synthesized and emitted into the atmosphere recently, and the growth of emissions in some areas during certain periods could be quadratic. Important anthropogenic sources that contribute to trace chemicals in the troposphere include on-road and off-road vehicles over the land, consumer products, power plants, refineries, industrial boilers, ships, smelters, specialty plants, as well as traditional sources such as dairies and biomass burning. Depending on location, time, and chemicals, some sources may dominate over others. For example, methyl-t-butyl-ether (MTBE) was used as an additive to gasoline fuel in California for a short period of time, and was then recalled due to its toxicity in water. To substitute MTBE, ethanol has become a common additive. As a result, ethanol emission from vehicles has recently increased, as has the emission of propane, due to the increasing use of liquefied propane fuel. In developing nations, social advancement has driven dramatic increases in anthropogenic emissions; for example, anthropogenic emissions of NH_3, N_2O, and NO_x increased by 2–4 times in China from 1980 to 2010 (Gu et al., 2012). Due to dramatic reductions in anthropogenic volatile organic compound (VOC) emissions, natural sources have begun to exceed anthropogenic emissions in developed nations. Table 1.5 lists types of

Table 1.5 Emissions of criteria air pollutants in the USA in 2008 (US EPA, in 1000 tons)

Sector	CO	VOCs	NO$_x$	SO$_2$	PM$_{2.5}$	PM$_{10}$	NH$_3$	Pb	HAPs
Agriculture – Crops and livestock dust	0	0	0	0	923	4650	0	0	0
Agriculture – Fertilizer application	0	0	0	0	0	0	1190	0	0
Agriculture – Livestock waste	0	92	0	0	7	24	2448	0	0
Bulk gasoline terminals	1	93	0	0	0	0	0	0	5
Commercial cooking	30	13	0	0	86	89	0	0	5
Dust – Construction dust	0	0	0	0	220	2115	0	0	0
Dust – Paved road dust	0	0	0	0	280	1539	0	0	0
Dust – Unpaved road dust	0	0	0	0	812	8104	0	0	0
Fires – Agric. field burning	576	53	25	3	66	68	4	0	6
Fires – Prescribed fires	8176	1697	139	65	701	824	119	0	240
Fires – Wildfires	12,203	2847	96	70	999	1178	198	0	1051
Fuel comb – Comm/institutional – Biomass	17	1	5	2	3	3	0	0	1
Fuel comb – Comm/institutional – Coal	15	0	20	88	2	5	0	0	2
Fuel comb – Comm/institutional – Natural gas	100	9	146	1	6	6	1	0	2
Fuel comb – Comm/institutional – Oil	18	3	60	66	4	5	1	0	1
Fuel comb – Comm/institutional – Other	12	1	9	1	1	1	0	0	0
Fuel comb – Electric Generation – Biomass	21	1	10	3	1	2	1	0	1
Fuel comb – Electric Generation – Coal	573	29	2813	7603	276	370	11	0	144
Fuel comb – Electric Generation – Natural gas	93	9	181	15	20	21	11	0	3
Fuel comb – Electric generation – Oil	13	2	82	142	9	11	2	0	1
Fuel comb – Electric generation – Other	27	2	27	15	2	2	2	0	1
Fuel comb – Industrial boilers, ICEs – Biomass	192	8	80	25	32	39	2	0	7
Fuel comb – Industrial boilers, ICEs – Coal	59	2	208	663	24	51	0	0	15
Fuel comb – Industrial boilers, ICEs – Natural gas	367	60	777	38	29	32	7	0	22
Fuel comb – Industrial boilers, ICEs – Oil	26	4	93	135	9	13	1	0	1

(Continued)

Table 1.5 Emissions of criteria air pollutants in the USA in 2008 (US EPA, in 1000 tons)—cont'd

Sector	CO	VOCs	NO$_x$	SO$_2$	PM$_{2.5}$	PM$_{10}$	NH$_3$	Pb	HAPs
Fuel comb – Industrial boilers, ICEs – Other	121	6	72	67	31	33	1	0	2
Fuel comb – Residential – Natural gas	94	13	230	1	5	5	38	0	1
Fuel comb – Residential – Oil	14	2	57	121	6	7	3	0	0
Fuel comb – Residential – Other	49	3	38	9	1	2	0	0	0
Fuel comb – Residential – Wood	2378	350	35	10	344	345	20	0	70
Gas stations	0	643	0	0	0	0	0	0	28
Industrial processes – Cement manuf	100	9	186	106	13	24	1	0	4
Industrial processes – Chemical manuf	205	99	77	196	23	30	19	0	31
Industrial processes – Ferrous metals	467	19	63	33	36	44	1	0	2
Industrial processes – Mining	29	2	6	4	106	748	0	0	1
Industrial processes – NEC	260	217	198	158	120	187	51	0	58
Industrial processes – Non-ferrous metals	328	16	16	132	20	25	1	0	10
Industrial processes – Oil and gas production	219	1688	409	61	7	10	0	0	8
Industrial processes – Petroleum refineries	84	68	93	144	24	27	3	0	6
Industrial processes – Pulp and paper	132	130	74	40	40	49	6	0	54
Industrial processes – Storage and transfer	17	238	7	6	23	53	5	0	15
Miscellaneous non-industrial NEC	29	227	2	0	3	4	11	0	25
Mobile – Aircraft	458	35	121	13	4	10	0	1	8
Mobile – Commercial marine vessels	143	21	827	150	35	37	0	0	3
Mobile – Locomotives	120	44	846	11	25	28	0	0	4
Mobile – Non-road equipment – Diesel	870	166	1551	32	122	127	1	0	41
Mobile – Non-road equipment – Gasoline	15,462	2281	250	2	55	61	1	0	539
Mobile – Non-road equipment – Other	1015	47	187	1	2	2	1	0	0
Mobile – On-road – Diesel heavy duty vehicles	981	238	3353	85	204	224	6	0	45
Mobile – On-road – Diesel light duty vehicles	47	11	75	3	6	6	0	0	2

Sector	CO	VOCs	NOx	SO2	PM2.5	PM10	NH3	Pb	HAPs
Mobile – On-road – Gasoline HD vehicles	2530	166	251	1	4	7	5	0	45
Mobile – On-road – Gasoline LD vehicles	32,492	2640	3455	29	82	138	128	0	733
Solvent – Consumer and commercial solvent use	0	1619	0	0	0	0	0	0	174
Solvent – Degreasing	0	198	0	0	0	0	0	0	28
Solvent – Dry cleaning	0	49	0	0	0	0	0	0	4
Solvent – Graphic arts	3	356	4	0	0	0	0	0	24
Solvent – Industrial surface coating and solvent use	3	648	3	0	3	4	0	0	75
Solvent – Non-industrial surface coating	0	429	0	0	0	0	0	0	68
Waste disposal	1384	180	97	21	205	240	66	0	42
Subtotal (no federal waters)	**82,552**	**17,784**	**17,356**	**10,369**	**6066**	**21,631**	**4366**	**1**	**3659**
Fuel comb – Industrial boilers, ICEs – Natural gas	78	1	64	0	0	0	0	0	0
Fuel comb – Industrial boilers, ICEs – Oil	2	0	8	1	0	0	0	0	0
Fuel comb – Industrial boilers, ICEs – Other	0	0	0	0	0	0	0	0	0
Industrial processes – Oil and gas production	2	58	2	0	0	0	0	0	0
Industrial processes – Storage and transfer	0	1	0	0	0	0	0	0	0
Mobile – Commercial marine vessels	61	26	738	457	56	61	0	0	1
Subtotal (federal waters)	**143**	**87**	**813**	**458**	**57**	**62**	**0**	**0**	**1**
Subtotal (all but vegetation and soil)	**82,696**	**17,871**	**18,168**	**10,827**	**6123**	**21,693**	**4367**	**1**	**3660**
Biogenics – Veg and soil	6474	31,744	1078	0	0	0	0	0	4322
Total	**89,170**	**49,615**	**19,246**	**10,827**	**6123**	**21,693**	**4367**	**1**	**7981**

HAP, hazardous air pollutant; NOx, nitrogen oxides (NO + NO2); PM2.5, fine particles in the (ambient) air 2.5 μm or less in size; PM10, particles in the atmosphere with a diameter of less than or equal to 10 μm; VOC, volatile organic compound.

anthropogenic and natural emissions responsible for reactive organic gases (ROGs) and other air pollutants in the USA in 2008.

1.2.3   Anthropogenic organic chemicals in the regional troposphere

As fossil fuel and biomass contain many organic compounds, it may be unsurprising that hundreds of organic chemicals have been detected in air samples worldwide. Scientists in China detected hundreds of organic acids in acid rain samples in the late 1980s. Partial speciation of primary particulate matter in California detected several hundred semi-volatile organic compounds. Using improved analytical methods, scientists in Australia were able to identify many dozens of organic compounds in addition to the ~ 600 organic compounds identified in the USA. Currently, vehicle exhausts are routinely analyzed for several hundreds of reactive organic gases in California, USA. Using a subset of these compounds, atmospheric chemists in the UK have built a nearly explicit photochemical mechanism that describes photochemical reactions of several thousand chemical species in the gas phase. With a number of educated assumptions, scientists have generated over a million photochemical reactions to forecast aerosol formation from a single organic compound with eight carbons. In fact, if analytical methods are available, a single compound could have many isomers detected. In sum, organic chemicals in the modern atmosphere mainly contain C, H, and O atoms; some of them, such as HCHO and C_6H_6, are carcinogens, while others, such as 1,3-butadiene and isoprene, may participate in photochemical reactions to produce harmful O_3 efficiently near the surface. Table 1.6 lists selected organic chemicals and their concentration levels identified in air samples collected in Asian cities. In rural areas, concentration levels of organic chemicals may be orders of magnitudes lower (Guo et al., 2004).

1.2.4   Trace elements in the regional troposphere

Besides C, H, and O atoms, about 35 other atoms and isotopes have been identified in air samples. Among them are particle-bound Hg, Cd, Cr, Pb, and As, which are toxic to humans, animals, and plants. Atmospheric deposition of these chemicals and subsequent runoff may contribute significantly to water pollution in certain areas, such as ponds near metropolitan areas where runoff water is not treated and ponds that accumulate polluted runoff water. The N and S atoms are the two most abundant elements found in tropospheric aerosols besides C, H, and O, and they are in the form of ammonium, nitrate and sulfate in aerosols and may be in the form of NH_3, N_2O, NO, HNO_2, NO_2, HNO_3, HNO_4, NO_3, N_2O_5, $R-C(O)OONO_2$, H_2S, CH_3SCH_3, OCS, CS_2, SO_2, H_2SO_4, $R-OSO_3$, etc., in the gas phase. Other relatively abundant elements in the air include Fe, Ca, Mn, and Mg, which may participate in catalytic and neutralizing reactions in wet aerosols as well as in fog and rain droplets. However, the exact form of many identified elements present in the air is uncertain due to the limitations of instrumentation and analytical methods. Toxic elements and concentration levels identified in air samples will be discussed in Chapter 10.

Table 1.6 Peak mixing ratios of selected organics in Asian cities

AVOC	ppb	AVOC	ppb
CH$_4$	6000	2-Methylpentane	16.1
Ethane	93	3-Methylpentane	13.4
Ethene	48.4	n-Hexane	13.8
Ethyne	58.3	n-Heptane	3.9
Propane	35	n-Octane	1.3
Propene	12.8	n-Nonane	1.4
i-Butane	16.4	n-Decane	0.7
n-Butane	42.2	Benzene	10.4
1-Butene	2.5	Toluene	17.5
i-Butene	4	Ethylbenzene	2.9
trans-2-Butene	3.4	*m*-Xylene	10.1
cis-2-Butene	2.7	*p*-Xylene	5.2
i-Pentane	25.9	*o*-Xylene	6.9
n-Pentane	24.8	i-Propylbenzene	0.2
1,3-Butadiene	2.5	n-Propylbenzene	0.3
Isoprene	1.7	3-Ethlytoluene	1.7
trans-2-Pentene	5.3	4-Ethyltoluene	0.7
cis-2-Pentene	9.4	2-Ethyltoluene	0.5
2,2-Dimethylbutane	0.6	1,3,5-Trimethylbenzene	1.1
2,3-Dimethylbutane	5	1,2,4-Trimethylbenzene	2.9

AVOC, anthropogenic volatile organic compound.

1.2.5 Trace chemicals in the global troposphere

As most chemicals identified in the troposphere are fairly reactive and their emission sources are near the surface, mixing ratios of trace chemicals in the free troposphere are often much lower than near the surface. A number of flight campaigns have been conducted over global oceans. In contrast to the atmospheric boundary layer over populated areas, the mixing ratio of NO$_x$ (defined as NO + NO$_2$) in the marine boundary layer and free troposphere is of the order of ~100 ppt, except in areas affected by plumes from ocean-going ships and outflows from the continental boundary layer. CH$_4$, C$_2$H$_6$, HCHO, CH$_3$CHO, CH$_3$COCH$_3$, and CH$_3$C(O)OONO$_2$ (PAN) are major reactive organic compounds in the free troposphere.

1.2.6 Isotopic tracers in the troposphere

Isotopes may differentiate in geochemical reservoirs, such as atmosphere, ocean, plants, fossil fuels, and volcanoes, due to physical, chemical, or biological processes, and these differentiations may sometimes be used to derive unambiguous origins or formation pathways for related chemicals in the atmosphere as well as in bodies of water.

Measurements of isotopic compositions of C (^{13}C, ^{14}C), O (^{17}O, ^{18}O), and ^{34}S in air samples and corresponding reservoirs have been reported to apportion relative contributions to atmospheric CO and CO_2 from fossil fuels and biomass burning and to derive relative importance of homogeneous versus heterogeneous reaction pathways for the formation of atmospheric nitrate and sulfate. For example, as the isotopic composition of C (^{13}C, ^{14}C) in fossil fuels and biomass is distinct, and the uncertainty due to stratosphere–troposphere exchange is considered negligible, the apportioning of atmospheric CO and CO_2 with regard to their fossil fuel origin or biomass burning would be straightforward if no other processes were involved. In that case, the major burden would be the acquisition of the isotopic composition data representing various conditions. Of course, in reality, atmospheric photochemical reactions may play a significant role in air samples. Another example is the application of isotopic compositions in deriving the relative importance of atmospheric formation pathways of nitrate and sulfate using O (^{17}O, ^{18}O). At a coastal pier downwind of Los Angeles, California, USA, nitrate in outflowing metropolitan plumes was suggested to have been formed more via the oxidation of NO by O_3 than by radicals, and non-sea-salt, secondary sulfate was suggested to be formed mainly from the oxidation of ship-emitted SO_2 by O_3. On one hand, it is hard to determine the actual ratio of various formation pathways accurately in certain atmospheric conditions via other methods, which makes the isotopic method a vital alternative. On the other hand, stringent requirements regarding the representation of isotopic composition data are necessary, as the uncertainty is sometimes large. Nevertheless, results from isotopic methods of non-sea-salt-sulfate formation pathways may constrain atmospheric photochemical model simulations in the future, as cloud volume in current atmospheric photochemical models has no counterpart in measurements.

1.3   Trace chemicals observed in the stratosphere

The number of trace chemicals in the stable stratosphere is much less than in the troposphere, and so is the mixing ratio of anthropogenic trace chemicals. Besides dry air and a few chemicals (CO_2, H_2O, O_3) with mixing ratios at ppm level, trace chemicals with sub-ppm level in the stratosphere include chemicals transported from the troposphere (CH_4, CO, N_2O, NH_3, SO_2, OCS, CS_2, CH_3Cl, CF_2Cl_2, HCF_2Cl, $CFCl_3$, $CFCl_2CF_2Cl$, CH_3CCl_3, CCl_4, CH_3Br, and other CFCs and hydrofluorocarbons (HFCs)) and chemicals emitted and produced in the stratosphere (NO, NO_2, $ClNO_3$, NO_3, N_2O_5, HNO_3, H_2O_2, HCl, HOCl, $BrNO_3$, HBr, HOBr, BrCl, O, O^1D, H, OH, HO_2, Cl, ClO, OClO, Cl_2O_2, Cl_2, S, SO, HS, SO_3, H_2SO_4).

Measurements via balloons, flights, rockets and satellites as well as ground spectroscopy were conducted to understand trace chemicals in the stratosphere in the last two decades of the twentieth century. Reactive nitrogen species ($NO_y = NO + NO_2 + HNO_3 + ClNO_3 + HNO_4 + N_2O_5$) were found to range from 10 to 15 ppb collectively above 30 km ASL, with NO, NO_2, and HNO_3 being major components, over mid-latitudes; compared with N_2O whose mixing ratio ranges from 10 to 250 ppb, the accumulation of reactive nitrogen species in the upper stratosphere is evident, possibly due to in-situ emissions. Mixing ratios of OH, HO_2, NO, NO_2,

ClO, BrO, HF, HCl, and ^{14}C in the stratosphere were found to be larger than in the troposphere, due to their origin from photochemical reactions. The mixing ratio of H_2O_2 was found to be significant up to 1 ppb. Trace chemicals, such as CO, NH_3, SO_2, OCS, and CS_2, have also been measured in the stratosphere, and their mixing ratios are lower than in the troposphere due to their origins from the surface. For the remaining trace chemicals, numerical modeling may provide reasonable estimates of their mixing ratios in the stratosphere using observational and laboratory constraints, before measurements become possible.

1.4   Greenhouse chemicals in the atmosphere

Greenhouse gases include H_2O, CO_2, CH_4, N_2O, CFCs and their substitutes, aerosols and O_3. To describe greenhouse gases collectively, CO_2 (eq) is used to represent the sum of six anthropogenic greenhouse gases (CO_2, CH_4, N_2O, SF_6, HFCs, perfluorocarbons (PFCs)) times their 100-year global warming potentials (1, 21, 310, 23,900, 650–11,700, 6500) respectively, following IPCC convention.

The emission of anthropogenic CO_2, as an end product of carbon-fuel combustion, is closely related to consumption of fossil fuels, construction materials, minerals and industrial materials. On a per-capita basis, sparsely populated developed countries such as Australia, Finland, Canada, and the USA consumed the largest amount of materials. Populated developed countries such as Germany, Austria, Japan, and the UK consumed similar amounts of materials to sparsely populated developing countries such as Argentina, Brazil, South Africa, and Saudi Arabia on a per-capita basis. Also on a per-capita basis, populated developing countries such as Nigeria, India, China, Cuba, and Mexico consumed the least amounts of materials. However, due to population, the total amount of materials consumed in China has exceeded the USA since 2005. As a result, anthropogenic emission of CO_2 (eq) from China reached 7.5 billion tons in 2005; by comparison, the USA emitted 7.3 billion tons CO_2 (eq) in 2005, and global emission was 45 billion tons CO_2 (eq) in that year. According to a survey in California, USA, CO_2 emission was 393.2 million tons (Tg) in the state, and accounted for 86% of total CO_2 (eq) emission in 2009. The corresponding percentages were 7% for CH_4, 3.3% for N_2O, 3.3% for HFCs, and 0.24% for SF_6 in California, USA in 2009, and greenhouse gases CF_4 and NF_3 were also emitted in small amounts.

The six greenhouse gases are rather stable in the troposphere, and their mixing ratios in the free troposphere are determined by accumulated emissions except for CO_2, which interacts at the Earth's surface. For example, $[CO_2] = 380$ ppmv, $[CH_4] = 1.8$ ppmv, and $[N_2O] = 315$ ppbv. CFCs have been present in the atmosphere only since early last century, when they were synthesized, and their mixing ratios are declining due to their role in destroying the stratospheric O_3 layer and the resulting ban of their use; meanwhile, their substitutes HCFCs are now present in the troposphere. Nevertheless, the mixing ratios of CFCs and HCFCs are below ~1 ppb in the free troposphere, and international treaties have been negotiated to control their adverse effects in the stratosphere.

Aerosols and O_3 may have greenhouse effects especially near the tropopause. Radiative effects of aerosols depend on their chemical composition, especially the content of black carbon. Without black carbon in the atmosphere, global temperature would be significantly cooler. As the lifetime of black carbon is of the order of a week in the troposphere, reducing black carbon emission is effective in mitigating global warming in addition to yielding health benefits. O_3 near the tropopause is mainly affected by NO_x level and downward transport from the stratosphere. Lightning and aircraft emissions of NO_x may elevate the NO_x level there, and subsequently increase O_3 near the tropopause. However, both sources are difficult to reduce, as the mechanism of formation of lightning is not sufficiently understood and aviation is expected to rise.

1.5 Toxic chemicals in breathing zones

When a chemical is toxic to humans, its dose–response curve takes the form shown in Figure 1.3. In the figure, the horizontal axis denotes various doses applied to experimental animals, and the vertical axis denotes the percentage of experimental animals that died or suffered certain damage such as cancer. To determine if a chemical is a toxin, sophisticated toxicological studies have to be conducted; the latter is desirable for new active ingredients of medicines, new synthesized chemicals in consumer products, and newly identified chemicals in ecological and environmental systems.

In breathing zones, 0.5–1.5 m above ground level, the air that humans inhale often contains many chemicals known to be toxic to humans, especially in populated areas. Identified toxic chemicals include inorganic gases (CO, O_3, NH_3, Hg, Rn, Cl_2, etc.), organic gases (aldehydes, aromatics, etc.), and particulate matter components (Cd, Cr, Pb, As, Ni, V, Co, Cu, Sb, Sn, Zn, H_2SO_4, NH_4NO_3, polycyclic aromatic hydrocarbons, etc.). In the USA, nearly 200 chemicals with noticeable toxicity are monitored

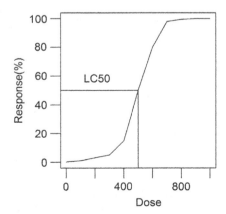

Figure 1.3 Dose–response curve of a fictitious air toxin similar to the lung cancer risk of inhaling cadmium. The unit of dose is neglected for generality.

and regulated in ambient and indoor air, due to concerns from the public, research and industrial communities, and civil servants. Details will be described in Chapter 10.

Air toxins in breathing zones traditionally originated from emissions from industries, power plants, smelters, biomass burning, building and paving materials, and natural sources. Various health standards have been developed and are enforced in many countries, including the USA and China, for professional workers who are routinely exposed to much higher risk levels in built environments than the general public in an ambient environment, and recommendations for chemicals such as Rn, CO, and HCHO in residential, commercial, and congregational indoor air have been made in California, USA and other regions.

As urbanization progresses from the regional to the global scale, vehicle exhausts and consumer products become more and more important. In fact, they are dominant sources for air toxins in many areas of California, USA. Smoking used to be the highest risk of personal exposure in indoor air. However, a calculation of accumulated risk on highways due to vehicle exhausts and evaporations within vehicles shows that riding a vehicle on California highways could result in similar personal risk exposure to smoking cigars continuously. Thus, clean driving may significantly reduce human exposure to toxic chemicals in breathing zones, as vehicle transportation is indispensable for humans nowadays.

Summary

In this chapter, major and minor chemicals observed in the modern atmosphere, which consists of the troposphere, stratosphere, mesosphere and above, have been discussed. Then, trace chemicals observed in the troposphere and stratosphere have been described. Finally, greenhouse gases and toxic chemicals have been discussed. Typical values have been tabulated for atmospheric chemical modeling purposes.

Table 1.A1 Sample vertical profiles of temperature, pressure, and O_3 (0–46 km ASL)

Z (km)	T (K)	P (hPa)	O_3 (ppmv)	Z (km)	T (K)	P (hPa)	O_3 (ppmv)
0	301	1013	0.04	24	220	30	4.86
2	288	800	0.04	26	225	22	6.26
4	277	626	0.05	28	229	16	7.32
6	266	485	0.05	30	234	12	7.61
8	252	371	0.06	32	238	9	7.97
10	238	279	0.06	34	242	7	8.01
12	224	206	0.08	36	247	5	7.65
14	210	149	0.11	38	251	4	6.94
16	203	107	0.22	40	256	3	5.98
18	207	77	0.67	42	261	2	5.03
20	212	56	1.74	44	264	2	3.99
22	216	41	3.29	46	267	1	3.17

Exercises

1. Rn is an air toxin due to its radioactive property. Survey the literature to find its concentration levels in breathing zones, and calculate the corresponding personal lifetime risk exposure to Rn in units of kg Rn per kg of human body, for a baby girl born in a coastal area and a baby boy born in Tibet, China ($\sim$4 km ASL).

2. Assume a 1-km column of standard air over an ocean surface initially had an adiabatic lapse rate of 10 K km^{-1} without H_2O. Then, a certain amount of H_2O vapor was added to the column, so that the lapse rate was 7.0 K km^{-1}, and the relative humidity was constant throughout the column. Survey the literature to estimate the amount of H_2O contained in the column, using the units kg H_2O per km^2 of air column.

3. The CO_2 mixing ratio was 280 ppm in 1750, 310 ppm in 1950, and 380 ppm in 2010 at a mountain site (3 km ASL) in Hawaii. (a) Survey the literature to estimate how much H_2O, in kg, is needed to convert all atmospheric CO_2 in 1750, 1950, and 2010 respectively into plant tissues. (b) Assume the increase of atmospheric CO_2 from 1750 to 1950 was due to fossil fuel combustion only, and all geochemical reservoirs other than the atmosphere absorb half of CO_2 emissions from fossil fuel combustion. Calculate how much fossil fuel (in kg) was combusted during the 200 years. (c) Do the same calculation for the period 1950–2010.

4. Suppose fertilizer $C(O)(NH_2)_2$ is responsible for the presence of 0.2 ppbv N_2O and 0.2 ppbv CH_4 in the troposphere. Survey the literature to find how much energy may be saved, if soil management were perfect.

5. Aviation emits NO_x, like other combustions, but injects pollutants at much higher altitudes than others. Assuming it takes 20 years for a molecule in the troposphere to get into the stratosphere, survey the literature and find how much N_2O emission has to be curbed in the troposphere to compensate aviation emission of NO_y with 1 Tg N per year in the stratosphere. Make reasonable assumptions as necessary.

6. Hg and BaP are both potent toxins for humans. Survey the literature to estimate their atmospheric burdens, and identify areas with high risks for children and the elderly.

2 Chemical reactions in the atmosphere

Atmospheric chemicals react with each other, transfer among various phases, and are transformed via photolysis, photoionization, aqueous dissociation and near interfaces. In laboratories, reaction rate parameters have been measured for nearly 1000 elementary reactions involving inorganic and organic chemicals. Using smog chambers, many thousands of experiments have been conducted to investigate formation mechanisms of O_3 and secondary organic aerosols from degradations of organic chemicals. While the exact number of chemical reactions occurring in the atmosphere may be overwhelming, known inorganic, organic, and heterogeneous reactions have been compiled and updated regularly for several decades.

2.1 Inorganic reactions

Inorganic chemicals in the atmosphere include five important groups, namely odd oxygen species, odd hydrogen species, reactive nitrogen species, reactive sulfur

Chemical Modeling for Air Resources. http://dx.doi.org/10.1016/B978-0-12-408135-2.00002-1

species, and reactive halogen species. Odd oxygen species consists of O_3, O, O^1D, and odd hydrogen species consists of OH, HO_2, H. Reactive nitrogen species consists of NO, HNO_2, NO_2, NO_3, HNO_3, N_2O_5, HNO_4, NH_3, N_2O, and reactive sulfur species consists of SO_2, H_2SO_3, SO_3, H_2SO_4, H_2S, OCS, CS_2. Reactive halogen species consists of F, FO, Cl, ClO, Br, BrO, I, IO, HCl, $ClNO_3$, Cl_2, etc. The relative importance of group members depends on photochemical regimes and altitudes. For example, reactive sulfur species mainly includes SO_2, H_2SO_3, SO_3, and H_2SO_4 in the troposphere, but also includes H_2S, OCS and CS_2 in the stratosphere. Inorganic reactions in the atmosphere include first-order, second-order, and third-order elementary reactions, with reactants from the five important groups as well as a few other chemicals such as CO, O_2, and H_2O.

2.1.1 First-order reactions

First-order reactions include photolysis, chemical transfer at the interface of two phases, and pseudo-first-order thermal dissociation.

Important photolysis reactions in the troposphere include:

$$O_3 + hv \rightarrow O^1D + O_2 \qquad\qquad\qquad (R2.1.1)$$

$$NO_2 + hv \rightarrow O + NO \qquad\qquad\qquad (R2.1.2)$$

Reaction (R2.1.1) provides O^1D for the formation of OH, via reaction (R2.1.3), in the troposphere and stratosphere, and OH is the most important oxidant in the Earth's atmosphere:

$$O^1D + H_2O \rightarrow 2OH \qquad\qquad\qquad (R2.1.3)$$

Reaction (R2.1.2) is the major pathway in the troposphere for providing O for O_3 formation via reaction (R2.1.4), and the latter is the dominant formation pathway for O_3 in the atmosphere:

$$O + O_2 + M \rightarrow O_3 + M \qquad\qquad\qquad (R2.1.4)$$

Unique photolysis reactions in the stratosphere include:

$$O_2 + hv \rightarrow O + O \qquad\qquad\qquad (R2.1.5)$$

$$N_2O + hv \rightarrow N_2 + O^1D \qquad\qquad\qquad (R2.1.6)$$

Reaction (R2.1.5) provides O for the formation of the "O_3 layer" in the stratosphere via reaction (R2.1.4), and reaction (R2.1.6) decomposes N_2O that is emitted from the surface and is rather stable in the troposphere.

Unique photolysis reactions in the mesosphere and thermosphere include:

$$O_2 + hv \rightarrow O + O^1D \tag{R2.1.7}$$

$$N_2 + hv \rightarrow N + N \tag{R2.1.8}$$

Chemical transfer across the interface of two phases in the atmosphere occurs between gas–liquid phases, gas–solid phases, and liquid–solid phases. For example, during rainfall in summer, HNO_3 may be washed out by raindrops (R2.1.9). During cold winter, $ClNO_3$ may be collected by HCl particles in the stratospheric vortex over Antarctica (R2.1.10). In an internally mixed wet aerosol particle containing ammonium sulfate in the aqueous phase and PbO in the solid core, as illustrated in Figure 2.1, sulfate may enter the core (R2.1.11):

$$HNO_3(g) \xrightarrow{H_2O\,(l)} NO_3^- + H_3O^+ \tag{R2.1.9}$$

$$ClNO_3(g) \xrightarrow{HCl\,(s)} Cl_2 + HNO_3(s) \tag{R2.1.10}$$

$$SO_4{}^{2-}(aq) \xrightarrow{PbO\,(s)} PbSO_4(s) \tag{R2.1.11}$$

The last type of first-order elementary reaction is pseudo-first-order thermal dissociation in high-pressure conditions, such as the dissociation of N_2O_5 in the planetary boundary layer (PBL) at dawn:

$$N_2O_5 \xrightarrow{M} NO_2 + NO_3 \tag{R2.1.12}$$

Calculations of first-order elementary reaction rate coefficients are rather complex, due to the nature of the processes involved. For example, the photolysis rate coefficient of reaction (R2.1.1), commonly denoted as $J(O_3)$, is the sum of the products of

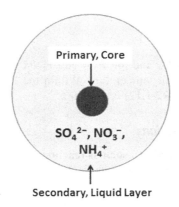

Figure 2.1 An internally mixed wet aerosol particle.

the absorption cross-section of O_3, the corresponding quantum yield for O^1D, and the actinic flux over the relevant solar spectrum:

$$J(O_3) = \sum C(i) \cdot Q(i) \cdot F(i), \quad i = 1, 2, ..., n \tag{F2.1.1}$$

In formula (F2.1.1), $C(i)$ denotes the cross-section of O_3 at wavelength interval i, $Q(i)$ denotes quantum yield of O^1D from O_3 at wavelength interval i, $F(i)$ denotes the actinic flux of wavelength interval i, and n denotes the total number of wavelength intervals relevant to the reaction. While cross-sections and quantum yields may be determined via laboratory experiments, actinic fluxes depend on atmospheric conditions, such as sunlight intensity, absorption and scattering along the path from the Sun to the location of interest, albedos of the Earth's surface, absorption and scattering along the path from the Earth's surface to the location of interest. Thus, calculating photolysis rate coefficients requires radiative transfer modeling.

The rate coefficient of reaction (R2.1.9) depends on transport processes near the surface of rain droplets. Assume all rain droplets have the same geometry, an ideal spherical shape with radius r. In addition, assume reaction (R2.1.9) is limited by the diffusion of HNO_3 in interstitial air. Then, the rate coefficient of reaction (R2.1.9), denoted by $k(g \rightarrow aq.)$, may be calculated as a function of the radius of rain droplets (R), Henry's constant of HNO_3 (H), the gas-phase diffusion coefficient of HNO_3 near rain droplets (D_g), and the uptake coefficient of HNO_3 by rain droplets (γ):

$$k(g \rightarrow aq.) = \text{Function}(R, H, D_g, \gamma) \tag{F2.1.2}$$

The exact formulation is listed in footnote "g" of Table 6.3 in Chapter 6 for rain droplets and in formula (F9.3) for smaller particles. The rate coefficient of N_2O_5 thermal dissociation reaction (R2.1.12), $k_1(\text{thermal})$, depends on air pressure as well as air temperature, activation energy of the reaction, and other factors. Numerically, it is a function of four variables, namely air mass ([M]), the rate coefficient when air pressure is approaching zero, $k_0([M])$, the rate coefficient when air pressure is approaching infinity (k_∞), and a so-called broadening factor that is a function of F_c as well as the ratio of $k_0([M])$ to k_∞:

$$k_1(\text{thermal}) = F \cdot k_0/(1 + r) \tag{F2.1.3}$$

In formula (F2.1.3), $r = k_0/k_\infty$ and $\log F \approx \log F_c /[1 + (\log r)^2]$; all independent variables above depend on air temperature. Within the PBL, the pseudo-first-order rate coefficient of reaction (R2.1.12) is ~ 0.06 s^{-1}.

2.1.2 Second-order reactions

Second-order elementary reactions, such as reaction (R2.1.3), are the most common form of inorganic reactions in the atmosphere. Examples of second-order elementary reactions in the troposphere include:

$$OH + CO \rightarrow CO_2 + H \tag{R2.1.13}$$

$$H + O_2 \xrightarrow{M} HO_2 \qquad\qquad (R2.1.14)$$

$$HO_2 + NO \rightarrow NO_2 + OH \qquad\qquad (R2.1.15)$$

Adding reactions $(R2.1.1) + (R2.1.3) + 2(R2.1.13 + R2.1.14 + R2.1.15 + R2.1.2 + R2.1.4)$ yields the following effective pathway:

$$2CO + 3O_2 + H_2O \xrightarrow{hv} 2CO_2 + O_3 + 2OH \qquad\qquad (P2.1.1)$$

The above pathway is not an elementary reaction, but it denotes that, in certain conditions, CO may be effectively oxidized by O_2 in the presence of H_2O and sunlight to form CO_2 and oxidants O_3 and OH. As CO is relatively abundant in the PBL, e.g. the 1-hour ambient air quality standard is over 40 mg m^{-3} in the USA and 10 mg m^{-3} in China, the above pathway may be important for oxidant budgets in certain conditions. When global, annual CO emission is 350 million tons, the above pathway leads to the production of 600 million tons of O_3 and 212.5 million tons of OH.

Examples of second-order elementary reactions in the stratosphere are:

$$O^1D + O_2 \rightarrow O + O_2 \qquad\qquad (R2.1.16)$$

$$O^1D + N_2 \rightarrow O + N_2 \qquad\qquad (R2.1.17)$$

By adding reactions $(R2.1.1)$, $(R2.1.4)$, $(R2.1.16)$, and $(R2.1.17)$, a null cycle is formed. Within the null cycle, no chemical is formed or lost except that sunlight is absorbed and air molecules are heated. Adding reactions $2 \times (R2.1.4) + (R2.1.5)$, a pathway to form the "O_3 layer" in the stratosphere is formed:

$$3O_2 \xrightarrow{hv} 2O_3 \qquad\qquad (P2.1.2)$$

Second-order elementary reactions occurring in the modern stratosphere also include reactions, such as $(R2.1.18)$ and $(R.2.1.19)$, which involve fragments of anthropogenic chemicals such as CFCs. CFCs such as CF_2Cl_2 are rather stable in the troposphere after emission from the surface, but dissociate rapidly in the stratosphere via photolysis. After dissociation, CFCs produce halogen radicals such as F, FO, Cl, and ClO. These radicals participate in chemical reactions and are partly responsible for the "O_3 hole".

$$O + FO \rightarrow F + O_2 \qquad\qquad (R2.1.18)$$

$$F + O_3 \rightarrow FO + O_2 \qquad\qquad (R2.1.19)$$

Second-order elementary reaction $(R2.1.20)$ mainly takes place when solar radiation with very short ultraviolet wavelength is significant:

$$N + O_2 \rightarrow NO + O \qquad\qquad (R2.1.20)$$

Adding reactions $(R2.1.8) + 2(R2.1.20)$ yields the following pathway:

$$N_2 + 2O_2 \xrightarrow{hv} 2NO + 2O \qquad\qquad (P2.1.3)$$

The pathway (P2.1.3) illustrates productions of reactive NO and O from inert N_2 and O_2 in the presence of very short ultraviolet radiation of the Sun, which may occur near the top of the atmosphere.

2.1.3   Third-order reactions

Third-order elementary reactions occur less frequently in the atmosphere, partly explained by collision theory and transition-state theory. Reaction (R2.1.4) is a third-order elementary reaction. However, at a fixed location in the atmosphere where the air pressure is almost constant, this reaction may be approximated as a pseudo-second-order elementary reaction when "air molecules" are abundant. Most third-order elementary reactions in the atmosphere occur at similar conditions to reaction (R2.1.4), and their reaction rate coefficients may be calculated from formula (F2.1.3), similar to the last type of first-order reactions.

Reactions of fourth order are rare in the atmosphere. Reaction (R2.1.21) is an example:

$$HO_2 + HO_2 + H_2O + M \rightarrow H_2O_2 + O_2 + H_2O + M \qquad\qquad (R2.1.21)$$

In the above reaction, M denotes a bulk air molecule $(0.79N_2 + 0.21O_2)$, and the role of M in the reaction is to dissipate collision energy according to transition-state theory. The rate coefficient of the above reaction may be calculated using formula (F2.1.3). Important inorganic species and their reactions in the lower atmosphere are listed in Tables 6.2 and 6.3 in Chapter 6, together with a coupled gas–aqueous photochemical mechanism for the O_3–CH_4–C_2H_6–NO_x–SO_x–Cl^- system, which may be used to simulate atmospheric chemistry in the global troposphere for general purposes.

2.2   Organic reactions

Chemical reactions describing the transformation of many organic chemicals in the atmosphere are numerous, especially near emission sources, e.g. in PBL or in various plumes. In principle, calculations of rate coefficients for organic reactions are similar to those for inorganic reactions. However, for practical purposes, elementary reactions are rarely used to represent organic reactions in atmospheric chemical modeling. Instead, effective pathways are often used to condense organic reactions.

2.2.1   Oxidation and formation of CH₄

CH_4 is the simplest hydrocarbon compound but has the highest mixing ratio in the atmosphere. Its oxidation pathway to CO_2 and H_2O in the gas phase of the atmosphere may be outlined by the following organic, elementary reactions (R2.2.1)–(R2.2.17), besides inorganic reactions such as (R2.1.13):

$$CH_4 + OH \rightarrow CH_3 + H_2O \tag{R2.2.1}$$

$$CH_3 + O_2 + M \rightarrow CH_3O_2 + M \tag{R2.2.2}$$

$$CH_3O_2 + NO \rightarrow NO_2 + CH_3O \tag{R2.2.3}$$

$$CH_3O_2 + CH_3O_2 \rightarrow 2CH_3O + O_2 \tag{R2.2.4}$$

$$CH_3O_2 + HO_2 \rightarrow CH_3OOH + O_2 \tag{R2.2.5}$$

$$CH_3O + O_2 \rightarrow HCHO + HO_2 \tag{R2.2.6}$$

$$CH_3O + HO_2 \rightarrow CH_3OH + O_2 \tag{R2.2.7}$$

$$CH_3OH + OH \rightarrow H_2O + HCHO + H \tag{R2.2.8}$$

$$HCHO + h\nu \rightarrow 2H + CO \tag{R2.2.9}$$

$$HCHO + h\nu \rightarrow H_2 + CO \tag{R2.2.10}$$

$$HCHO + OH \rightarrow H_2O + CO + H \tag{R2.2.11}$$

$$CH_3OOH + OH \rightarrow H_2O + CH_2OOH \tag{R2.2.12}$$

$$CH_3OOH + OH \rightarrow H_2O + CH_3OO \tag{R2.2.13}$$

$$CH_2OOH + M \rightarrow HCHO + OH + M \tag{R2.2.14}$$

$$CH_2OOH + O_2 \rightarrow HCOOH + HO_2 \tag{R2.2.15}$$

$$CH_3OOH + h\nu \rightarrow OH + CH_3O \tag{R2.2.16}$$

$$HCOOH + OH \rightarrow H_2O + CO_2 + H \tag{R2.2.17}$$

CH_4 oxidation reactions in the gas phase of the atmosphere may be classified into four types from a radical perspective: (a) radical generation reactions, (b) radical propagation reactions, (c) radical termination reactions, and (d) others. For example, photolysis reactions (R2.2.9) and (R2.2.16) generate radicals, and reactions (R2.1.13), (R2.2.1)–(R2.2.4), (R2.2.6), (R2.2.8), (R2.2.11)–(R2.2.15),and (R2.2.17) propagate radicals. Reactions (R2.2.5) and (R2.2.7) terminate radicals, and reaction

(R2.2.10) proceeds without involving radicals. When CH_4 is oxidized into CO_2 in the gas phase of the atmosphere, intermediate products CH_3OH, CH_3OOH, HCHO, HCOOH, and CO are formed. It is noteworthy that, in the gas phase of the atmosphere, HCHO is oxidized into CO, and there is no pathway for the conversion of HCHO to HCOOH.

CH_4 may be produced in the atmosphere during the oxidation of 2-alkenes, such as 2-butene, by O_3:

$$CH_3HC{=}CHCH_3 + O_3 \rightarrow CH_3CHOO^* + CH_3CHO \qquad \text{(R2.2.18)}$$

$$CH_3CHOO^* \rightarrow CH_4 + CO_2, \sim 14\% \qquad \text{(R2.2.19a)}$$

$$CH_3CHOO^* \rightarrow CH_3 + CO + OH, \sim 26\% \qquad \text{(R2.2.19b)}$$

$$CH_3CHOO^* \rightarrow CH_3 + CO_2 + H, \sim 28\% \qquad \text{(R2.2.19c)}$$

$$CH_3CHOO^* \rightarrow CH_3O + HCO, \sim 17\% \qquad \text{(R2.2.19d)}$$

$$CH_3CHOO^* + M \rightarrow CH_3CHOO + M, \sim 15\% \qquad \text{(R2.2.19e)}$$

In the above reactions, CH_3CHOO^* denotes a peroxyethyl biradical in an excited state. Note that the peroxyethyl biradical CH_3CHOO^* has the same formula as acetic acid, but is much more reactive than the latter due to its different structure. In the atmosphere, CH_3CHOO^* oxidizes NO, NO_2, SO_2, and CO into NO_2, NO_3, H_2SO_4, and CO_2 respectively to form CH_3CHO; only a negligible fraction of CH_3CHOO^* is transformed to CH_3COOH. While the rate coefficient of elementary reaction (R2.2.18) is measurable, it is conceivable that reaction (R2.2.19) is difficult to measure for many possible branches. For atmospheric chemical modeling purposes, elementary reactions (R2.2.18) and (R2.2.19) may be denoted by the following effective reaction:

$$CH_3HC{=}CHCH_3 + O_3 \rightarrow CH_3CHO + 0.14CH_4 + 0.14CO_2 + 0.54CH_3$$
$$+ 0.26CO + 0.26OH + 0.28CO_2 + 0.28H$$
$$+ 0.17CH_3O + 0.17HCO + 0.15CH_3CHOO$$
$$\text{(R2.2.20)}$$

From now on, "reactions" will be used to refer to "effective reactions" and "effective pathways" as well as "elementary reactions" for clarity.

2.2.2 Degradation of $\geq C_2$ organics

According to the USA National Institute of Standards and Technology, there were 38,000 reaction constants measured for 11,700 distinct reactant pairs in $\sim 15,000$ reactions of 9000 chemical compounds by 2000. Despite this large database, kinetic data for atmospheric reactions, especially those occurring in PBL, can only provide a full description of photochemical reactions of $\leq C_3$ organic compounds in the gas phase so far, according to the International Union of Pure and Applied Chemistry (IUPAC). For larger organic compounds, statistical and modeling methods in combination with hypotheses have to be used to close the gap between experimental data and needs by atmospheric chemical models for a systematic set of chemical reactions.

Atmospheric hydrocarbons may contain one to over a dozen carbons in the gas phase, and may be oxidized into oxygenated hydrocarbons, such as alcohols, organic nitrates, hydroperoxides, carbonyls, carboxylic acids, and peroxyacylnitrates, before their transformation into CO, CO_2 and H_2O in the atmosphere. While it is impossible to list all those reactions explicitly, the degradation of a generic alkane molecule to an alkyl radical with at least one less carbon is elaborated below.

Let RCH_3, $R_1CH_2R_2$, and $R_1CH(R_2)R_3$ denote chain alkanes with an H to be abstracted by an oxidant from a primary, secondary, and tertiary carbon respectively. The corresponding reactions in the atmosphere proceed as follows:

$$RCH_3 + OH \rightarrow H_2O + RCH_2 \tag{R2.2.21a}$$

$$RCH_3 + NO_3 \rightarrow HNO_3 + RCH_2 \tag{R2.2.21b}$$

$$R_1CH_2R_2 + OH \rightarrow H_2O + R_1CHR_2 \tag{R2.2.22a}$$

$$R_1CH_2R_2 + NO_3 \rightarrow HNO_3 + R_1CHR_2 \tag{R2.2.22b}$$

$$R_1CH(R_2)R_3 + OH \rightarrow H_2O + R_1C(R_2)R_3 \tag{R2.2.23a}$$

$$R_1CH(R_2)R_3 + NO_3 \rightarrow HNO_3 + R_1C(R_2)R_3 \tag{R2.2.23b}$$

Rate constants have been measured for oxidations of CH_4, C_2H_6, C_3H_8, n-C_4H_{10}, n-C_5H_{12}, ..., n-C_{16}, and some of their isomers by OH. For unmeasured chain alkanes, their OH oxidation rate constants may be estimated as follows:

$$k(R2.2.21a) = F(R) \cdot 4.47 \times 10^{-18} \cdot T^2 \cdot \exp(-303/T) \tag{F2.2.1}$$

$$k(R2.2.22a) = F(R_1) \cdot F(R_2) \cdot 4.32 \times 10^{-18} \cdot T^2 \cdot \exp(233/T) \tag{F2.2.2}$$

$$k(R2.2.23a) = F(R_1) \cdot F(R_2) \cdot F(R_3) \cdot 1.89 \times 10^{-18} \cdot T^2 \cdot \exp(711/T) \tag{F2.2.3}$$

where $F(-CH_3) = 1$ and $F(-CH_2-) = F(-CH<) = F(>C<) = \exp(76/T)$. The unit of T is K and the units of the rate constant k are cm^3 molecule^{-1} s^{-1}. At 300 K, the ratio of the rate constants of the OH abstraction of primary, secondary, and tertiary H from chain alkanes via reactions (R2.2.21a), (R2.2.22a), and (R2.2.23a) is about 1:10:30.

Alkanes (CH_4, C_2H_6, C_3H_8, n-C_4H_{10}, n-C_5H_{12}, ..., n-C_9) and a few isomers have been measured for rate constants of their oxidations by NO_3. For unmeasured chain alkanes, their NO_3 oxidation rate constants may be estimated as follows:

$$k(\text{R2.2.21b}) = F(R) \cdot 7 \times 10^{-19} \tag{F2.2.4}$$

$$k(\text{R2.2.22b}) = F(R_1) \cdot F(R_2) \cdot 1.5 \times 10^{-17} \tag{F2.2.5}$$

$$k(\text{R2.2.23b}) = F(R_1) \cdot F(R_2) \cdot F(R_3) \cdot 8.2 \times 10^{-17} \tag{F2.2.6}$$

where $F(-CH_3) = 1$ and $F(-CH_2-) = F(-CH<) = F(>C<) = 1.5$, while $T = 298$ K. The units of the rate constant k are cm^3 molecule^{-1} s^{-1}. Based on measurements, reactions (R2.2.21)–(R2.2.23) have OH:NO_3 branching ratios of $\sim 100{:}1$, with the OH branch dominant during daytime and the NO_3 branch important at night.

The alkyl radicals formed in reactions (R2.2.21)–(R2.2.23) react with O_2 to form peroxyalkyl radicals in an excited state, and the latter either collides with air molecules to form RO_2 [$=RCH_2O_2$, $R_1CHO_2R_2$, or $R_1CO_2(R_2)R_3$] or decomposes into HO_2 + alkene. As chain alkanes larger than CH_4 mainly stay in the troposphere, the collision branch dominates, and the decomposition branch is negligible. RO_2 may react with NO, HO_2, and NO_2 as follows:

$$RO_2 + NO \rightarrow NO_2 + RO \tag{R2.2.24a}$$

$$RO_2 + NO + M \rightarrow RONO_2 + M \tag{R2.2.24b}$$

$$RO_2 + HO_2 \rightarrow ROOH + O_2 \tag{R2.2.25}$$

$$RO_2 + NO_2 + M \rightarrow RO_2NO_2 + M \tag{R2.2.26a}$$

Among reactions (R2.2.24)–(R2.2.26), the NO branch dominates in PBL, and the NO_2 branch has a rapid reverse reaction:

$$RO_2NO_2 + M \rightarrow RO_2 + NO_2 + M \tag{R2.2.26b}$$

It is noteworthy that reaction (R2.2.26) is negligible for RO_2 radicals from an atmospheric chemical modeling perspective, contrary to similar reactions for RCO_3 radicals where peroxyacyl nitrates (PANs) may be present significantly at low temperature. The RO_2 radical may also react with another RO_2 or RCO_3 radical, though no measurements have been reported for the rate constants of these reactions. Constraints from smog chamber data and chemical kinetics suggest that reactions between identical RO_2 radicals may proceed as follows:

$$RCH_2O_2 + RCH_2O_2 \rightarrow RCH_2O + RCH_2O + O_2 \tag{R2.2.27a}$$

$$RCH_2O_2 + RCH_2O_2 \rightarrow RCH_2OH + RCHO + O_2 \tag{R2.2.27b}$$

$$R_1CHO_2R_2 + R_1CHO_2R_2 \rightarrow R_1CHOR_2 + R_1CHOR_2 + O_2 \tag{R2.2.28a}$$

$$R_1CHO_2R_2 + R_1CHO_2R_2 \rightarrow R_1CH(OH)R_2 + R_1C(O)R_2 + O_2 \tag{R2.2.28b}$$

$$R_1CO_2(R_2)R_3 + R_1CO_2(R_2)R_3 \rightarrow R_1CO(R_2)R_3 + R_1CO(R_2)R_3 + O_2 \tag{R2.2.29}$$

The bulk rate constants of reactions (R2.2.27) and (R2.2.28) are estimated to be 2.5×10^{-13} and 5×10^{-15} cm^3 molecule^{-1} s^{-1} respectively, and that of reaction (R2.2.29) is 2×10^{-17} cm^3 molecule^{-1} s^{-1}, with an uncertainty of at least a factor of 5. Branching ratios of the radical (a) versus neutral (b) branches in reactions (R2.2.27) and (R2.2.28) are estimated to be 0.45:0.55.

Reactions between non-identical RO$_2$ radicals may proceed similarly. The (bulk) rate constant of a reaction between non-identical RO$_2$ radicals, e.g. reaction (R2.2.30), is estimated to be twice the geometric mean of rate constants of corresponding reactions between identical RO$_2$ radicals:

$$R_1O_2 + R_2O_2 \rightarrow \text{products} \tag{R2.2.30}$$

For example, if R_1O_2 denotes RCH_2O_2 and R_2O_2 denotes $R_1CHO_2R_2$ in reaction (R2.2.30), then $k(\text{R2.2.30}) \approx 2\sqrt{k(\text{R2.2.27}) \cdot k(\text{R2.2.28})} = 7.1 \times 10^{-14}$ cm^3 molecule^{-1} s^{-1}.

Reactions of RO$_2$ radicals [RCH_2O_2, $R_1CHO_2R_2$, or $R_1CO_2(R_2)R_3$] with peroxyacyl radicals $R'C(O)O_2$ in the atmosphere result in the preferential formation of carboxylic acids:

$$RCH_2O_2 + R'C(O)O_2 \rightarrow RCH_2O + R'C(O)O + O_2 \tag{R2.2.31a}$$

$$RCH_2O_2 + R'C(O)O_2 \rightarrow RCHO + R'C(O)OH + O_2 \tag{R2.2.31b}$$

$$R_1CHO_2R_2 + R'C(O)O_2 \rightarrow R_1CHOR_2 + R'C(O)O + O_2 \tag{R2.2.32a}$$

$$R_1CHO_2R_2 + R'C(O)O_2 \rightarrow R_1C(O)R_2 + R'C(O)OH + O_2 \tag{R2.2.32b}$$

$$R_1CO_2(R_2)R_3 + R'C(O)O_2 \rightarrow R_1CO(R_2)R_3 + R'C(O)O + O_2 \tag{R2.2.33}$$

A unique case here is that HO$_2$ may react with CH$_3$C(O)O$_2$ to form O$_3$ with a branching ratio of $\sim 20\%$:

$$HO_2 + CH_3C(O)O_2 \rightarrow O_2 + CH_3C(O)OOH \tag{R2.2.34a}$$

$$HO_2 + CH_3C(O)O_2 \rightarrow O_3 + CH_3C(O)OH \tag{R2.2.34b}$$

Alkyloxy radicals [RCH_2O, R_1CHOR_2, $R_1CO(R_2)R_3$] formed from previous reactions in the atmosphere rapidly decompose, via unimolecular reactions, to form alkyl radicals with at least one less carbon than the original alkanes. The decomposition propagates radicals and results in the formation of HCHO and R radical for RCH_2O, the formation of two aldehydes (R_1CHO, R_2CHO) and corresponding (R_2, R_1) radicals for R_1CHOR_2 with larger branching ratios for smaller aldehyde, and the formation of three ketones [$R_1C(O)R_2$, $R_1C(O)R_3$, $R_2C(O)R_3$] and corresponding (R_3, R_2, R_1) radicals for $R_1CO(R_2)R_3$ with larger branching ratios for smaller ketones. Besides decomposition, primary and secondary alkyloxy radicals RCH_2O and R_1CHOR_2 may also react with O_2 via reaction (R2.2.35) to form HO_2 radical and carbonyl compounds in the atmosphere:

$$RCH_2O + O_2 \rightarrow HO_2 + RCHO \tag{R2.2.35a}$$

$$R_1CHOR_2 + O_2 \rightarrow HO_2 + R_1C(O)R_2 \tag{R2.2.35b}$$

Rate constants have been measured for $\leq C_4$ alkyloxy radicals. Based on a few data points, for an unmeasured primary alkyloxy radical RCH_2O with R consisting of four or more carbons, the rate constant of reaction (R2.2.35a) may be estimated by the following formula:

$$k_n = k_{n-1} + 2^{n-2} \times 10^{-15}, \quad n \geq 5 \tag{F2.2.7}$$

where n denotes the number of carbons in the RCH_2O radical, and the units of k are cm^3 molecule^{-1} s^{-1}; $k_4 = 14 \times 10^{-15}$ cm^3 molecule^{-1} s^{-1}. For an unmeasured secondary alkyloxy radical R_1CHOR_2 with R_1 and R_2 consisting of two or more carbons, the rate constant of reaction (R2.2.35b) may be assumed as follows:

$$k_n = k_{n-1} + (n - 4) \times 0.6 \times 10^{-15}, \quad n \geq 5 \tag{F2.2.8}$$

where n denotes the number of carbons in the R_1CHOR_2 radical, and the units of k are cm^3 molecule^{-1} s^{-1}; $k_4 = 7.6 \times 10^{-15}$ cm^3 molecule^{-1} s^{-1}. The uncertainty in estimated rate constants is expected to be of the order of $\sim 10\%$ in the atmosphere.

The decomposition rate constant for an alkyloxy radical is $\sim 2 \times 10^{20}$ times that of its reaction with O_2. Though of less importance in the atmosphere, alkyloxy radicals may also react with NO_2 and NO, similar to RO_2, and possible isomerization of alkyloxy radicals to form alcohols has also been hypothesized in the literature.

Besides chain alkanes, cyclic alkanes, unsaturated hydrocarbons such as isoprene, monoterpenes, and aromatic compounds may also be degraded in the atmosphere. For unsaturated hydrocarbons, the addition of atmospheric oxidants (OH, O_3, NO_3, NO_2, O) to the double or triple bonds initiates degradation processes for most such molecules, similar to reactions (R2.2.18) and (R2.2.19). Though the database of kinetic reactions for unsaturated hydrocarbons is no better than for chain alkanes, the decomposition in the atmosphere is much faster for the former. Reaction (R2.2.36) illustrates the oxidation of propene by OH:

$$OH + CH_3CH{=}CH_2 \rightarrow (f)CH_3CHCH_2OH + (1 - f)CH_3CH(OH)CH_2 \tag{R2.2.36}$$

where the branching ratio (f) is estimated to range from 0.65 to 0.87 in the literature. Reaction (R2.2.36) proceeds ~ 10 times faster than propane in PBL, and the two β-hydroxyalkyl radicals formed in the reaction are at least as active as the propyl radical. In fact, unsaturated hydrocarbons are mostly decomposed within PBL, and intermediate products from their decomposition, namely oxygenated hydrocarbons, may be transported to the global atmosphere.

2.2.3 Oxygenated hydrocarbons

Oxygenated hydrocarbons in the atmosphere are an important group of chemicals. In the gas phase, this group includes carbonyls, organic nitrates, peroxyacyl nitrates, organic hydroperoxides, alcohols, and carboxylic acids. In the aqueous phase, organic sulfate and nitrogen compounds have been reported. In the aerosol phase, this group is collectively termed as OC, which is often twice the amount of carbon mass without any oxygen. Some of the oxygenated hydrocarbons are produced as intermediates from the decomposition of hydrocarbons, and the rest may be directly emitted from natural sources and anthropogenic activities. Their transformations in the atmosphere are outlined below.

2.2.3.1 Carbonyls

Among oxygenated hydrocarbons in the atmosphere, carbonyls [RCHO, $R_1C(O)R_2$] may be rapidly photolyzed to form alkyl and acyl radicals with at least one carbon less than the original molecules, such as HCHO in reactions (R2.2.9) and (R2.2.10). Photolysis of CH_3CHO, which may be formed from radical products of reaction (R2.2.36), proceeds similarly to HCHO with two branches – one produces two radicals, and the other produces "stable" compounds:

$$CH_3CHO + hv \rightarrow CH_3 + CHO \qquad\qquad (R2.2.37a)$$

$$CH_3CHO + hv \rightarrow CH_4 + CO \qquad\qquad (R2.2.37b)$$

In general, photolysis of aldehyde proceeds via reaction (R2.2.38) and photolysis of ketone proceeds via reaction (R2.2.39):

$$RCHO + hv \rightarrow R + CHO \qquad\qquad (R2.2.38a)$$

$$RCHO + hv \rightarrow RH + CO \qquad\qquad (R2.2.38b)$$

$$R_1C(O)R_2 + hv \rightarrow R_1C(O) + R_2 \qquad\qquad (R2.2.39a)$$

$$R_1C(O)R_2 + hv \rightarrow R_1 + R_2C(O) \qquad\qquad (R2.2.39b)$$

$$R_1C(O)R_2 + hv \rightarrow R_1 + R_2 + CO \qquad\qquad (R2.2.39c)$$

There are other possible branches for photolysis of larger aldehydes reported in the literature, such as the production of ethene from butanal, but details remain to be explored.

The oxidation of carbonyls in the atmosphere is mainly initiated by OH abstraction of α-H, followed by subsequent dissociations. Reaction (R2.2.40) shows OH degradation of a general aldehyde and reaction (R2.2.41) outlines a pathway for the degradation of a general ketone:

$$RCHO + OH \rightarrow RC(O) + H_2O \tag{R2.2.40}$$

$$R_1CH(R_2)C(O)R_3 + OH \rightarrow R_1C(R_2)C(O)R_3 + H_2O \tag{R2.2.41a}$$

$$R_1C(R_2)C(O)R_3 + O_2 + M \rightarrow R_1CO_2(R_2)C(O)R_3 + M \tag{R2.2.41b}$$

$$R_1CO_2(R_2)C(O)R_3 + NO \rightarrow NO_2 + R_1CO(R_2)C(O)R_3 \tag{R2.2.41c}$$

$$R_1CO(R_2)C(O)R_3 \rightarrow R_1C(O)R_2 + CO + R_3 \tag{R2.2.41d}$$

$$R_1CO(R_2)C(O)R_3 \rightarrow R_1 + R_2C(O)C(O)R_3 \tag{R2.2.41e}$$

$$R_1CO(R_2)C(O)R_3 \rightarrow R_2 + R_1C(O)C(O)R_3 \tag{R2.2.41f}$$

The alkyl glyoxal may dissociate under sunlight to form alkyl radicals and CO, and so on.

2.2.3.2 Organic nitrates

Organic nitrates ($RONO_2$) may be formed via reaction (R2.2.24b), where the branching ratio is small for CH_3ONO_2 and $C_2H_5ONO_2$, but significant when R contains three or more carbons. The yield of $RONO_2$ increases with the number of carbons in R in reaction (R2.2.24). For example, during propene oxidation, the yield of hydroxyl organic nitrates with three carbons is $\sim 2\%$ of nitrate; for isoprene oxidation, the corresponding yield is estimated to be $\sim 12\%$ when R contains five carbons. Empirical formulae may be fitted from measured pairs of carbon numbers in R and the yield of $RONO_2$ to estimate the yield of an unmeasured organic nitrate.

Organic nitrates may be oxidized by OH and other oxidants, and may decompose via photolysis. Rate constants of unmeasured $RONO_2$ may be estimated from measured compounds with similar structures. For example, based on measured rate constants of the oxidation of C_1–C_4 unsubstituted n-alkyl nitrates by OH, as shown in reaction (R2.2.42), an empirical formula (F2.2.9) may be obtained for estimating the rate constant of the oxidation of an unmeasured alkyl nitrate by OH:

$$OH + RCH_2ONO_2 \rightarrow H_2O + RCHO + NO_2 \tag{R2.2.42}$$

$$k(R2.2.36) = 2.3 \times 10^{-14} \cdot [1 + 1.06456 \cdot (n^3 - 1)] \tag{F2.2.9}$$

In formula (F2.2.9), n denotes the number of carbons in n-alkyl nitrates, and the units of k are cm^3 molecule^{-1} s^{-1}. Available data suggest that 2-alkyl nitrates and β-oxo-n-alkyl nitrates are oxidized by OH at about half the speed compared with corresponding n-alkyl nitrates, and Cl oxidizes alkyl nitrates over an order of magnitude faster than OH. The aldehyde RCHO formed in reaction (R2.2.42) may be decomposed in the atmosphere, as discussed earlier.

2.2.3.3 Peroxyacyl nitrates

Peroxyacyl nitrates (PANs) represent a special reservoir for reactive nitrogen in the global atmosphere. Their formation may be initiated by the OH oxidation of aldehydes (R2.2.40), and realized by:

$$RC(O) + O_2 + M \rightarrow RC(O)O_2 + M \qquad (R2.2.43)$$

$$RC(O)O_2 + NO_2 \leftrightarrow RC(O)O_2NO_2 \qquad (R2.2.44)$$

In summer, PANs rapidly dissociate back to peroxyacyl radicals in the PBL. When air temperature is low, the equilibrium moves towards the formation of PANs, which may then be transported to remote oceanic areas. PANs may be photolyzed to form NO_2 and NO_3, and oxidized by OH, which abstracts a β-H if no unsaturated bond is available for addition. The latter may proceed similarly to ketone oxidation (R2.2.41).

2.2.3.4 Organic hydroperoxides

Organic hydroperoxides with two or more carbons (ROOH) are formed via reaction (R2.2.25) in the atmosphere. Rate constants of the formation reactions are estimated to be $\sim 1.5 \times 10^{-11}$ cm^3 molecule^{-1} s^{-1} when no measurement is available. In the atmosphere, ROOH may degrade via a photolysis reaction:

$$ROOH + hv \rightarrow RO + OH \qquad (R2.2.45)$$

The photolysis reaction (R2.2.45) produces RO, which decomposes rapidly to carbonyls with at least one carbon less than ROOH. ROOH is also oxidized by OH via reaction (R2.2.46):

$$ROOH + OH \rightarrow H_2O + RO_2 \qquad (R2.2.46)$$

Combining reactions (R2.2.25) and (R2.2.46), an effective pathway is formed: $OH + HO_2 \rightarrow H_2O + O_2$. Along this pathway, two HO_x radicals are terminated, and ROOH has no change. Thus, ROOH formation is a partial sink for oxidants.

2.2.3.5 Alcohols

Alcohols may be emitted by anthropogenic activities and are also produced in the atmosphere via reactions (R2.2.27b), (R2.2.28b). For a radical $RR'CHO_2$, the yield of alcohol $RR'CHOH$ is lower than carbonyl $RR'C(O)$, due to the branch that forms alkyloxy radical $RR'CHO$; the latter may also form $RR'C(O)$. Measurements indicate that, in reactions between RO_2 radicals, while CH_3O_2 yields $\sim 1/3CH_3OH$ and $\sim 2/3HCHO$, when R contains 2–3 carbons, corresponding yields decrease for alcohols, e.g. 1/4–1/8 alcohols and 3/4–7/8 aldehydes.

Alcohols are oxidized by OH in the gas phase via:

$$R_1R_2CHOH + OH \rightarrow R_1R_2CHO + H_2O \tag{R2.2.47a}$$

$$R_1R_2C(R_3)OH + OH \rightarrow H_2O + R_1R_2C(R_3)O \tag{R2.2.47b}$$

The OH oxidation of a primary or secondary alcohol yields a corresponding aldehyde, as shown in reaction (R2.2.47a). For a tertiary alcohol, a tertiary alkyloxy radical is formed, as shown in reaction (R2.2.47b). The tertiary alkyloxy radical, also a product of reaction (R2.2.29), decomposes rapidly to form carbonyl molecules (R_1R_2CO, R_1R_3CO, R_2R_3CO) and alkyl radicals (R_3, R_2, R_1). The rate constants of OH oxidation reactions have been measured for some alcohols, and an empirical formula (F2.2.10) is fitted for estimating rate constants of unmeasured n-alkyl alcohols (ROH) with five or more carbons:

$$k(ROH + OH) = 9 \times 10^{-13} \cdot (-0.646 + 2.03 \cdot n^{1.5} - 0.384 \cdot n^2) \tag{F2.2.10}$$

where n denotes the number of carbons in ROH, and the units of k are cm^3 molecule^{-1} s^{-1}. Alcohols are water soluble, so may also be oxidized in liquid water in the atmosphere. Besides the –OH functional group, other parts of alcohols also react in the atmosphere, similar to hydrocarbons discussed earlier.

2.2.3.6 Carboxylic acids

Carboxylic acids are formed via reactions (R2.2.31b), (R2.2.32b), and (R2.2.34b) in the atmosphere, from photo-oxidation intermediates of hydrocarbons and carbonyls. In the gas phase, the –C(O)OH functional group of a carboxylic acid may be stripped by the OH abstraction of the α-H followed by the departure of CO_2:

$$RC(O)OH + OH \rightarrow H_2O + CO_2 + R \tag{R2.2.48}$$

The rate constants of reaction (R2.2.48) have been measured for some carboxylic acids, with values ranging from 4.5×10^{-13} for formic acid to 1.2×10^{-12} for propionic acid at 298 K. Thus, the chemical lifetime of organic acids is of the order of a week in the gas phase. As organic acids are intermixable with water, hundreds of carboxylic acids were detected in rain samples in China in the late 1980s. Thus, heterogeneous loss may be important for organic acids in the atmosphere, especially over areas with significant rainfall.

2.2.4 Reactivities

Reactivities are defined as extra pollutants, such as O_3, particulate matter (PM), acidity, CO_2 (eq), and toxicity, formed within a certain time period per unit of an organic compound RH emitted to the atmosphere. O_3 reactivity has been studied extensively for several decades in the USA, especially in California, where hundreds of hydrocarbons have been regularly modeled for their potential to produce extra O_3 in photochemical conditions conducive for short-term, severe O_3 pollution. It has been found that, in general, O_3 reactivity is smallest for small alkanes, such as CH_4 and C_2H_6, which are larger than CO. O_3 reactivity is greater for larger alkanes, such as C_3H_8 and C_8H_{18}, and largest for alkenes and aromatics, such as 1,3-butadiene, isoprene, toluene, and xylenes. Aldehydes usually have higher O_3 reactivity than (other oxygenated) hydrocarbons with the same number of carbons.

As anthropogenic combustion becomes fairly complete ($\sim$99%) nowadays, emissions of organic compounds result in little enhancement of CO_2 in the atmosphere. However, some chemicals have been found to absorb spectral windows, left by CO_2, efficiently near the tropopause where H_2O is scarce, and thus to serve as greenhouse gases together with H_2O and CO_2. Global warming potentials of greenhouse gases have been thoroughly studied since 1990, when the first scientific assessment on climate change was reported by the IPCC, and corresponding research led to the Nobel Peace Prize being shared by the IPCC in 2007. Dominant greenhouse gases in the modern atmosphere are CO_2, CH_4, N_2O, chlorofluorocarbons (CFCs), hydrochlorofluorocarbons (HCFCs), SF_6, and other perfluorocarbons (PFCs), and their 100-year global warming potentials compared with CO_2 are 1, 21, 310, 3500 (CFC_{11}), $\sim$7300 (CFC_{12}), 650 (HFC_{32}), $\sim$11,700 (HFC_{23}), and 23,900 (SF6). In California, NF_3 was recently identified as a greenhouse gas with significant emissions. Due to the difference in timescale, only CH_4 and HCFCs participate in tropospheric chemistry, while others are active in stratospheric chemistry only.

Reactivities for secondary organic aerosols (SOAs), acidity buffering, and toxicity could be assessed similar to O_3 and CO_2 (eq), when corresponding information is available. Currently, SOA chemical mechanisms are being actively researched. Compared with the O_3 chemical mechanism that is constrained by thousands of smog chamber experiments, the SOA chemical mechanism was recently updated in California for aromatics based on hundreds of chamber experiments specially designed for aerosol mechanisms. Research on aerosol and rain acidities buffered by organic acids is related to the SOA mechanism, as is the research on toxicity.

2.3 Heterogeneous reactions

Heterogeneous reactions are important in both the troposphere and stratosphere. In the stratosphere, polar stratospheric clouds are responsible for the formation of the Antarctic "O_3 hole". In the troposphere, heterogeneous reactions involving gas–liquid–solid phases are closely related to air quality problems. Despite the aqueous and solid phases occupying less than 10^{-6} and 10^{-10} respectively of atmospheric volume fractions on average, there are many chemical reactions occurring in these media.

2.3.1 Gas–aqueous reactions

Gas–aqueous reactions may occur when liquid water is present in the atmosphere, such as in clouds or aerosols. In the modern atmosphere, a number of chemicals, such as OH and HO_2 radicals, H_2O_2, SO_2, O_3, N_2O_5, HNO_3, NH_3, alcohols, aldehydes, and carboxylic acids, may transfer between the gas phase and liquid droplets.

Gas–aqueous transfer in the atmosphere is illustrated in Figure 2.2. The transfer consists of three steps: (1) diffusion from gas phase to the gas–droplet interface; (2) equilibrium between the gas and liquid phases; and (3) diffusion within droplets. For example, OH radical and HCOOH may be homogeneous in the gas phase beyond a distance from the center of a cloud droplet. The distance is denoted as R in the figure. Mixing ratios of OH radical and HCOOH decrease with R until the radius of the droplet (R_p) is reached, due to diffusion from the gas phase and absorption by the liquid phase. At the surface of the droplet, concentrations of OH (aq) radical and HCOOH (aq) are governed by Henry's law, with respect to corresponding gas-phase mixing ratios. It is seen that the ratio is much higher for HCOOH than for OH. From the surface to the center of the droplet, OH (aq) and HCOOH (aq) concentrations decrease and then may be constant due to chemical reactions within the droplet. It needs to be clarified that the aqueous phase is sometimes a net source for HCOOH; in that case, trends are the opposite to what is depicted in the figure.

Heterogeneous reactions proceed faster than gas-phase reactions for S(IV), HO_x, H_2O_2, aldehydes, and organic acids. For example, SO_2 may be oxidized into sulfate in the aqueous phase within an hour when enough oxidants are available; in the gas phase, its lifetime is several days. With wet aerosols, the SO_2 oxidation rate may be doubled compared with the case without wet aerosols. The presence of aqueous phase in the atmosphere may deplete HO_2 and HCHO rapidly from the gas phase. However, if an air parcel with abundant pollutants passes through a cloud within a few hours, the air parcel may eventually produce nearly as much O_3 as if it did not encounter clouds, since NO_x species are insoluble and may be the limiting precursor for O_3 formation.

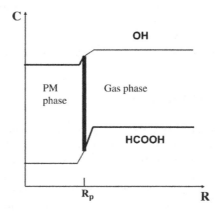

Figure 2.2 Gas–droplet transfer processes.

2.3.2 Gas–solid reactions

Gas–solid reactions are believed to be responsible for the "O_3 hole" observed in the stratospheric vortex over the Antarctic continent in early spring.

In the stratosphere, O_3 is formed via reactions (R2.1.4) and (R2.1.5). The normal loss reactions that keep O_3 at a healthy level, ~ 300 Dobson units, include reactions (R2.3.1)–(R2.3.7), as shown below, besides reactions (R2.1.1), (R2.1.18), and (R2.1.19):

$$O_3 + O \rightarrow 2O_2 \qquad\qquad (R2.3.1)$$

$$O_3 + OH \rightarrow HO_2 + O_2 \qquad\qquad (R2.3.2)$$

$$O + HO_2 \rightarrow OH + O_2 \qquad\qquad (R2.3.3)$$

$$O_3 + NO \rightarrow NO_2 + O_2 \qquad\qquad (R2.3.4)$$

$$O + NO_2 \rightarrow NO + O_2 \qquad\qquad (R2.3.5)$$

$$Cl + O_3 \rightarrow ClO + O_2 \qquad\qquad (R2.3.6)$$

$$ClO + O \rightarrow Cl + O_2 \qquad\qquad (R2.3.7)$$

During winter, air temperature in the stratospheric vortex over the Antarctic continent is low (~ 180 K). In the absence of sunlight, O_3 there is transported from the tropical stratosphere, and reactive nitrogen and chlorine species are converted into corresponding reservoirs HNO_3 (s), HCl (s), and $ClONO_2$. Together with H_2SO_4 (s) and H_2O (s), polar stratospheric clouds are formed in the vortex during winter. At the surface of polar stratospheric cloud (PSC) particles, reaction (R2.3.8) is believed to proceed to accumulate Cl_2, while solid particles precipitate down to the troposphere:

$$HCl(s) + ClONO_2 \xrightarrow{\text{psc}} Cl_2 + HNO_3(s) \qquad\qquad (R2.3.8)$$

In the first few weeks of spring, Cl_2 is rapidly photolyzed to form abnormally high amounts of Cl radicals that may destroy most O_3 in the vortex to form the observed O_3 hole over the Antarctic continent. This finding resulted from numerous laboratory, field, and theoretical studies, which were cited for the Nobel Prize in Chemistry in 1995.

2.4 Ozone photochemical mechanisms

Tropospheric O_3 is formed mainly from photochemical reactions of hydrocarbons and NO_x, as illustrated in Figure 2.3. Photolysis of O_3 produces O^1D, which is converted to OH by H_2O; this step serves as the radical initiating step. After initiation, OH

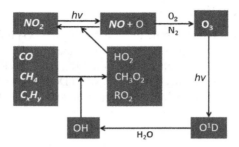

Figure 2.3 Formation pathway of bad ozone near the surface.

radicals oxidize CO and organics, which propagate and multiply radicals to form peroxyl radicals. Peroxyl radicals oxidize NO into NO_2 and are transformed into OH and alkyloxy radicals, which propagate and multiply radicals. NO_2 photolyzes back to NO and produces O, and the latter combines with O_2 to form O_3. The cycling between NO_2 and NO is rapid in the troposphere, compared with the formation of nitric acid. Thus, for each NO_x molecule converted to nitric acid, dozens of O_3 molecules may be formed in the troposphere.

Systematic photochemical mechanisms for simulating O_3 formation in the urban, regional, and global atmosphere emerged in the late 1980s, and representative sets are Carbon-Bond Mechanisms (CBM series), California State Air Pollution Research Center Photochemical Mechanisms (SAPRC series), Regional Acid Deposition Mechanisms (RADM family), and Master Chemical Mechanisms (MCM series), amongst others. The inorganic reaction sets included in these mechanisms are similar to those listed in Chapter 6, Table 6.3; but degradations of hundreds of organic compounds observed in the PBL are represented differently.

Among the four O_3 mechanisms, the CBM chemical mechanism groups most organic compounds according to their structures, and is the most condensed photochemical mechanism. In this mechanism, alkanes and most of the alkyl groups in other organics are represented by a single model species PAR ($\sim CH_3$). The alkene functional group is represented by a model species OLE ($\sim C_2H_4$) for 1-alkenes, and with a model species ALD2 ($\sim CH_3CHO$) for 2-alkenes, assuming immediate transformation of $CH_3CH=$ to ALD2. Model species (PAR, OLE, ALD2) have been called "carbon-bond surrogates" by various authors, but are commonly termed "structural surrogates". All mono-alkylbenzenes are represented by a model species TOL similar to toluene, and di- and tri-alkylbenzenes are represented by a model species XYL similar to m-xylene. Model species (TOL, XYL) are termed "molecular surrogates". Only three organics, namely HCHO, C_2H_4, and isoprene, are represented explicitly for their degradations. Other organics, such as CH_4, ketones formed from PAR, and dicarbonyls formed from aromatics, were treated in the extended version of the CBM mechanism, but were unnecessary for simulating data acquired from smog chamber experiments.

The SAPRC mechanism represents alkanes, biogenic and anthropogenic alkenes, and aromatics more explicitly than the CBM mechanism, using the "lumped molecules" method. In a recent version implemented for regulatory air quality modeling, five model species are used to represent alkanes according to their reaction rates with OH, and each model species represents a weighted average of alkanes with OH reaction rates within its range ("lumped molecules"). Besides isoprene and monoterpenes, model species are used to represent 1-alkenes, internal alkenes, mono- and multi-substituted benzenes. The numbers of model species and reactions in the SAPRC mechanism are about twice that in the CBM mechanism. Both mechanisms have been evaluated with available data from thousands of indoor and outdoor smog chamber experiments, and have been used for regulatory O_3 air quality simulations in the USA.

The RADM mechanism is similar to the SAPRC mechanism in the number of model species and reactions, but use "molecular surrogates" to represent most organics. Emissions of organics need to be converted into that of corresponding "molecular surrogates". For example, hexane emission needs to be converted into pentane emission with a conversion factor. Atmospheric chemical models using the RADM mechanism simulate photochemical reactions of surrogate molecules, as if converted molecules do not exist. The RADM mechanism has also been evaluated with smog-chamber experimental data.

The MCM mechanism is the largest among the four. It represents over 100 organics nearly explicitly, which results in ~ 5000 model species and $\sim 14,000$ reactions. Compared with condensed mechanisms, the MCM mechanism describes many intermediate and terminal chemicals, and is especially suitable for modeling other air pollutants, such as secondary organic aerosols and toxicity, besides O_3. The MCM mechanism has been compared with condensed mechanisms, and has been implemented in regional and global models to simulate air quality and atmospheric chemistry (Jacobson and Ginnebaugh, 2010).

Summary

In this chapter, chemical reactions in the atmosphere have been discussed for important inorganic and organic compounds. Degradations of organic compounds in the troposphere have been elaborated for CH_4, higher hydrocarbons, and oxygenated organics. Heterogeneous reactions have been described for gas–aqueous reactions in the troposphere and gas–solid reactions responsible for the "O_3 hole". Systematic photochemical mechanisms have been described, especially for regulatory O_3 air quality modeling.

Exercises

1. Historically, the stratosphere only contained O_2, O_3, O, and O^1D besides inert N_2, and the null cycle mentioned in this chapter and reaction (R2.1.5) were all the reactions in the stratosphere. Survey the literature, and estimate the "O_3 layer" in Dobson units during that era. Using reactions (R2.1.18) and (R2.1.19) to estimate how much

halogen is needed to bring down the historical O_3 layer to the level in the modern stratosphere.

2. An effective pathway for CH_4 oxidation in the atmosphere may be expressed as: $CH_4 + 2O_2 \xrightarrow{hv} CO_2 + 2H_2O$. Survey the literature, and write elementary reactions responsible for the effective pathway in typical PBL conditions.

3. Methylglyoxal may be photolyzed under visible light. Survey the literature, and write its reactions in a zero-pollution atmosphere where the air contains the "dry air" chemicals and H_2O only.

4. Assume a fictitious chemical molecule AB absorbs 10 times the amount of energy as a CO_2 molecule from Earth's long-wave radiation, but its lifetime in the atmosphere is the same as CO_2. Answer without calculation what the global warming potential of AB is, if scaled to that of CO_2. Now, assume that climate change has prolonged the lifetime of AB by 10 times, but the lifetime of CO_2 stays at ~ 100 years in the atmosphere. Estimate the global warming potentials of AB relative to CO_2 at 100 and 1000 years respectively.

5. Find the most recent version of the SAPRC photochemical mechanism, and analyze its strengths and weaknesses in representing O_3 reactivities of hundreds of organics emitted from anthropogenic activities on urban, regional, and global scales.

3 Radiation in the atmosphere

Radiation is one of the most important constituents in the atmosphere. In the sky, it is responsible for wonderful visual images such as rainbows, coronas, auroras, and mirages. Near the surface, it maintains comfortable temperature ranges for plants, animals, and humans. It also provides photons to initiate photochemical reactions in the atmosphere.

3.1 Distribution of the solar spectrum

The Sun emits electromagnetic waves that center at visible light and range from infrared radiation to γ rays. From the ground, visible light is observable in many ways. At an altitude, towers, balloons, aircrafts, and rockets have been used for over a century to observe various parts of the solar spectrum at limited locations for a limited time. In the last 50 years, satellites have been employed to provide long-term observations of the solar spectrum in space.

Chemical Modeling for Air Resources. http://dx.doi.org/10.1016/B978-0-12-408135-2.00003-3

3.1.1 Black-body radiation

Black-body radiation, also termed Planck's law, determines the intensity of a radiation (I_e) at a wavelength (λ) from the temperature (T) of the emitter, if the latter is a perfect absorber and emitter (black body):

$$I_e(\nu,\ T) \ = \ 2h\nu^3 c^{-2}/\{\exp[h\nu/k(B)T] - 1\} \tag{F3.1.1}$$

where h and $k(B)$ are Planck's and Boltzmann's constants respectively, and c is the speed of light, while $\nu = c/\lambda$. Note that if the units of I_e are W m^{-2} nm^{-1} sr^{-1}, then the unit of λ is nm. The "sr" denotes steradian, the unit of solid angle.

The Sun is very similar to a black body. Despite significant variations of the solar spectrum near the surface due to many atmospheric processes, especially clouds, total solar radiation in space has been found to be fairly constant. For example, at 645 km ASL, the daily average total solar radiation ranged from 1357 to 1362 W m^{-2}, with an average of 1361 W m^{-2}, from March 2003 to July 2012, according to satellite measurements in a recent mission, the Solar Radiation and Climate Experiment (SORCE) in the USA. Long-term satellite measurements in earlier missions recorded a similar value (1360 W m^{-2}). Satellite-observed solar spectrum during the SORCE mission is illustrated (solid line) in Figure 3.1 for X-ray–ultraviolet–visible–infrared wavelengths (0.1–40 nm, 115–2400 nm). For comparison, the theoretical spectrum from Planck's law is also shown (dashed line) at $T = 5750$ K. The correlation coefficient between observed spectrum and theoretical curve was 0.96 over the observed range.

In the 220–800 nm range, the theoretical spectrum, denoted as "Planck's law 1", does not capture observations well, as shown in Figure 3.2. The observed solar radiation may be fitted against wavelength using the following formula using the least-squares constraint:

$$\ln (I_e(\lambda)) \ = \ 114.7918 + 0.0092\lambda - 17.0772 \ln \lambda - 6291.2504/\lambda \tag{F3.1.2}$$

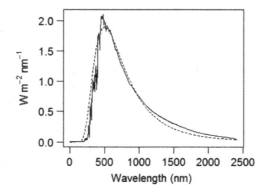

Figure 3.1 The observed and theoretical solar spectrum.

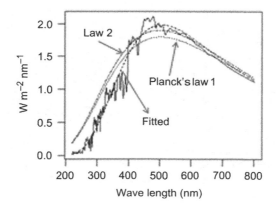

Figure 3.2 Observed, theoretical, and fitted solar spectrum from 220 to 800 nm.

The fitted curve is also shown in Figure 3.2, together with a curve estimated using the formula (F3.1.3), denoted as "Law 2", which performs better than Planck's law but worse than the fitted curve:

$$I_e(\lambda, T) = 2h\nu^3 c^{-2}/\{\exp[h\nu/k(B)T] - 1\} \qquad (F3.1.3)$$

It is shown that the fitted curve is closer to observations, especially for wavelengths <400 nm. Of course, formulae (F3.1.2) and (F3.1.3) are only suitable for the observations described above.

3.1.2 Atmospheric protection

The modern atmosphere protects humans at the surface from harmful electromagnetic waves with wavelengths <300 nm. As ultraviolet–visible–infrared photons emitted from the Sun enter the top of the atmosphere at ∼100 km ASL, N_2 absorbs most harmful photons with wavelength <100 nm within the thermosphere. Meanwhile, O_2 absorbs harmful photons with wavelengths <245 nm from the top of the atmosphere down to the stratosphere. For harmful photons with wavelengths >245 nm, O_3 is the most important absorber. Figure 3.3 shows the first group of photographs that recorded solar UV spectrum from a rocket rising up to 55 km ASL in the USA on October 10, 1946 (USA National Aeronautics and Space Administration). From the pictures, the shortest UV wavelength decreased from >300 nm at 2 km to ∼300 nm at 8 and 17 km. At 34 km, the shortest UV wavelength decreased to ∼280 nm. UV radiation with wavelengths <240 nm was captured at 55 km ASL. Thus, it is shown that significant harmful UV was present in the stratosphere and upwards, but UV with wavelengths <300 nm was undetected near the surface.

With solar radiation, the modern atmosphere also serves as a natural photochemical reactor for the transformation of anthropogenic emissions. For example,

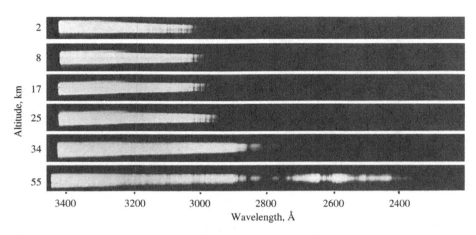

Figure 3.3 The first group of photographs that recorded the solar UV spectrum from a rocket rising up to 55 km ASL in the USA on October 10, 1946. The horizontal axis is wavelength in Å and the vertical axis is altitude in km.
Photographs are from NOAA, USA.

anthropogenic activities injected large amounts of oxidized N and S compounds and organic halogens into the atmosphere in the last century. Without transformations, these compounds could have accumulated to dangerous levels in the atmosphere. In the case of organic halogens, photochemical reactions in the stratosphere transformed them into inorganic halogens, but the latter were found to be responsible for the stratospheric "O_3 hole" observed over the Antarctic continent in the first few weeks of local spring in the last few decades. The finding of the O_3 hole and its causes alarmed global communities to "repair" the O_3 hole by reducing anthropogenic emissions of organic halogens; this effort is similar to an ancient Chinese legend "Lady Nü-wa patches the sky". In the case of oxidized N and S compounds, photochemical reactions in the troposphere have converted them into soluble products (NO_3^-, SO_4^{2-}) with a time scale of days. With proper amounts of NH_3, NO_3^- and SO_4^{2-} may be further transformed into fertilizers and rapidly deposited on to the surface. Of course, there are concerns for public health on human exposure to fertilizers in breathing zones.

3.1.3 Sunlight at the surface

Complex atmospheric processes may dramatically alter the solar spectrum that reaches the Earth's surface. For example, gas-phase H_2O and CO_2 are strong absorbers of solar infrared radiation. In the visible range, precipitation and sandstorms reduce solar radiation. Human activities used to emit all incompletely combusted black plumes into industrial regions, and are still responsible for unhealthy air pollutions, such as brown haze, over many metropolitan areas. In the ultraviolet range, O_3 destructive substances invented half a century ago have caused the formation of a polar stratospheric O_3 hole over the Antarctic and Australian continents.

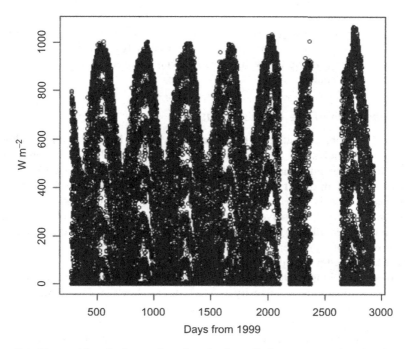

Figure 3.4 Observed hourly time series of total solar radiation at a ground station in Fresno city, California, USA from September 30, 1999 to December 31, 2006.
Data are from California Air Resources Board, USA.

Figure 3.4 illustrates the solar spectrum observed at the surface of a PM supersite for air quality studies in Fresno city, California, USA from late September 1999 to December 2006. In the 87-month hourly time series, total solar radiation at the ground peaked at 1063 W m^{-2} on an hourly averaged basis. Monthly averages of total solar radiation ranged from 75 W m^{-2} in January to 333 W m^{-2} in July, and the average diurnal cycle peaked at noon with 683 W m^{-2}. The average total solar radiation reaching the ground of this city during the 8-year observational period was 211–215 W m^{-2} depending on the averaging methods. As the solar energy at the top of the atmosphere was ∼340 W m^{-2}, 62–63% reached the ground during the observational period. This value is close to the global average, where the Earth's surface absorbs ∼50% of solar radiation.

3.2 Long-wave radiation

The Earth emits long-wave radiation constantly, similar to the Sun that emits short-wave radiation. After emission from the condensed Earth, most long-wave photons are first absorbed by H_2O, CO_2 and other greenhouse gases in the troposphere.

Then, these greenhouse gases emit long-wave photons at different temperatures aloft. On an annual average basis, the energy of the long-wave radiation that escapes from the top of the atmosphere equals the energy of the short-wave radiation received from the Sun; any significant deviation from the balance may result in global climate change.

3.2.1 Spectrum of the Earth's radiation

As almost every object emits electromagnetic waves, the Earth also emits radiation. According to Planck's law, the radiation spectrum of a black body depends on its temperature. As the temperature of the Earth's surface changes from <220 K in polar winter to >320 K in desert summer, the spectral distribution of Earth's long-wave radiation at source is expected to change significantly, in contrast to the nearly constant solar spectrum observed by satellites in space. Of course, observations on solar activities are still very limited in terms of spatial coverage for the Sun, despite the involvement of a number of satellites, while the spatial coverage for the Earth is relatively much better now due to satellites.

The theoretical long-wave spectrum of the Earth's radiation at the surface centers at 9–13 μm with significant intensity at 5–30 μm, as shown in Figure 3.5. It is seen that the peak intensity differs by an order of magnitude between polar winter and desert summer. Global observations of long-wave radiation from Earth have been conducted from the ground throughout the top of the atmosphere by satellites, e.g. Nimbus, and the data support Planck's law. Satellite data also include strong absorptions of the long-wave radiation by greenhouse gases, which is in contrast to short-wave radiation observed in space.

Despite significant differences in the long-wave radiation emitted from polar winter and desert summer, the energy absorbed by the Earth at the top of the atmosphere must be in close balance with the energy escaping from the top of Earth's atmosphere on a multi-year average basis. Using the solar constant/average albedo pair of 1360 W m^{-2}/0.3, the Earth needs to release energy amounting to 238 W m^{-2}

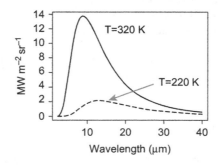

Figure 3.5 Theoretical spectrum of long-wave radiation from a black body at temperatures of 220 K (dashed line) and 320 K (solid line).

to space on a global, multi-year average basis, which corresponds to the emission of a black body at ~ 255 K:

$$E(T) = k(\text{SB}) \cdot T^4 \qquad \text{(F3.2.1)}$$

where E denotes the energy flux, in units of W m^{-2}, from a black body with temperature T (K), and $k(\text{SB})$ is the Stefan–Boltzmann constant (5.6705×10^{-8} W m^{-2} K^{-4}).

3.2.2 Greenhouse gases

Greenhouse gases in the atmosphere increase the temperature at the surface. Photons of the infrared radiation emitted from the Earth's surface have wavelengths ranging from 4 to 100 μm. The corresponding energy levels are in the range of the vibrational–rotational spectrum for a number of minor and trace chemicals, termed greenhouse gases, in the atmosphere, such as H_2O, CO_2, CFCs and their substitutes, CH_4, N_2O, and O_3. These chemicals firstly absorb long-wave radiation emitted from the Earth's surface, and convert it into vibrational–rotational energy at a molecular level. Secondly, molecules in vibrationally or rotationally excited states emit thermal long-wave photons back to the atmosphere in all directions. About half of the photons thermally emitted by excited molecules return to the Earth's surface, and the other half are emitted upwards. Thirdly, the upwardly mobile photons are absorbed by greenhouse gases above, and these three steps may iterate if there are extra greenhouse gases further above.

For example, if only one iteration described above occurs in the atmosphere, with the constraint of energy balance, as illustrated in Figure 3.6, the temperature at the

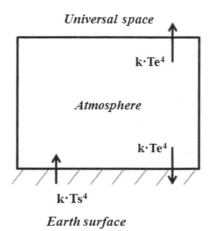

Figure 3.6 Hypothetical one-box, well-mixed atmosphere with greenhouse gases in energy balance with the Earth and universal space.

surface (T_s) will be related to the effective temperature of the Earth ($T_e = 255$ K) as follows:

$$k(\text{SB}) \cdot T_s^4 = 2k(\text{SB}) \cdot T_e^4 \tag{F3.2.2}$$

Solving the above equation yields $T_s = T_e \cdot 2^{0.25} = 303$ K. When extra greenhouse gases are present in the atmosphere to warrant the addition of the second iteration described in the last section, two vertical layers may be needed to solve for the surface temperature. In that case, $T_s = T_e \cdot 3^{0.25} = 336$ K!

There are several points to consider when looking at the effects of greenhouse gases. Firstly, there are significant differences in the concentration levels of greenhouse gases in the atmosphere. For example, H_2O is of the order of 0.1–1% in the PBL, and concentration of CO_2 is ~ 400 ppmv. Thus, H_2O is more important than CO_2 in the PBL from the perspective of greenhouse gas effects. However, in the tropopause, the CO_2 concentration is about the same as in the PBL, while H_2O is scarce due to precipitation and condensation (Marcy et al., 2007). Thus, the relative importance of CO_2 versus H_2O changes with altitude. Secondly, there are significant differences among greenhouse gases in strengths of absorption bands. For example, a CO_2 molecule absorbs photons with less energy than a molecule of O_3, CH_4 or other greenhouse gases. Thirdly, some chemicals without greenhouse effects may affect greenhouse gases via photochemical reactions. For example, an NO molecule emitted from the surface may facilitate the formation of several O_3 molecules in the tropopause. The former does not have a unique absorption spectrum important for a global warming effect, but the latter does. Lastly, besides greenhouse gases, aerosols (especially black carbon) have been demonstrated to play an important role in global warming at the surface. Mitigation measures that reduce black carbon concentration in the atmosphere may also reduce other air pollutants.

3.2.3 Climate change

On a global scale, climate changed rapidly in the last few decades, which coincides with a rapid increase in anthropogenic emissions of air pollutants, especially greenhouse gases, due to the globalization of industrialization and urbanization. While it is still a hot topic for research with regard to the exact causes of climate change, changes in radiation budgets could account for much of the recent temperature rise.

The total solar energy that reaches the top of the atmosphere is 1365 W m^{-2}, with oscillations of 5 W m^{-2}. This data originated from satellite observations (1370 W m^{-2}) in combination with estimates inferred from earlier sunspot number counts and cosmogenic isotopes, which provide estimates of solar radiation over 1000 years. The data suggests that solar radiation could have changed quasi-periodically with an amplitude of 0.24–0.30% per 100 years. Thus, total solar energy could have changed by 1.6–2.0 W m^{-2} at the top of the atmosphere in the last 50 years, which corresponds to a quasi-radiative force of 0.4–0.5 W m^{-2}.

Mixing ratios of the important greenhouse gases CO_2, CH_4, N_2O, and halocarbons are inferred to have been 280, 0.7, 0.18–0.26, and 0 ppm respectively in 1880, and were measured as being 379, 1.77, 0.32, and ~ 0.005 ppm respectively in 2005. The significant increase has been attributed to fossil fuel combustion since industrialization, which also emits black carbon, SO_2, NO_x, and ROGs. The latter chemicals produce sulfate aerosol and O_3 in the troposphere. Together, these greenhouse gases trap significant amounts of long-wave radiation from the surface, and could provide radiative force that is an order of magnitude larger than solar cycles, according to the IPCC.

Urbanization and associated changes in land surface and organisms over the land and in surface water have important implications for radiation budgets. Firstly, the surface albedo has been changed. For example, black carbon in snow and ice increases absorption efficiencies and reduces albedos. It is conceivable that a small change in albedo on a global scale, e.g. by 0.01, which corresponds to a radiative force of 3.4 W m^{-2}, may result in the same effect on radiative force as doubling the CO_2 mixing ratio in the atmosphere. Secondly, the albedo of the Earth may also be changed by altered properties of clouds and aerosols and increased level of black carbon particles. Thirdly, deforestation that changes surface albedo may increase the CO_2 mixing ratio in the atmosphere, as does anthropogenic pollution of surface waters.

3.3 Radiative transfer measures

Atmospheric radiation has been measured worldwide for over a century, and the spatial and temporal coverage of observations has expanded greatly since the invention of satellites about 50 years ago. Observable quantities include spectral light intensity and energy flux, which may be used to calculate heating rate and actinic flux.

3.3.1 The Beer–Lambert law

Most readers are assumed to be familiar with this law, which relates the decrease of the intensity of light along its path through a column of an absorbing agent. For example, if the attenuation of a direct beam from light intensity I_0 to I through a particle with diameter D_p, as shown in Figure 3.7(a), is due to absorption by its sole component chemical (i), then the Beer–Lambert law becomes:

$$I = I_0 \cdot \exp\left[-k(i) \cdot D_p \right], \quad k(i) = 4\pi A_i(\lambda)/\lambda \tag{F3.3.1a}$$

where $A_i(\lambda)$, commonly termed the imaginary index of the refraction of light at wavelength λ by chemical i, is close to 1 for black carbon particles and close to 0 for pure water droplets over the solar spectrum.

For atmospheric chemical modeling purposes, the Beer–Lambert law is written as follows:

$$dI(\lambda, \theta, \phi)/dZ_p = -I(\lambda, \theta, \phi) \cdot \sum_{i=1}^{n} [\sigma_{ext}(i, \lambda) \cdot C(i, Z_p)] \tag{F3.3.1b}$$

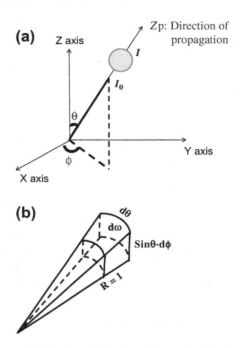

Figure 3.7 Illustrations of a light beam (a) and a solid angle (b) in three-dimensional coordinates.

where $I(\lambda, \theta, \phi)$ is the intensity of a light wave with wavelength λ from direction (θ, ϕ) in units of W m^{-2} sr^{-1}, as shown in Figure 3.7(a), and Z_p is the distance along the path of light propagation in meters. The extinction coefficient (cross-section) of $I(\lambda, \theta, \phi)$ by a chemical i is denoted by $\sigma_{ext}(i, \lambda)$ in units of m^2 mol^{-1}, and the concentration of chemical i at position Z_p is denoted by $C(i, Z_p)$ in units of mol m^{-3}. The "solid angle" ω, in units of steradians (sr), refers to an area on the surface of a sphere with unit radius, and its maximum value is the surface area of the sphere (4π). A small solid angle with the value of dω ($= \sin \theta \cdot \mathrm{d}\theta \cdot \mathrm{d}\phi$) is illustrated in Figure 3.7(b). In fact, the direct solar beam to the Earth's surface is within a very small solid angle (10^{-4}) due to the relative size of the Earth and its distance from the Sun. However, as scattered light exceeds direct solar beams near the surface, the solid angle is an important measure for calculating radiative transfer.

3.3.2 Actinic flux

In the atmosphere, some chemicals may be transformed via photolysis by absorbing photons from solar radiation. The first-order rate constant of a photolysis reaction,

$J(i)$, may be calculated as the sum of products of actinic fluxes and their absorption parameters over relevant wavelengths:

$$J(i) = \sum_{\lambda} \sigma_{abs}(i, \lambda) \cdot q(i, \lambda) \cdot F(\lambda) \tag{F3.3.2a}$$

$$F(\lambda) = \oint I(\lambda, \omega) d\omega = \int_{\theta=0}^{\pi} \int_{\phi=0}^{2\pi} I(\lambda, \theta, \phi) d\phi \sin\theta \, d\theta \tag{F3.3.2b}$$

where $\sigma_{abs}(i, \lambda)$ denotes the extinction coefficient (cross-section) of $I(\lambda)$ due to absorption by a chemical i, and $q(i, \lambda)$ denotes the corresponding quantum yield that represents the ratio of the number of molecules dissociated to the number of photons with wavelength λ absorbed by chemical i. $F(\lambda)$ is the actinic flux of photons with wavelength λ in units of W m^{-2}, and the notation is the same as in formula (F3.3.1). Note that actinic flux is non-negative by definition, and is not a vector.

3.3.3 Energy flux

An important measurable quantity that describes radiation fields in the atmosphere is "energy flux", also called "net flux" or "irradiance". The irradiance is defined as all radiation passing through a unit plane surface, e.g. the X–Y plane in Figure 3.7(a), along its normal direction, e.g. the Z axis, per unit time:

$$\overrightarrow{E(\lambda)} = E(\lambda) \cdot \overrightarrow{k} \tag{F3.3.3a}$$

where $\overrightarrow{k}$ denotes the unit vector along the vertical axis in Figure 3.7(a), and

$$E(\lambda) = \oint I(\lambda) \cos\theta \, d\omega = \int_{\theta=0}^{\pi} \int_{\phi=0}^{2\pi} I(\lambda, \theta, \phi) d\phi \cos\theta \sin\theta \, d\theta \tag{F3.3.3b}$$

If $\mu = \cos\theta$, then the above formula becomes

$$E(\lambda) = \int_{\mu=-1}^{1} \int_{\phi=0}^{2\pi} I(\lambda, \mu, \phi) d\phi\mu \, d\mu \tag{F3.3.3c}$$

The units of irradiance are W m^{-2}. According to the sign of μ, the irradiance may be split into upward (positive μ) and downward (negative μ) terms:

$$\overrightarrow{E(\lambda)} = \left[\int_{\mu=0}^{1} \int_{\phi=0}^{2\pi} I(\lambda, \mu, \phi) d\phi\mu \, d\mu + \int_{\mu=-1}^{0} \int_{\phi=0}^{2\pi} I(\lambda, \mu, \phi) d\phi\mu \, d\mu \right] \cdot \overrightarrow{k}$$

$$= \left[\int_{\mu=0}^{1} \int_{\phi=0}^{2\pi} I(\lambda, \mu, \phi) d\phi\mu \, d\mu \right] \cdot \overrightarrow{k} - \left[\int_{\mu'=0}^{1} \int_{\phi=0}^{2\pi} I(\lambda, \mu', \phi) d\phi\mu' \, d\mu' \right] \cdot \overrightarrow{k}$$

$$= E^{\uparrow}(\lambda) - E_{\downarrow}(\lambda)$$

$$\tag{F3.3.4}$$

where $\mu' = -\mu$ and $\overrightarrow{k'} = -\overrightarrow{k}$. The upward irradiance is denoted by $E^{\uparrow}(\lambda)$, and $E_{\downarrow}(\lambda)$ denotes downward irradiance.

Measurements of irradiance at two altitudes may be used to estimate the photolysis rate constant $J(i)$, if the change in irradiance is due to the absorption by chemical i only. In that case, formula (F3.3.1) becomes:

$$dI(\lambda)/dZ_p = -I(\lambda) \cdot \sigma_{abs} \cdot C(Z_p) \tag{F3.3.5a}$$

Differentiating the first two terms of formula (F3.3.3b) against a short vertical distance (dZ) yields

$$dE(\lambda)/dZ = \oint dI(\lambda)/dZ \cdot \cos\theta \cdot d\omega = \oint dI(\lambda)/dZ_p \cdot d\omega \tag{F3.3.5b}$$

Combining (F3.3.5a, b) and using the definition of actinic flux in (F3.3.2b) yields:

$$dE(\lambda)/dZ = -\left[\oint I(\lambda) \cdot d\omega \right] \cdot \sigma_{abs}(\lambda) \cdot C(Z_p) = -F(\lambda) \cdot \sigma_{abs}(\lambda) \cdot C(Z_p) \tag{F3.3.5c}$$

Combining the above formula with formula (F3.3.2a) yields a conditional relationship for photolysis rate constant $J(i)$:

$$J(i) = -\frac{1}{C(i, Z)} \cdot \sum_{\lambda} \left[q(i, \lambda) \cdot \frac{dE(\lambda)}{dZ} \right] \tag{F3.3.6}$$

3.3.4   Heating rate

Radiative heating is responsible for the increase of air temperature with height in the thermosphere and stratosphere. The corresponding heating rate may be calculated from the partial derivative of irradiance against height:

$$(\partial T/\partial t)_r = k(air) \cdot [(\partial E/\partial Z) + E(J)], \quad k(air) = 1/(C_p \cdot \rho) \tag{F3.3.7}$$

where $(\partial T/\partial t)_r$ denotes radiative heating rate and $\partial E/\partial Z$ denotes the absorption rate of radiative energy. $E(J)$ is the release rate of reaction energy from corresponding photolysis. C_p and ρ are heat capacity and density of air respectively.

3.4   Simulating radiation fields in the atmosphere

The spatial distributions of light intensity, energy, and actinic fluxes are fundamental quantities for simulating atmospheric chemical reactions. The fields of these quantities may be simulated for spectral intervals at each time step. Examples are given below, and then complete equations are provided.

3.4.1 A Beer–Lambert experiment with a reflective cell

In a hypothetical Beer–Lambert experiment with a reflective cell at the bottom, a direct light beam passes through a thin column of absorbent and partially reflects back, as shown in Figure 3.8. The incident beam intensity I_1 is 244 units at line A, and the optical depth is 0.00667 and 0.0333 for paths A–B and B–C respectively. In the experiment, the extinction was purely due to absorption, and the albedo at line C was 0.2431402.

As the radiative transfer processes in this experiment only involved absorption and reflection, the light intensity may be calculated along the trajectory of the light (A–B–C–B–A). After the light intensity is calculated at lines A, B, C in the forward and backward directions, irradiance (EF$\downarrow$) and actinic flux (AF) may be calculated. The results are shown in Figure 3.8. It is seen that, in the experiment, actinic flux is over 50% larger than irradiance in quantity.

3.4.2 A more realistic example: two-stream transfer in three layers

In the atmosphere, photons from solar radiation are not only absorbed by chemicals in photolysis reactions, but also scattered by air molecules due to processes involving the oscillation of non-forbidden vibrational–rotational energy levels. Thus, radiative transfer calculations need to include the attenuation of a direct beam due to absorption and scattering along its path, distribution of scattered lights, and the attenuation of the reflected beam.

Figure 3.9 illustrates radiative transfer processes of a single line of solar radiation in three layers when the solar zenith angle is 0, using two streams to represent scattered light and assuming that the light intensity is uniform in all azimuthal angles. Incident light intensity of the single line ($I[1]$), optical depths of three layers (Tau),

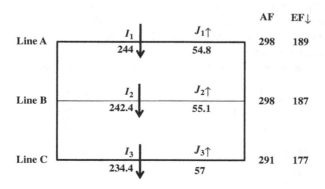

Figure 3.8 A hypothetical, reflective Beer–Lambert experiment.

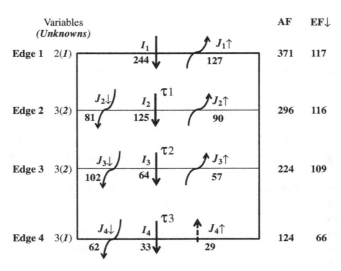

Figure 3.9 An example of atmospheric radiative transfer processes related to a single line of solar radiation with two streams in three layers.

surface albedo (Alpha), and single scattering albedo in three layers (A) are listed in (E3.4.1). Etau is used to denote $\exp(-Tau)$ in subsequent formulae.

$$
\begin{aligned}
I[1] &\quad 2.44 \times 10^{15} \\
Tau &\quad c(1, 1, 1)\cdot 2/3 \\
Etau &= \exp(-Tau) \\
Alpha &\quad 0.3 \\
A &\quad c(0.99, 0.95, 0.90)
\end{aligned}
\tag{E3.4.1}
$$

In the example, the attenuation of the incident light beam by each layer may be calculated first, and the light intensity of the reflected beam is expressed as the product of surface albedo and the sum of attenuated intensities of the incident light beam and the downward scattered light beam. The number of total and unknown variables at the edges of each layer is denoted in the figure, and there are six unknown variables to represent the intensity of scattered light in the upward and downward directions in three layers. These six unknown variables are determined in (E3.4.2):

$$
\begin{bmatrix}
1 & M_{12} & 0 & 0 & 0 & 0 \\
0 & 1 & M_{23} & 0 & M_{25} & 0 \\
0 & M_{32} & 1 & 0 & 0 & 0 \\
0 & 0 & M_{43} & 1 & M_{45} & 0 \\
0 & 0 & 0 & M_{54} & 1 & M_{56} \\
0 & 0 & 0 & 0 & M_{65} & M_{66}
\end{bmatrix}
\cdot
\begin{bmatrix}
J_1\uparrow \\
J_2\uparrow \\
J_2\downarrow \\
J_3\downarrow \\
J_3\uparrow \\
J_4\downarrow
\end{bmatrix}
=
\begin{bmatrix}
RHS1 \\
RHS2 \\
RHS3 \\
RHS4 \\
RHS5 \\
RHS6
\end{bmatrix}
\tag{E3.4.2}
$$

where M_{ij} denotes a non-zero element in the coefficient matrix ($M_2[i,j]$) for the six unknown variables ($J_1\uparrow, J_2\uparrow, J_2\downarrow, J_3\downarrow, J_3\uparrow, J_4\downarrow$), and RHS$i$ denotes RHS[i]. Their expression is listed in (E3.4.3):

$$
\begin{aligned}
M_2[1,2] && -(\text{Etau}[1]\cdot(1-0.5\cdot A[1])+0.5\cdot A[1]) \\
M_2[2,3] && -0.5\cdot A[2]\cdot(1-\text{Etau}[2]) \\
M_2[2,5] && -(0.5\cdot A[2]+\text{Etau}[2]\cdot(1-0.5\cdot A[2])) \\
M_2[3,2] && -0.5\cdot A[1]\cdot(1-\text{Etau}[1]) \\
M_2[4,3] &= & -(\text{Etau}[2]\cdot(1-0.5\cdot A[2])+0.5\cdot A[2]) && \text{(E3.4.3a)} \\
M_2[4,5] && -0.5\cdot A[2]\cdot(1-\text{Etau}[2]) \\
M_2[5,4] && -0.5\cdot A[3]\cdot(1-\text{Etau}[3]) \\
M_2[5,6] && -\text{Alpha}\cdot(\text{Etau}[3]+0.5\cdot A[3]\cdot(1-\text{Etau}[3])) \\
M_2[6,5] && -(0.5\cdot A[3]+\text{Etau}[3]\cdot(1-0.5\cdot A[3]))
\end{aligned}
$$

$$
\begin{aligned}
\text{RHS}[1] && 0.5\cdot A[1]\cdot(I[1]-I[2]) \\
\text{RHS}[2] && 0.5\cdot A[2]\cdot(I[2]-I[3]) \\
\text{RHS}[3] &= & 0.5\cdot A[1]\cdot(I[1]-I[2]) \\
\text{RHS}[4] && 0.5\cdot A[2]\cdot(I[2]-I[3]) \\
\text{RHS}[5] && 0.5\cdot A[3]\cdot(I[3]-I[4])+\text{Alpha}\cdot I[4]\cdot\{\text{Etau}[3]+0.5\cdot A[3]\cdot(1-\text{Etau}[3])\} \\
\text{RHS}[6] && 0.5\cdot A[3]\cdot(I[3]-I[4])+\text{Alpha}\cdot I[4]\cdot 0.5\cdot A[3]\cdot(1-\text{Etau}[3])
\end{aligned}
$$

$$\text{(E3.4.3b)}$$

Thus, equation (E3.4.2) becomes (E3.4.4), and the solution is listed in Figure 3.9.

1	−0.754	0	0	0	0		$J_1\uparrow$		58.8
0	1	−0.231	0	−0.745	0		$J_2\uparrow$		29
0	−0.241	1	0	0	0	·	$J_2\downarrow$	=	58.8
0	0	−0.745	1	−0.231	0		$J_3\downarrow$		29
0	0	0	−0.219	1	−0.22		$J_3\uparrow$		21.3
0	0	0	0	−0.732	0.934		$J_4\downarrow$		16.3

$$\text{(E3.4.4)}$$

After the intensity of scattered light is solved, actinic flux (AF) and energy flux (EF) may be calculated at each of the four edges. These results are also listed in the figure. It is seen that actinic flux is 2–3 times the value of energy flux.

Comparing the top two layers of this example with the previous example, the addition of the strong scattering process reduced the intensity of the direct beam by 2–4 times and reduced the value of energy flux by ∼60%. However, changes of actinic flux were only ~20% at the top and bottom edges with little change in the middle edge.

3.4.3 A six-stream radiative transfer model

To simulate atmospheric phenomena, dozens of vertical layers with multiple reflection levels are often used nowadays. A popular six-stream radiative transfer model used in many atmospheric simulations is described below, assuming vertical layers are evenly spaced for optical depth (dO) and the incident light intensity is 1.

In the six-stream model, scattered lights are assumed to distribute among six angles in each layer using a phase function for distribution. Climatological cloud data were used to extract approximate cloud levels and effective reflectivities, to avoid large uncertainties from model assumptions. Corresponding parameters, formulae, and solution pathways are as follows.

When optical depth is <0.0001, the attenuation of light intensity for the O_2 absorption band (180–200 nm) is neglected. For other spectra with little optical depth, the actinic flux is assumed to be the incident light density multiplied by $(1 + 2\alpha\mu)$. The α denotes albedo and μ denotes cosine of solar zenith angle; the factor "2" denotes the multiple nature of reflection. When optical depth is significant, light attenuation due to O_2 absorption is calculated. The phase matrix for the six-stream model is a 3×3 matrix, denoted as **Ph**[3,3], with the value of [0.3099, 0.2861, 0.2323, 0.4578, 0.4479, 0.4255, 0.2323, 0.2660, 0.3422] in Fortran convention. The phase functions corresponding to the three distinct scattering angles are:

$$Ph0(J) = 0.375 \cdot [3.0 - U_2(J) - \mu^2 + 3.0 \cdot U_2(J) \cdot \mu^2]; \quad J = 1, 2, 3 \quad \text{(E3.4.5)}$$

where $U_2 = U \cdot U$, and $U = [0.1127, 0.5, 0.8873]$. If absorption accounts for all extinction, then the incident and reflected light beams may be solved at each level and actinic flux at each level is simply the sum of both beams. When a single scattering albedo is not trivial, the scattering process is solved from the top to the bottom. For example, at the top edge of the top layer, the scattered light beams are denoted as $4 \cdot W$ $[1:3] \cdot C[1:3]$ (top), and C is governed by:

$$B[3 \times 3] \cdot C[3] = Q[3] \quad \text{(E3.4.6)}$$

where $W = [0.2778, 0.4444, 0.2778]$ and $B[3 \times 3]$ is the phase function with diagonal elements:

$$B[i, i] = 1 + U[i]/dO + 0.5 \cdot dO/U[i] \quad \text{(E3.4.7)}$$

and non-diagonal elements:

$$B[i, j] = -0.5 \cdot A[1] \cdot Ph[i, j] \cdot dO/U[i] \quad \text{(E3.4.8)}$$

$A[1]$ denotes the single scattering albedo at the upper edge of the top layer. The $Q[3]$ term in equation (E3.4.6) is given below:

$$Q(i) = 0.125 \cdot S(1) \cdot dO \cdot Ph0(i)/U(i)$$ (E3.4.9)

Figure 3.10 shows simulated results from the six-stream radiative transfer model for the solar spectrum near Los Angeles, California, USA, at 6 a.m. on August 26, 1987. Calculated actinic flux in units of incident light intensity is shown in Figure 3.10(a) at 80, 50, 10, and 0 km ASL for the solar spectrum within 200–500 nm with intervals of 5 nm. At 80 km ASL, the top of the vertical domain of the model, simulated actinic flux is found to exceed incident light for all wavelengths. At 50 km ASL, the top of the stratosphere, simulated actinic flux is seen to be larger than incident light for wavelengths longer than ~ 310 nm; for shorter ultraviolet wavelengths, actinic flux is seen to be reduced to 10% of incident light intensity with a minimum at ~ 250 nm. At 10 km ASL, actinic flux is seen to be nearly 0 for wavelengths shorter than 300 nm; at longer wavelengths, actinic flux is observed to be up to 65% of incident intensity at ~ 500 nm. On the surface, actinic flux is seen to be further reduced by tropospheric air mass; the maximum ratio of actinic flux to the incident light is found to be only 22%.

The intensity of actinic flux and contributions from combined direct and reflected beams versus scattered beams are shown in Figure 3.10(b) for wavelength intervals of 300–305 and 500–505 nm. It is seen that the actinic flux of the wavelength interval of 300–305 nm is dominated by the combined direct and reflected beams above ~ 10 km ASL; below this altitude, scattered light beams dominated the actinic flux. For the wavelength interval of 500–505 nm, it is observed that the combined direct and reflected light beams dominate the actinic flux. It must be noted that these values represent a condition with large optical depth for solar light, as the solar zenith angle in the calculation was 85.3°. The surface albedo was assumed to be 0.1, as was the total aerosol extinction coefficient at 310 nm. In polluted air, the total aerosol extinction coefficient could be an order of magnitude larger.

3.4.4 Generalized equations of radiative transfer

In numerical simulations, both the atmosphere and oceans of the Earth are split into a large number of "thin" layers, characterized by small variations in the vertical direction and relatively much larger variations in the horizontal direction. The radiative transfer through a thin layer with a vertical optical depth of τ_v, decreasing downwards, at a solar zenith angle (θ), characterized by U ($\cos \theta$), is described in general terms by:

$$U \cdot dI/d\tau_v = I - S$$ (E3.4.10)

where S represents added intensity from scattering and thermal emission, and $dI/d\tau$ is always negative. The thermal emission part of S was given in formula (F3.1.1) as $I_e(\nu, T)$, and the scattering part of S is commonly expressed as the product of a single scattering albedo and incident light intensity, $A \cdot I$. The single scattering albedo (A) is

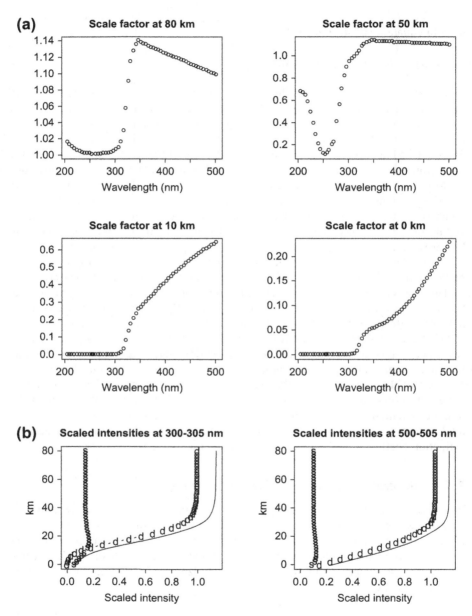

Figure 3.10 Solar spectrum resolved from a six-stream radiative transfer model ($U = 0.1$). (a) Actinic flux at 80, 50, 10, and 0 km ASL. (b) Intensity of actinic flux and contributions from combined direct and reflected beams versus scattered beams at wavelength intervals of 300–305 and 500–505 nm.

an average quantity calculated over all scattered directions in space, characterized by the angle (α) between incident light and scattered light beams and the corresponding phase function (P):

$$A = \frac{1}{4\pi} \int_0^{2\pi} d\phi' \int_0^{\pi} P(\cos\alpha) \cdot \sin\theta' \cdot d\theta' \qquad (E3.4.11)$$

If $P(\cos\alpha)$ is independent of scattering direction, i.e. it is isotropic, then the single scattering albedo $A = P$. In the solar spectrum, thermal emission is unimportant, so equation (E3.4.10) becomes:

$$\frac{dI}{d\tau} = I - \frac{1}{4\pi} \int_0^{2\pi} d\phi' \int_0^{\pi} I(\theta', \phi') \cdot P(\cos\alpha) \cdot \sin\theta' \cdot d\theta' \qquad (E3.4.12)$$

where $\tau = \tau_v/U$. If scattering is isotropic, then equation (E3.4.12) simplifies to:

$$\frac{dI}{d\tau} = I - A \cdot \bar{I} \qquad (E3.4.13)$$

where

$$\bar{I} = \frac{1}{4\pi} \int_0^{2\pi} d\phi' \int_0^{\pi} I(\theta', \phi') \cdot \sin\theta' \cdot d\theta'$$

If the scattered light intensity is independent of ϕ', then equation (E3.4.12) simplifies to:

$$\frac{dI}{d\tau} = I - \frac{1}{2} \int_{-1}^{1} I(U) \cdot P(U) \cdot dU \qquad (E3.4.14)$$

The second term on the right-hand side of equation (E3.4.14) may be approximated numerically by Gaussian quadrature, so that:

$$U_i \frac{dI(\tau_v, U_i)}{d\tau_v} = I(\tau_v, U_i) - 0.5 \sum_{j-1}^{N} a_j \cdot I(\tau_v, U_i) \cdot P(U_i, U_j) \qquad (E3.4.15a)$$

Thus, for N quadratures, there are N equations in the above form, with $i = 1:N$. Commonly used quadrature parameters are:

$$N = 2: a_j = 1, \ U_j = \pm 1/\sqrt{3};$$

$$N = 4{:}a[1{:}4] = [0.35,\ 0.65,\ 0.65,\ 0.35],$$
$$U[1{:}4] = [-0.8611, -0.34,\ 0.34,\ 0.8611]$$

(E3.4.15b)

When $N = 2$, two streams of scattered light may be denoted as $I^+(\tau_v) = I(\tau_v, +1/\sqrt{3})$ and $I^-(\tau_v) = I(\tau_v, -1/\sqrt{3})$ for the upward and downward directions of scattered fields respectively. Given direct beam I_0 at the top, $I_{direct} = I_0 \cdot \exp(-\tau_v/U)$. If boundary conditions for the two equations are (1) $\Gamma^-(0) = 0$ and (2) $I^+ + I^- \to$ constant (i.e. 0) as $\tau_v \to \infty$, then solutions to the two-stream equations are:

$$I^+ = (I_0/4\pi) \cdot \left(\sqrt{3}AU\right) \Big/$$
$$\left(1 - U^2\beta^2\right)\left[\frac{3A\left(1 + \sqrt{3}U\right)}{\left(\beta + \sqrt{3}\right)^2}\exp(-\beta\tau_v) + \left(1 - \sqrt{3}U\right)\exp(-\tau)\right]$$

$$I^- = (I_0/4\pi) \cdot \left(1 + \sqrt{3}U\right) \cdot \left(\sqrt{3}AU\right) \Big/ \left(1 - U^2\beta^2\right)[\exp(-\beta\tau_v) - \exp(-\tau)]$$

(E3.4.16)

where $\beta = \sqrt{3(1-A)}$, which is related to the classic random walk effect according to the original author of this solution (McElroy, 1971). The corresponding actinic flux (F) and energy flux (E) are:

$$F(\tau_v, U) = 4\pi\left(\frac{I^+ + I^-}{2}\right) + I_0 \exp(-\tau)$$

(E3.4.17a)

$$E(\tau_v, U) = \left(2\pi/\sqrt{3}\right) \cdot \left(I^+ - I^-\right) - |U|I_0\exp(-\tau)$$

(E3.4.17b)

Summary

In this chapter, solar radiation observed from recent satellite campaigns was discussed first, together with the protective role of the atmosphere for people living on the surface. Secondly, the theoretical spectrum of thermal radiation from the Earth was discussed together with greenhouse effects. Thirdly, radiative transfer measures including light intensity, actinic and energy fluxes, and heating rate were defined. Finally, modeling examples for radiative transfer were discussed, and general equations for radiative transfer simulation were presented.

Exercises

Using formula (F3.1.1), conduct the following exercises:

1. Given $\int_0^\infty \dfrac{x^3 dx}{e^x - 1} = \dfrac{\pi^4}{15}$ and $\sigma = \dfrac{2\pi^5 k^4}{15h^3 C^2}$, prove that the energy flux of a black body over all spectra is $(\sigma/\pi)T^4$.

2. Derive the relationship between peak wavelength and temperature of a black body.

3. Providing that the Earth's radiation has peak wavelength at 10 μm and solar radiation at 500 nm, survey the literature to estimate the effective albedo of the Earth by assuming both the Earth and Sun are perfect black bodies and are in energy balance with greenhouse effects.

4. Assuming anthropogenic emissions will reduce the effective solar albedo of the Earth by 0.1, from the above, calculate the resulting change in the effective temperature of the Earth as a black body and its peak wavelength.

5. Using equation (E3.4.10) and the two-stream approximation for thermal emission, derive a formula for energy flux in similar form to (E3.4.17b).

4 Modeling chemical changes in the atmosphere

Chapter Outline

Chemicals in the atmosphere change constantly with time and space, partly due to numerous photochemical reactions occurring simultaneously in the air. To understand the temporal and spatial distributions of important chemicals, such as air pollutants, it is imperative to simulate chemical changes numerically in the atmosphere as the problem usually cannot be solved analytically.

4.1 Chemical ordinary differential equations

4.1.1 General formula

The concentration of a chemical in a volume of the atmosphere ($C[i]$) changes due to production and loss processes, such as chemical reactions, emissions, depositions, and transport processes. This phenomenon is described by:

$$\mathrm{d}C[i]/\mathrm{d}t = P - L \cdot C[i] \tag{E4.1.1}$$

Chemical Modeling for Air Resources. http://dx.doi.org/10.1016/B978-0-12-408135-2.00004-5

where $dC[i]/dt$ denotes the rate of change, P denotes the production rate, and L denotes the pseudo-first-order loss rate constant. In the atmosphere, P is often a combination of many reaction rates involving many other chemicals, and so is L. Thus, equation (E4.1.1) must be expanded to include other chemicals important for the production and loss of the chemical concerned:

$$dC[i]/dt = P(C[j]) - L\,(C[j]) \cdot C[i]; \quad i = 1, 2, \ldots, N \qquad (E4.1.2)$$

where N denotes the number of chemicals included in the system of ordinary differential equations. Usually, the indices i and j are different, but they could be the same for some elementary reactions or effective reaction pathways used in compressed photochemical mechanisms. Depending on the target chemical, the system of ordinary differential equations represented by equation (E4.1.2) ranges from about a dozen to many thousands. Typical examples in important atmospheric regions are discussed below.

4.1.2 Example for mesospheric OH

In the modern mesosphere, the OH mixing ratio is believed to be mainly determined by photochemical reactions of a handful of short-lived radicals (O, O^1D, H, OH, HO_2), besides O_2, O_3, and H_2O. Corresponding photochemical reactions are listed in Table 4.1.

In the table, three photolysis reactions are shown in bold and there are six pressure-dependent reactions, while the other 11 reactions are all second-order elementary reactions. It can be seen that, without H_2O, only the first six reactions would occur. The addition of H_2O results in 14 more reactions. The system of ordinary differential equations, according to equation (E4.1.2), is listed in Table 4.2.

It is seen in Table 4.2 that OH has complex production and loss terms, both involving radicals and relatively stable chemicals. For comparison, both O_3 and O^1D have a single production term, and H_2O has a single loss term. In particular, radical O^1D only involves reactions with relatively stable chemicals.

Table 4.1 OH photochemical mechanisms in the mesosphere

Species	Reactions	Reactions
O_2	1. $\mathbf{O_2 + h\nu \rightarrow 2O}$	11. $H + HO_2 \rightarrow H_2O + O$
O_3	2. $\mathbf{O_3 + h\nu \rightarrow O_2 + O^1D}$	12. $OH + OH \rightarrow O + H_2O$
O^1D	3. $O + O_3 \rightarrow 2O_2$	13. $OH + HO_2 \rightarrow O_2 + H_2O$
O	4. $O^1D + M \rightarrow O + M$	14. $H_2O + O^1D \rightarrow 2OH$
H_2O	5. $O + O + M \rightarrow O_2 + M$	15. $O_3 + H \rightarrow O_2 + OH$
H	6. $O + O_2 + M \rightarrow O_3 + M$	16. $O + OH \rightarrow O_2 + H$
OH	7. $\mathbf{H_2O + h\nu \rightarrow H + OH}$	17. $O + HO_2 \rightarrow O_2 + OH$
HO_2	8. $O + OH + M \rightarrow HO_2 + M$	18. $H + HO_2 \rightarrow O_2 + H_2$
	9. $O_2 + H + M \rightarrow HO_2 + M$	19. $H + HO_2 \rightarrow 2OH$
	10. $OH + H + M \rightarrow H_2O + M$	20. $O_3 + OH \rightarrow O_2 + HO_2$

Table 4.2 Terms of ordinary differential equations for mesospheric OH

Species	Production terms	Loss terms
O_3	$k_6[O][O_2]$	$[O_3](J_2 + k_3[O] + k_{15}[H] + k_{20}[OH])$
O^1D	$J_2[O_3]$	$[O^1D](k_4 + k_{14}[H_2O])$
O	$2J_1[O_2] + k_4[O^1D] + k_{11}[H][HO_2] +$ $k_{12}[OH][OH]$	$[O](k_3[O_3] + k_5[O] + k_6[O_2] + k_8[OH] +$ $k_{16}[OH] + k_{17}[HO_2])$
H_2O	$k_{10}[OH][H] + k_{11}[H][HO_2] +$ $k_{12}[OH][OH] + k_{13}[OH][HO_2]$	$J_7[H_2O]$
H	$J_7[H_2O] + k_{16}[O][OH]$	$[H](k_9[O_2] + k_{10}[OH] + k_{11}[HO_2] +$ $k_{15}[O_3] + k_{18}[HO_2] + k_{19}[HO_2])$
OH	$J[H_2O] + 2k_{14}[H_2O][O^1D] +$ $k_{15}[O_3][H] + k_{17}[O][HO_2] +$ $2k_{19}[H][HO_2]$	$[OH](k_8[O] + k_{10}[H] + 2k_{12}[OH][OH] +$ $k_{13}[OH][HO_2] + k_{16}[O][OH] +$ $k_{20}[O_3][OH])$
HO_2	$k_8[O][OH] + k_9[O_2][H] + k_{20}[O_3][OH]$	$[HO_2](k_{11}[H] + k_{13}[OH] + k_{17}[O] +$ $k_{18}[H] + k_{19}[H])$

4.1.3 Example for stratospheric O_3

In the modern stratosphere, O_3 closely interacts with a small number of chemicals, namely odd oxygen chemicals (O, O^1D), odd hydrogen chemicals (OH, HO_2, H_2O_2), reactive nitrogen chemicals (N_2O, NO, NO_2, NO_3, N_2O_5, HNO_3), reactive chlorine chemicals (CF_2Cl_2, Cl_2, Cl, ClO, HCl, $ClONO_2$), and simple carbon chemicals (CH_4, CH_3O_2, HCHO, CO) since most hydrocarbons cannot escape from the troposphere. If CF_2Cl_2 represents all halocarbons in the stratosphere and H_2O is constant, then the size of the system of ordinary differential equations for stratospheric O_3 is $N \approx 22$, as shown in Table 4.3. In the table, reaction rate constants are denoted as k for gas-phase thermal reactions, as k_h for heterogeneous reactions, and as J for photolysis reactions for convenience, but in reality these constants change with reactions, product branches, temperature, surface area of particles, and sometimes pressure.

It is noteworthy that the complexity of stratospheric O_3 originates from eight loss reactions that involve odd oxygen species, odd hydrogen species, reactive nitrogen species, and chlorine radicals. The odd hydrogen species are important for the oxidation of CH_4 with a lifetime of 10 years, while reactive nitrogen species and chlorine radicals originate from dissociations of N_2O and CF_2Cl_2 with lifetimes of ~100 years. In the stratosphere, CH_4, N_2O, and CF_2Cl_2 have a single production term, transport from the troposphere. The dissociation of N_2O has two pathways, producing NO and N_2 respectively. The latter pathway links the biogeochemical cycle of nitrogen from the biosphere to the major reservoir of nitrogen in the atmosphere, though N_2 is constant in the timescale relevant to the O_3 system, which could collapse in a few weeks. The dissociation of CF_2Cl_2 provides chlorine radicals, which are responsible for the formation of the "O_3 hole" observable over Antarctica in early spring.

Table 4.3 Important terms of ordinary differential equations for stratospheric O_3

Species	Production terms	Loss terms
O_3	$k[O][O_2]$	$[O_3](J + k[O] + k[O^1D] +$ $k[NO] + k[NO_3] + k[OH] +$ $k[HO_2] + k[Cl])$
O	$2J[O_2] + J[O_3] + J[NO_2] +$ $k[O^1D] + J[NO_3]$	$[O](k[O_2] + k[O_3] + k[ClO])$
O^1D	$J[O_3] + J[N_2O]$	$[O^1D](k + k[H_2O] + k[N_2O] +$ $k[H_2] + k[CF_2Cl_2])$
OH	$k[HO_2][O_3] + h[HO_2][NO] +$ $2J[O^1D][H_2O] + J[HNO_3] + 2J[H_2O_2]$	$[OH](k[O_3] + k[NO_2] +$ $k[HNO_3] + k[HCl] + k[H_2O_2] +$ $k[HCHO] + k[CH_4] + k[CO])$
HO_2	$k[OH][O_3] + k[NO][CH_3O_2] +$ $2J[HCHO] + k[OH][HCHO] +$ $k[CO][OH] + k[OH][H_2O_2]$	$[HO_2](k[O_3] + k[NO] + 2k[HO_2])$
H_2O_2	$2k[HO_2][HO_2]$	$[H_2O_2](J + k[OH])$
N_2O	*Transport from troposphere*	$[N_2O](J + k[O^1D])$
NO	$J[NO_2] + k[O^1D][N_2O]$	$[NO](k[O_3] + k[HO_2] +$ $k[CH_3O_2] + k[NO_3] + k[ClO])$
NO_2	$k[O_3][NO] + k[HO_2][NO] +$ $k[NO][CH_3O_2] + 2k[NO_3][NO] +$ $J[HNO_3] + J[NO_3] + J[N_2O_5] +$ $k[ClO][NO]$	$[NO_2](J + k[OH] + k[O_3] +$ $k[NO_3] + k[ClO])$
NO_3	$k[O_3][NO_2] + k[OH][HNO_3]$	$[NO_3](J + k[NO_2])$
N_2O_5	$k[NO_3][NO_2]$	$[N_2O_5](k + J + k_h)$
HNO_3	$k[NO_2][OH] + 2k[N_2O_5] +$ $k_h[ClONO_2][HCl]$	$[HNO_3](J + k[OH] + k_h)$
CF_2Cl_2	*Transport from troposphere*	$[CF_2Cl_2](J + k[O^1D])$
Cl_2	$k_h[HCl][ClONO_2]$	$J[Cl_2]$
Cl	$2J[Cl_2] + k[OH][HCl] + J[CF_2Cl_2] +$ $k[ClO][NO] + k[ClO][O]$	$k[O_3][Cl]$
ClO	$k[O^1D][CF_2Cl_2] + k[Cl][O_3]$	$[ClO](k[O] + k[NO])$
HCl	$k[Cl][CH_4]$	$[HCl](k[OH] + k_h[ClONO_2])$
$ClONO_2$	$k[ClO][NO_2]$	$[ClONO_2](k + J + k_h[HCl])$
CH_4	*Transport from troposphere*	$k[OH][CH_4]$
CH_3O_2	$k[OH][CH_4]$	$k[NO][CH_3O_2]$
HCHO	$k[NO][CH_3O_2]$	$[HCHO](J + k[OH])$
CO	$J[HCHO] + k[OH][HCHO]$	$k[CO][OH]$

4.1.4 Example for tropospheric O_3

In the troposphere, a large number of reactive hydrocarbon compounds are emitted from anthropogenic activities and natural sources, and decomposed via photochemical reactions. Abundant reactive hydrocarbon compounds are mainly alkanes (C_2H_6,

C_3H_8, and higher alkanes from fossil fuel evaporation and combustion), as unsaturated hydrocarbon compounds (C_2H_2, C_2H_4, C_3H_6, 1,3-butadiene, aromatic compounds, biogenic isoprene, and terpenes) are more reactive. Organic solvents for coating materials are a significant group of chemicals in the troposphere, as they may accumulate in the built environment and impose health risks. Due to the complex structure of reactive hydrocarbon compounds, especially those with large molecular weights, many intermediate chemicals may be produced from their oxidation into CO_2 and H_2O. As a result, it is extremely challenging to represent photochemical degradation reactions of reactive organic compounds in the troposphere at the level of elementary reactions, due to the limitations in detection limits of analytical instruments and the resulting lack of data necessary for clarifying reaction pathways and for determining rate parameters. The current understanding of quantum chemistry does not allow for reliable estimates of rate parameters for reactions involving reactive organic compounds, though clarification of reaction pathways is sometimes possible. To circumvent those limitations, degradation processes of complex organic compounds are often simplified in model representations.

The most simplified representation of complex organic compounds in the troposphere is currently the Carbon-Bond Mechanism, as briefly discussed in Chapter 2. Using the Carbon-Bond Mechanism, only 17 more model species are needed to simulate tropospheric O_3, besides those listed in Table 4.3. Table 4.4 lists the additional model species and their major production and loss terms. Of course, these terms affect the existing chemicals listed in Table 4.3, but the adjustment is omitted for clarity. It is shown that reactive organic gases with anthropogenic emissions are represented by only six model species, namely PAR, ETH, ALD2, OLE, TOL, and XYL, and biogenic emissions are represented by only a single model species, ISOP. More specifically, PAR represents primary alkanes as well as alkane-like intermediates produced from the oxidation of unsaturated hydrocarbons. ETH denotes C_2H_4. ALD2 represents CH_3CHO and higher aldehydes, as well as the functional group of 2-alkenes, and OLE denotes the functional group of terminal alkenes. TOL represents toluene and other monosubstituted aromatics with a single benzene ring, and XYL represents m-xylene and other similar aromatics. ISOP denotes isoprene. Though the degradation of the seven primary model species may involve a large number of intermediates, only a handful of intermediate chemicals have been detected in environmental smog chamber experiments. Based on these laboratory data, the Carbon-Bond Mechanism uses 10 model species to represent important stable and radical intermediates from the degradation of reactive organic compounds for simulating tropospheric O_3. Namely, important stable intermediates are MGLY (methylglyoxal), CRES (cresol), PAN (peroxyacetyl nitrate), and important radicals are CRO, OPEN, TO2, XO2, C_2O_3, and ROR. The "XO$_2$N" corresponds to the portion of organic peroxy radicals converted to organic nitrates, and other radicals may be characterized by their production and loss terms in Table 4.4. For example, it is shown that TO2 represents benzylperoxy radicals, and C_2O_3 denotes $CH_3C(O)O_2$, etc. As the Carbon-Bond Mechanism is highly optimized for representing photochemical degradation of reactive organic gases, it is widely applied in regulatory simulations of O_3 pollution events and

Table 4.4 Troposphere-specific CBM species and terms for O_3

Species	Production terms	Loss terms
PAR	Emission, $k[OLE][O] + k[ISOP][O] + k[ISOP][O_3] + k[XYL][OH]$	$k[OH][PAR]$
ETH	Emission, $k[ISOP][O] + k[ISOP][OH] + k[ISOP][O_3]$	$[ETH](k[O] + k[OH] + k[O_3])$
ALD$_2$	Emission, $k[OH][PAR] + k[ROR] + k[O][OLE] + k[OH][OLE] + k[O_3][OLE] + k[NO_3][OLE] + k[ETH][OH] + k[OPEN][O_3] + k[ISOP][O] + k[ISOP][OH] + k[ISOP][O_3]$	$[ALD_2](k[OH] + k[NO_3] + J)$
OLE	Emission, $k[ISOP][O]$	$[OLE](k[O] + k[OH] + k[O_3] + k[NO_3])$
TOL	Emission	$k[OH][TOL]$
XYL	Emission	$k[OH][XYL]$
ISOP	Emission	$[ISOP](k[O] + k[OH] + k[O_3] + k[NO_3])$
MGLY	$k[XYL][OH] + k[OPEN][O_3] + k[ISOP][OH] + k[ISOP][O_3]$	$[MGLY](k[OH] + J)$
CRES	$k[TOL][OH] + k[TO_2] + k[XYL][OH]$	$[CRES](k[OH] + k[NO_3])$
CRO	$k[CRES][OH] + k[CRES][NO_3]$	$k[NO_2][CRO]$
OPEN	$k[TO_2][NO] + k[CRES][OH]$	$[OPEN](k[OH] + k[O_3] + J)$
TO$_2$	$k[TOL][OH] + k[XYL][OH]$	$[TO_2](k[NO] + k)$
XO$_2$	$k[C_2O_3][NO] + 2k[C_2O_3][C_2O_3] + k[C_2O_3][HO_2] + k[OH][CH_4] + k[PAR][OH] + k[ROR] + k[O][OLE] + k[OH][OLE] + k[O_3][OLE] + k[NO_3][OLE] + k[O][ETH] + k[OH][ETH] + k[TOL][OH] + k[CRES][OH] + k[XYL][OH] + k[OPEN][OH] + k[OPEN][O_3] + k[MGLY][OH] + k[ISOP][O] + k[ISOP][OH] + J[ALD_2]$	$[XO_2](k[NO] + 2[XO_2])$
XO$_2$N	$k[PAR][OH] + k[ROR] + k[O][OLE] + k[OLE][NO_3] + k[ISOP][OH] + k[ISOP][NO_3]$	$k[XO_2N][NO]$
C$_2$O$_3$	$k[ALD_2][OH] + k[ALD_2][NO_3] + k[PAN] + k[OPEN][OH] + k[OPEN][O_3] + k[MGLY][OH] + k[ISOP][OH] + J[MGLY] + J[OPEN]$	$[C_2O_3](k[NO] + k[NO_2] + k[C_2O_3] + k[HO_2])$
PAN	$k[C_2O_3][NO_2]$	$k[PAN]$
ROR	$k[PAR][OH]$	$[ROR](k + k[NO_2])$

various methods have been developed to calculate changes of corresponding model species in a systematic way.

4.1.5 Example for aerosols

In the modern atmospheric boundary layer, the number of aerosols increases exponentially towards the surface, where most human activities occur. Typical aerosol types are sea-salt particles, dust particles of natural and anthropogenic origin, black carbon particles due to incomplete combustion, and internally mixed aerosols with significant amounts of secondary aerosol components. Primary aerosols, such as sea-salt particles, dust particles, and black carbon particles, contain unique chemicals by nature. After staying in the air for a couple of days, aerosols may become internally mixed due to physical processes and chemical reactions.

Figure 4.1 illustrates typical reactive chemicals in a wet aerosol droplet. The aerosol volume is circled, and the inner core is framed. In the inner core, a handful of ions, such as Ca^{2+}, Mg^{2+}, K^+, Na^+, Fe^{2+}, and Fe^{3+}, may be leached into aerosol solution, where a number of ions, radicals, and molecules are present. Radicals and molecules in aerosol solution interact with the gas phase. For example, NH_3, SO_3, SO_2, N_2O_5, and HNO_3 molecules in the gas phase may enter aerosol solution to form NH_4^+, NO_3^-, and SO_4^{2-}, which are responsible for most aerosol mass concentrations near the surface besides organic aerosol and black carbon.

Figure 4.2 illustrates organic aerosol species formed from the degradation of reactive organic gas molecules, as represented by the SAPRC photochemical mechanism implemented in the Community Multi-scale Air Quality (CMAQ) model by the US Environmental Protection Agency (EPA). It shows the six semi-volatile organic aerosol species that were used to represent observed aerosol mass produced from photochemical oxidations of six organic species in the model, namely toluene, m-xylene, cresol, terpene, internal alkenes, and large alkanes.

Table 4.5 lists 45 candidate species for simulating aerosol chemistry in the atmospheric boundary layer. Due to various limitations, H^+ and OH^- need to be resolved from the charge balance equation and the water dissociation constant. Major inorganic components are shown in bold, as are organic components expected to be major constituents of secondary organic aerosols. Not all production and loss terms are available for the species listed in the table. Decadal international research on acid deposition has produced fairly comprehensive kinetic data for the aqueous phase oxidation of S(IV) and dimethylsulfide (DMS) oxidation intermediates, represented

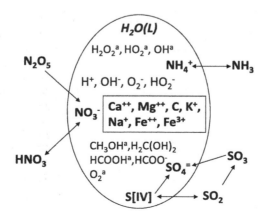

Figure 4.1 Reactive chemicals in a wet aerosol droplet.

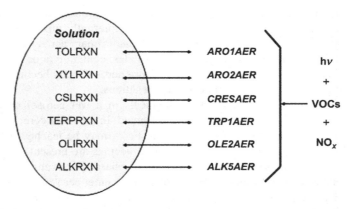

Figure 4.2 Organic aerosol species.

as SIV (a) in the table. Extensive environmental smog chamber experiments support the theory that organic aerosols with $-OH$, $-OOH$, $-C=O$, $-C(O)OH$, and $-ONO_2$ functional groups were produced from photochemical oxidations of both aromatic and non-aromatic hydrocarbons. Ambient measurements showed that total organic carbon could account for half of the aerosol mass in areas with heavy traffic, such as nearby highways, and the ratio of OC:EC is about 2. It follows that a significant fraction of the organic carbon present may have $HOC(O)-C-C(O)OH$ or a combination of $HOC(O)-C-C(O)OH$ and $HOC(O)-C-C-C(O)OH$ in its structure; this feature is represented by RCOOH and ACOOH for non-aromatic and aromatic carboxylic acids. Semi-volatile organic compounds that contain the structural component $C-C(OH)-C(ONO_2)$ may also have an OC:EC ratio of 2, though this type of

Table 4.5 Candidate model species in aerosols

1. NH_4^+	16. $RCOO^-$	31. $R_2C(OH)_2$
2. NO_3^-	17. $ACOO^-$	32. **RCOOH (a)**
3. SO_4^{2-}	18. HO_2^-	33. ROOH (a)
4. SIV (a)	19. O_2^-	34. AOH (a)
5. Cl^-	20. Cl_2^-	35. $A_2C(OH)_2$
6. Br^-	21. Cl (a)	36. **ACOOH (a)**
7. K^+	22. O_3 (a)	37. AOOH (a)
8. Na^+	23. OH (a)	38. ANO_3 (a)
9. Ca^{2+}	24. HO_2 (a)	39. RNO_3 (a)
10. Mg^{2+}	25. NO_3 (a)	40. A_2SO_4 (a)
11. Al^{3+}	26. H_2O_2 (a)	41. R_2SO_4 (a)
12. Mn^{2+}	27. CH_3OH (a)	42. **RPM (a)**
13. Fe^{2+}	28. $H_2C(OH)_2$	43. **APM (a)**
14. Fe^{3+}	29. HCOOH (a)	44. O_2 (a)
15. $HCOO^-$	30. ROH (a)	45. H_2O (a)

intermediate compound may be further oxidized in the atmosphere. Species RPM and APM are used to represent all other non-aromatic and aromatic aerosol organics listed in the table.

Some of the production and loss terms for the species in Table 4.5 have been reported in the literature. A full list could be derived theoretically, similar to the development of the Master Chemical Mechanism, with the foundation of the general chemical mechanism provided in Chapter 2, in spite of considerable uncertainty in reaction pathways, rate parameters, and activity coefficients. Ongoing and further studies in this area will certainly help reduce the gap between model needs and available data.

4.2 Accurate solvers

A system of ordinary differential equations for atmospheric chemical simulation involves radicals, stable intermediates, and trace gases, which have drastically different timescales. Thus, accurate solutions to the stiff system have become an engineering issue with the constraint of numerical round-off error. Two widely used stiff solvers are discussed below.

4.2.1 Stiff system

A stiff system is defined as a system that contains members with drastically different lifetimes. For example, in a fictitious environment, assume changes in concentrations of A and B are determined by a pseudo-first-order decay reaction (R4.2.1), and there is a pseudo-first-order reaction (R4.2.2) that converts C to B:

$$\text{AB} \xrightarrow{k_1} \text{loss} \tag{R4.2.1}$$

$$\text{C} \xrightarrow{k_2} \text{B} \tag{R4.2.2}$$

where $k_1 = 10^{-8}$ s^{-1} and $k_2 = 10^{-2}$ s^{-1}. Thus, the concentrations of A and C at an incremental time step (Δt seconds) after the initial time are given by:

$$[A] = [A]_0 \exp(-10^{-8}\Delta t); \quad [C] = [C]_0 \exp(-10^{-2}\Delta t) \tag{F4.2.1}$$

and the change in concentration of B, $d[B]/dt = d[A]/dt - d[C]/dt$, and the corresponding solution for B is given by:

$$[B] = [A]_0 \exp(-10^{-8}\Delta t) - [C]_0 \exp(-10^{-2}\Delta t) \tag{F4.2.2}$$

The time series of the concentration of B is plotted in Figure 4.3 for initial concentrations of A and C equal to 1. The top panel shows the time series of B in the first 400 seconds, mainly governed by the second term on the right-hand side of formula (F4.2.2). The bottom panel shows the time series of B over a much longer time period;

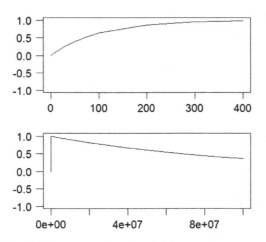

Figure 4.3 Time series of OH concentration in a fictitious environment.

besides the initial "stiff" change, the gradual decay is governed by the first term on the right-hand side of formula (F4.2.2).

Formula (F4.2.2) represents a typical stiff problem for atmospheric chemical modeling: the numerically optimal time increments for the two exponent terms on the right-hand side of formula (F4.2.2) differ by one million times. For example, if $\Delta t = 100$ seconds, the first exponent term will result in a small change from the initial condition, which will lead to an accurate solution. However, the second exponent term will result in a decrease of the initial condition by a factor of e, which will lead to an inaccurate solution considering the loss of curvature information between the two points with relatively large spans. As a result, the solution of OH will be inaccurate at the time step of 100 s in the example.

4.2.2 Newton's method

In order to solve a stiff system of photochemical ordinary differential equations, a number of numerical methods have been developed. A popular method is to use Newton's iteration, as described below.

For a system of chemical ordinary differential equations, defined in equation (E4.1.2), we may rearrange terms so that one side of the equation is 0 and the other side is a function of concentrations to be resolved, denoted as $F(C)$:

$$F(C) = \Delta t(R_p - R_l) - \Delta C \qquad\qquad (E4.2.1)$$

where Δt is the time step and ΔC is the change of concentration during the time step, while R_p and R_l denote the production rates and loss rates respectively. Taking derivatives of C on both sides of equation (E4.2.1) yields:

$$F'(C) = \Delta t \cdot \mathbf{J} - \mathbf{I} \qquad\qquad (E4.2.2a)$$

where J denotes a partial derivative matrix with element $[i, j]$ given by $\partial(R_p[i] - R_l[i])/\partial C[j]$. I denotes a unit matrix with 1 in diagonal terms and 0 elsewhere. The derivative of $F(C)$ against C at the $(n + 1)$th iteration is approximately expressed by:

$$F'(C) = [F(C_{n+1}) - F(C_n)]/(C_{n+1} - C_n) \qquad \text{(E4.2.2b)}$$

As $F(C_{n+1}) \equiv 0$, equations (E4.2.2a, b) yield the incremental change of concentrations at the $(n + 1)$th iteration:

$$C_{n+1} - C_n = [\Delta t(R_p - R_l) + C_0 - C_n]/(I - \Delta t J) \qquad \text{(F4.2.3)}$$

where C_0 denotes initial concentrations and C_n denotes concentrations at the nth iteration. The right-hand side of formula (F4.2.3) is evaluated using values at the nth iteration. The solution converges if $C_{n+1}/C_n - 1$ is within a small fraction, e.g. 0.001, for all chemicals in the system.

Figure 4.4 illustrates Newton's iteration from the nth iteration to the $(n + 1)$th iteration for an ideal one-chemical system. It is shown that, at the nth iteration, $F(C_n)$ was far from zero, so its derivative was used to update the value of C at the $(n + 1)$th iteration, as shown in formula (F4.2.3). It is seen that $F(C_{n+1})$ is smaller than $F(C_n)$. Following the trend, the solution would be reached in a few iterations.

In a real photochemical system, the change of $F(C)$ against C involves many chemicals. As the initial values of some chemicals are unknown, which may be unimportant for well-designed simulations, estimated initial values may be far away from the solution. In addition, updating chemicals in the initial steps may not bring all of them closer to the solution to similar extents. Moreover, the use of derivatives could direct some chemicals into a negative region in the first few steps. Hence, empirical constraints, such as bracketing chemical concentrations and time steps, are useful and sometimes very

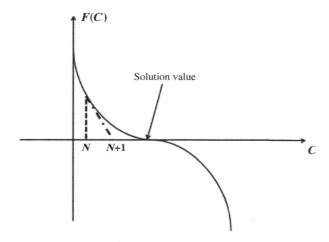

Figure 4.4 An illustration of Newton's iteration.

efficient in helping to find the solution. Nevertheless, the method described above is an implicit method with super-convergence for atmospheric chemical systems.

Figure 4.5 shows the convergence of an error criteria EPS, defined as the maximal, fractional, incremental change of a system during an iteration, and the curves of $F(C)$

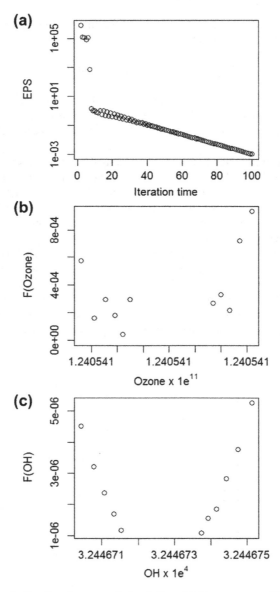

Figure 4.5 Newton's iteration from a real simulation. (a) Convergence of EPS with iteration times. (b) $F(O_3)$ versus O_3. (c) F(OH) versus OH. See text for details.

versus C for O_3 and OH in a real simulation. A system of 231 differential equations was solved using Newton's iteration to simulate coupled gas-phase photochemical reactions, vertical diffusions, emissions, and depositions of 33 species in seven layers of the atmospheric boundary layer over the Pacific Ocean offshore of Los Angeles, California, USA, in an early morning of mid-Summer. It is shown that Newton's iteration converged almost exponentially from a large initial error of $>100,000$ to a final error of 0.001. In the final 10 iterations, the relationship between $F(C)$ and C, as in equation (E4.2.1), was complex for both O_3 and OH.

4.2.3 Gear's method

Gear's method refers to a numerical method that solves for a general system of ordinary differential equations with automatic control of step size and order change using multistep information (Gear, 1971, pp. 158–166). The system of chemical ordinary differential equations described by equation (E4.1.2) or (E4.2.1) may be rewritten in the form:

$$C' = f(C) \tag{E4.2.3}$$

At an initial time t_0 (or "$n-1$"), information about chemical concentrations and their time derivatives in the system are stored in a matrix $A_{t0} = \{C_{t0}, \Delta t C_{t0}', ..., [\Delta t^{k-1} C_{t0}^{(k-1)}]/(k-1)!\}$, where $C_{t0}^{(k-1)}$ denotes the $(k-1)$th derivative of C_{t0}. At a later time t ($t_0 + \Delta t$, or "n"), the change of A is first predicted from information at time t_0 with explicit formula (F4.2.4) using chemical concentration information saved in previous k steps:

$$C_n = \sum_{i=1}^{k} (\alpha_i \cdot C_{n-1}) + \Delta t \cdot \beta_0 \cdot f(C_{n-1}) \tag{F4.2.4}$$

where parameters (α, β_0) are listed in Table 11.1–2 of Gear (1971) for orders 2–6, to be introduced shortly together with L. In the first step, the above formula is simply $C_n = C_{n-1} + \Delta t \cdot f(C_{n-1})$, or the first order. In later steps, the value of k may increase up to six times for chemical ordinary differential equations, in order to reach higher order solutions, namely to include more terms in the Taylor series using the same logic as the Runge–Kutta method. For example, when $k = 2$, $\alpha_i = (4/3, -1/3)$ and $\beta_0 = 2/3$, and when $k = 3$, $\alpha_i = (18/11, -9/11, 2/11)$ and $\beta_0 = 6/11$, which are the second- and third-order Runge–Kutta methods. Then, the roughly estimated values of A_t are corrected with a handful of iterations until convergence is reached to fulfill the implicit equation:

$$C_n = \sum_{i=1}^{k} (\alpha_i \cdot C_{n-1}) + \Delta t \cdot \beta_0 \cdot f(C_n) \tag{E4.2.4}$$

The solution of equation (E4.2.4) is given by formula (F4.2.5), which provides the corresponding corrector increment to update C_n from formula (F4.2.4) at the $(m+1)$

th iteration, using information of chemical ordinary differential equations at the mth iteration:

$$A_{t,(m+1)} = A_{t,(m)} - L \cdot W \cdot F(A_{t,m}) \tag{F4.2.5}$$

where L is a scalar corresponding to the order of the method and $W = [\partial F/\partial \mathbf{A} \cdot L]^{-1}$. For example, when $F(A) = hf(A_0) - A_1$, or $F(A_{t,m}) = [\Delta t(R_p - R_1) + C_0 - C_{n,m}]$ for chemical ordinary differential equations, $W = [-\mathbf{I} + \Delta t \cdot \beta_0 \cdot \mathbf{J}]^{-1}$.

As integration proceeds, the time step Δt from the input at the start is automatically adjusted in Gear's method by a multiplication factor that depends on the order of the method, desired accuracy, and relative changes of chemicals in the system.

Gear's method for stiff equations is accurate due to automatic controls of order and time step. However, as multistep information is used and the inversion of the partial derivative matrix is necessary, corresponding requirements of computer storage and time are large, which limited its application before a sparse-matrix, vectorized version became available in 1994 (Jacobson and Turco, 1994). Corresponding optimization techniques will be discussed later.

4.3   Empirical solvers

For a fixed set of chemical reactions, empirical methods may often be developed by classifying chemicals into slow reacting, short-lived, and exponential decaying categories.

4.3.1   Steady-state approximation

In a system of ordinary differential equations, some chemicals may have relatively very short chemical lifetimes compared with a target chemical. For example, in Table 4.3, the lifetime of O^1D against chemical production and loss is less than a second during daytime, much shorter than stratospheric O_3. At a reasonably short incremental time step for O_3 change, e.g. 1 minute, the change of the concentration of O^1D is approximately zero; this condition is termed "steady state" for O^1D. At steady state, the production and loss rates of O^1D are equal by definition. From equation (4.1.1) and Table 4.3, we have:

$$J_1[O_3] + J_2[N_2O] = [O^1D] (k_1 + k_2[H_2O] + k_3[N_2O] + k_4[H_2] + k_5[CF_2Cl_2])$$

and the steady-state concentration of O^1D is expressed as a function of concentrations of O_3, N_2O, H_2O, H_2, and CF_2Cl_2.

4.3.2   Family approximation

Some short-lived species may be combined together to form a family, which changes much more slowly than any member in the family. Then, the family instead of its members is included in the system of chemical ordinary differential equations, which results in a system of smaller size.

For example, in the upper mesosphere, with photochemical reactions listed in Table 4.1, the chemical lifetime of O is of the order of a day, while the lifetimes of O^1D, O_3, H, OH, and HO_2 are within 100 s. If combining odd hydrogen radicals into HO_x, then the lifetime of HO_x is of the order of a day. Assuming short-lived radicals are in a steady state, only O and HO_x are in the system of chemical ordinary differential equations.

Besides the HO_x family, reactive nitrogen species are often treated as a family in the troposphere. For example, NO_z may be defined to include $NO + NO_2 + HNO_2 + HNO_4 + NO_3 + 2N_2O_5$. In the gas phase, the loss of NO_z is via the formation of HNO_3 and organic nitrogen compounds. Thus, chemical production and loss of the NO_z family is much slower than the internal transformation between its members.

4.3.3 Separation of slow species

To reduce the size of a system of chemical ordinary differential equations, distinctly slow-reacting species may be solved separately from the coupled system. For example, in the troposphere, CH_4 and CO react with lifetimes of about 10 years and 1 month respectively, much slower than O_3, HO_x, NO_z, and their members. After steady-state and family approximations, concentrations of fast species may be solved before CO and CH_4. Then, concentrations of CO and CH_4 may be solved individually using equation (E4.1.1). In fact, some photochemical mechanisms, such as SAPRC, simply ignore the change of CH_4, as its change does not significantly affect their target chemicals in applied regions.

4.3.4 Polynomial fittings

In multidimensional atmospheric chemical modeling, many cells share the same system of chemical ordinary differential equations over many time steps. When computational limitations exist, detailed chemical mechanisms may not be directly implemented into multidimensional models. To represent target chemicals in such cases, the polynomial fitting method provides a vital alternative.

The polynomial fitting method consists of two steps. First, a chemical mechanism as detailed as possible is used in a box model to simulate a spectrum of chemical conditions covering the range expected in multidimensional modeling. The rates of chemical changes and other information are saved for the second step. Second, the net rates of changes of important chemicals that define the chemical environment for target chemicals are fitted as dependent variables against polynomial terms of important physical parameters, such as temperature and concentrations of important chemicals. The fitted formulae are then applied to calculate chemical changes in multidimensional atmospheric chemical modeling. This method has been success-fully applied in regional and global three-dimensional atmospheric chemical models to estimate O_3 budgets in North America and other regions (Jacob et al., 1993; Wang et al., 1998).

4.4   Optimization techniques

Among many techniques that may improve the accuracy or speed of a specific system of chemical ordinary differential equations, sparse-matrix, vectorized techniques improve the simulation speed of atmospheric chemical modeling without sacrificing accuracy.

4.4.1   Sparse-matrix techniques

To solve a system of chemical ordinary differential equations, accurate solvers that use Newton iteration need to inverse the partial derivative matrix, as shown in equations (E4.2.2a) and (E4.2.4). The inversion of a chemical partial derivative matrix, of size $N \times N$, includes decomposition and back substitution steps, which involves the number of multiplications of the order of N^3. For a typical atmospheric chemical system, most chemicals only interact with a few others. Thus, lots of elements in the partial derivative matrix are zero, and most of the N^3 multiplications are null.

Sparse-matrix techniques include two parts. First, the sequence of chemical ordinary differential equations may be rearranged to maximize the null multiplications. The null multiplications are maximized when chemicals with the fewest interactions with others in the system are listed at the top. Second, all non-null multiplications are identified for the decomposition and back substitution steps before chemical integrations. During chemical integrations, only non-null multiplications are conducted for each inversion. The sparser the partial derivative matrix, the more effective the sparse-matrix techniques. For example, in a simulation of boundary layer O_3 for a column of air over the Pacific Ocean offshore of Los Angeles, CA, USA, a total of 231 simultaneous differential equations were used to represent 33 model species in seven vertical layers. Of 53,361 elements of the partial derivative matrix, there are only 2335 non-zero elements. Without sparse-matrix techniques, the total number of floating-point operations for an inversion of the partial derivative matrix is 4,161,850; with the sparse-matrix techniques, the total number of floating-point operations is reduced to 205,570. Thus, sparse-matrix techniques saved over 95% of the floating-point operations for each inversion of the partial derivative matrix in the simulation. Most savings occurred in the decomposition step, as it accounted for almost 99% of the floating-point operations in the simulation.

4.4.2   Vectorization

While early computers calculated floating-point operations one by one, many modern computers conduct floating-point operations batch by batch. For example, assuming V_1 and V_2 are two one-dimensional arrays of size 128, the dot product of $V_1 \cdot V_2$ required 128 individual floating-point operations in the past but may only need one batch of operation nowadays in many modern vectorized computers. Usually, conducting one batch of N floating-point operations is many times faster than conducting N individual floating-point operations. Thus, vectorization may enhance the speed of floating-point operations by many times in modern computers.

For example, on an SGI origin 2000 workstation, a sparse-matrix, vectorized Gear solver, SMVGEAR (Jacobson and Turco, 1994; Jacobson, 1998a) could simulate 24-hour photochemical reactions in a cell using a version of the Carbon-Bond Mechanism in 0.536 seconds. For 500 cells, the computing time spent was 63.4 seconds. Thus, the saving of vectorization on the SGI workstation was 4.23 times. Using a version of the Master Chemical Mechanism, the corresponding computer time was 36.9 seconds for one cell and 4949 seconds for 500 cells; the saving of vectorization was 3.73 times. On a Cray J90 computer, which is a vectorized machine, the savings of vectorization amounted to 37.5 times for the Carbon-Bond Mechanism and 41.1 times for the Master Chemical Mechanism, an order of magnitude larger (Liang and Jacobson, 1999).

The significant benefit of vectorization in combination with sparse-matrix techniques described above has been successfully exploited by the SMVGEAR solver, which is faster than the original Gear code by two orders of magnitude. Before SMVGEAR became available in 1994, owing to the large number of chemical reactions occurring in the air, three-dimensional atmospheric chemical models either used empirical techniques to solve a fixed set of chemical reactions or applied numerical methods to solve comprehensive chemistry offline. Due to the accuracy and speed, SMVGEAR is widely applied in three-dimensional atmospheric chemical modeling nowadays, and may offer benchmark solutions for empirical methods.

4.5 Current issues

As chemicals change in the atmosphere due to simultaneous effects of transformation and transportation, simulation of chemical changes needs to carefully consider the time steps affordable by various chemicals and processes.

For example, most air pollution models integrate chemical changes due to transformation and transportation processes separately using a so-called operator-splitting technique. Namely, at a carefully chosen, common time step, chemical changes due to various processes are calculated one after the other. The accuracy of this method relies heavily on the understanding of the issues, which is reflected by the choice of the common time step.

Figure 4.6 illustrates the effect of using a time step that is probably too long for simulating regional O_3 responses to emission changes. To produce the two lines, two pairs of 3-day simulations were conducted over the greater Los Angeles area with horizontal resolution of 5 km × 5 km and bottom layer thickness of 30 m. The first pair of simulations was conducted using a coupled operator that includes chemical and vertical physical processes with sparse-matrix techniques, and the second pair of simulations used the operator-splitting technique for all processes with a time step of 5 minutes. In each pair of simulations, a baseline emission scenario and a hypothetic emission control scenario were used. The ratio of peak O_3 in each of the four counties in the greater Los Angeles area simulated from the two emission scenarios is plotted in the figure. The top line shows the result from the coupled operator, and the bottom line shows the result from the split operators with common time step of 5 minutes. It is

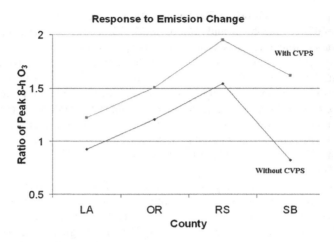

Figure 4.6 Simulated O_3 responses by the CVPS operator and split operators.

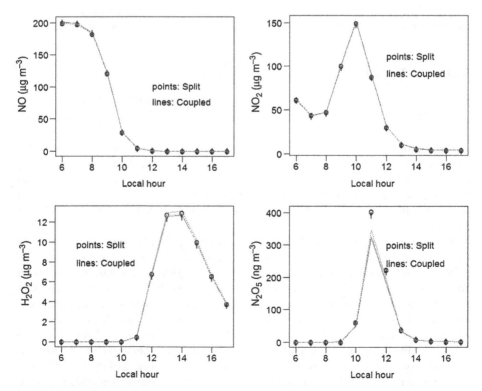

Figure 4.7 Simulated 12-hour time series of NO, NO_2, H_2O_2, and N_2O_5 over an isolated column of the atmospheric boundary layer near Los Angeles, CA, USA.

seen that the difference is significant, especially in Los Angeles and San Bernardino counties, where the sign of response differs between the coupled method and the split method with the common time step of 5 minutes.

Comparison between the two methods using a column of the atmospheric boundary layer is more optimistic. Figure 4.7 shows the 12-hour time series of NO, NO_2, H_2O_2, and N_2O_5 concentrations in the bottom layer simulated by the coupled operator and the split operators for an isolated column of the atmospheric boundary layer. It is seen that the two methods matched closely for the NO and NO_2 time series, while the peak values for H_2O_2 and N_2O_5 differed to some extent. Thus, it appears that horizontal advection in the three-dimensional model could amplify the difference from split operators in the simulation, which was known beforehand.

Summary

In this chapter, ordinary differential equations for atmospheric chemical systems in the mesosphere, stratosphere, troposphere, and atmospheric boundary layer have been elaborated. Accurate methods to systematically solve chemical changes in the above systems have been described, and empirical solution methods have also been described. For computational efficiency, general optimization techniques for matrix inversion have been presented. Finally, existing issues with simulating chemical changes in the atmosphere are discussed using an example for Los Angeles, California, USA.

Exercises

1. In the OH photochemical reaction system in the mesosphere, as defined in Table 4.1, (a) write the rate of change for $[HO_x]$ ($[H] + [OH] + [HO_2]$) and the rate of change for $[O]$ assuming $[O^1D]$ is at steady state, and (b) providing $[O_3]$ and $[OH]$ time series are measured, design a pathway for deriving $[H_2O]$ using the constraint of least squares error for $[HO_x]$ and $[O]$.
2. Survey the literature, and code a program to inverse a system of chemical ordinary differential equations. Print out the total number of floating-point operations for each inversion, with and without sparse-matrix techniques.
3. Apply a computer program, similar to the product of problem 2, to calculate the rates of change of an array of important chemicals that define the chemical environment for OH in the mesosphere in one day.
4. Under the same conditions as in problem 3, use the polynomial fitting method to represent the rates of chemical changes. Compare the time series from the polynomial fitting method and that from the accurate method.

Part Two

Applications

5 Ozone hole

O_3 is an interesting chemical in the modern atmosphere. In the natural atmosphere, O_3 is formed by photolysis of O_2 in the stratosphere. Without human perturbation, column O_3 could exceed 400 Dobson units (DU). From 1970 to 2010, the global long-term average column O_3 concentration was 270–310 DU. When photochemical smog pollution was prevalent in metropolitan Los Angeles areas during the 1960s, O_3 was the major pollutant of the smog and the measured peak mixing ratio reached 1 ppm. The origin of O_3 in the troposphere was initially thought to be mainly from the stratosphere via tropopause folding events until the mid-1990s, since 90% of global O_3 stays in the stratosphere. Though O_3 is a toxic gas in breathing zones, its presence further away from the Earth's surface acts as natural shield against harmful UV radiation.

5.1   Discovery of the stratospheric ozone hole

5.1.1   Measurement methods

Atmospheric O_3 concentration may be measured by two types of complementary methods. The first type is an electrochemical method that makes use of the oxidation

Chemical Modeling for Air Resources. http://dx.doi.org/10.1016/B978-0-12-408135-2.00005-7

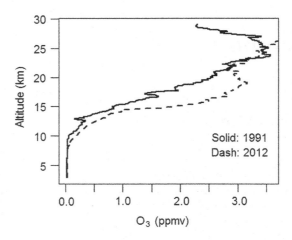

Figure 5.1 Vertical O_3 profiles from ozonesondes over Antarctica on September 4, 1991 and July 20, 2012.

of KI by O_3 in aqueous solution to produce a small electric current. This method is used in balloon-borne ozonesondes that can measure O_3 concentration together with air temperature, pressure, and relative humidity before balloons burst during ascent to about 30 km ASL. About 10% of the total O_3 above 30 km ASL is unmeasured by ozonesondes. The unique strength of this method is that vertical O_3 profiles from the surface to 30 km ASL may be measured day and night. In fact, ozonesonde data have been obtained over Antarctica since 1956 and are available at a network of ground stations around the globe (Logan, 1985; Chan et al., 1998; Lin et al., 2010). Figure 5.1 shows vertical O_3 profiles from ozonesondes launched over Antarctica on September 4, 1991 and July 20, 2012.

The second method involves spectroscopic measurement of atmospheric O_3 from its absorption spectrum in the ultraviolet, visible, and infrared regions (Lam et al., 2002). The spectrometers may be operated on the ground or mounted on to satellites. Satellite measurements have provided regional and global coverage during the daytime since the 1970s. Of course, satellite data need to be calibrated by ground observations.

5.1.2 Ozone over Antarctica

In 1985, scientists started to realize that stratospheric O_3 from ozonesonde measurements over Antarctica during austral spring had been significantly lower since 1980 than in the period from 1956 to 1970. Since 1986, year-round observations have been made over Antarctica with higher frequency during austral spring. In 1956–1970, column O_3 over Antarctica ranged from 300 to 330 DU; since 1980, column O_3 over Antarctica has decreased to below 100 DU. Complete depletion has often occurred in the range of the Antarctic polar vortex (12–20 km ASL), which is termed the "O_3 hole".

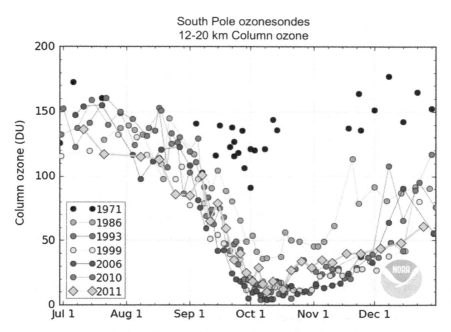

Figure 5.2 The column O_3 between 12 and 20 km ASL measured by ozonesondes launched at a station in Antarctica from July to December in selected years from 1971 to 2011. Obtained from NOAA, USA on August 8, 2012.

Figure 5.2 shows temporal variation of the column O_3 in the Antarctic vortex range measured by ozonesondes launched at a station in Antarctica from July to December in selected years from 1971 to 2011. It is seen that the column O_3 changed little in July and August during the polar night in all the years shown. During September and October of 1971, the column O_3 was only slightly lower than in other months. In 1986, the column O_3 decreased rapidly in September to fall below 50 DU in October. In 1993 and 2006, the column O_3 decreased to nearly 0 in October. In 2010 and 2011, the minimum column O_3 in the vortex was about 10 DU. By the end of December, the column O_3 recovered to the level found in early September in all the years shown. Thus, it is seen that the Antarctic O_3 hole has persisted for at least several weeks from late September to early October in recent years, based on ozonesonde observations.

While ozonesonde data clearly depicted the existence of the O_3 hole in the Antarctic vortex range in October in recent decades, the continental scale of stratospheric O_3 depletion has been mapped by satellite spectrometers. Figure 5.3 shows time series of total column O_3 averaged for all areas with latitudes from 70°S to 90°S in September and October from 1970 to 2011. No data are available from 1973 to 1978. It is seen that, on a continental scale, the 2-month average column O_3 for September and October has decreased from 310 to 270 DU in the last three and half decades. This trend is rather significant, considering that over 10% more harmful UV

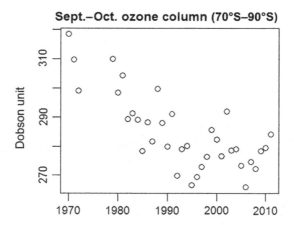

Figure 5.3 Time series 1970–2011 of average total column O_3 in September and October over Antarctic areas (70°S–90°S).

can reach 3% of the Earth's surface, despite the sparse population in the area, in 17% of the time as a result.

Observations also show that the minimum monthly total column O_3 averaged over a 5° × 10° area, i.e. not less than 27,630 km^2, in Antarctica decreased from over 250 DU in the early 1970s to below 150 DU in recent years (Figure 5.4), a 40% reduction. Meanwhile, areas with an O_3 hole have expanded significantly over Antarctica since 1979; in recent years, significant O_3 depletion occurred over 20 million km^2, according to data released by Nasa O_3 Watch, USA, as shown in Figure 5.5. It is seen that the minimum observed column O_3 in the Antarctic zone was below 100 DU in 1994, 1998, 2000, and 2006.

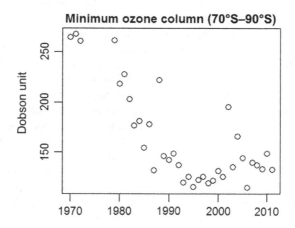

Figure 5.4 Time series 1970–2011 of minimum monthly total O_3 column in September and October over Antarctica (70°S–90°S) with horizontal area resolution of 5° × 10°.

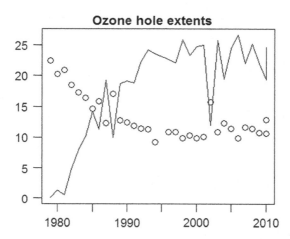

Figure 5.5 Extent of the ozone hole in terms of minimum O_3 (circles, 10 DU) and areas with low O_3 column (line, million km^2). Data from Nasa O_3 Watch, USA. The ozone hole area was averaged from September 7 to October 13, and the minimum O_3 was obtained from September 21 to October 16.

5.1.3 Global O_3 thinning

Besides the O_3 hole in the Antarctic vortex range, decadal observations from ground stations and satellites suggest that stratospheric O_3 has also changed in other areas (Lam et al., 2002). Figure 5.6 shows the distribution of minimal monthly column O_3 over a horizontal area of $5° \times 10°$ in all latitudinal intervals from 1970 to 2011. Empty areas in the figure indicate a lack of available data. It is shown that the minimal

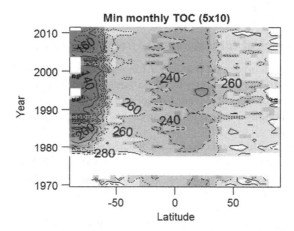

Figure 5.6 Minimal monthly total ozone columns (TOC) over a horizontal area of $5° \times 10°$ in all latitudinal intervals from 1970 to 2011.

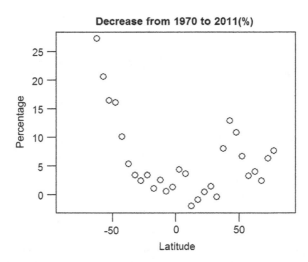

Figure 5.7 Global O_3 thinning from 1970 to 2011. Minimal monthly total O_3 column over a horizontal area of $5° \times 10°$ in 2011 was compared to that in 1970 at all latitudinal intervals.

monthly column O_3 in the Antarctic zone was lower than at other latitudes during the period, and values in mid-latitudes seem to decline too. The latter is clearly shown in Figure 5.7. By comparing minimal monthly column O_3 in all latitudinal intervals in 2011 with that in 1970, it is seen that the maximal depletion of column O_3 occurred in polar areas (27%) and mid-latitude zones (~10%). Near the Equator, the relative change was small, and the minimal monthly column O_3 even increased by 0.2–2% in latitudinal zones at 10–20°N and 30–35°N. The small rise at low latitudes in the northern hemisphere could be partly due to increased O_3 pollution in the troposphere as well as dynamical reasons.

5.1.4 Climatological total column O_3 from 2000 to 2011

Total column O_3 is an important parameter for atmospheric chemical modeling in the troposphere, as it affects photolysis rate constants of many reactions occurring near the surface. Based on recent decadal observations by satellites and at worldwide ground stations, monthly average total column O_3 may be obtained for representative latitudinal intervals.

Table 5.1 lists monthly total column O_3 at various latitudes averaged from 2000 to 2011. It is seen that total column O_3 ranged from 165 to 420 DU. Over equatorial areas, it changed little, with values of 250–270 DU. Over subtropical areas, total column O_3 peaked in spring and seasonal amplitudes were about 20%, ranging from 260 to 330 DU. At mid-latitudes, total column O_3 varied from 270 to 350 DU in the southern hemisphere, and from 290 to 380 DU in the northern hemisphere, with maximum in early spring. In northern high latitudes, total column O_3 peaked in spring (420 DU) and was minimal in fall (290 DU) based on available data; in southern high

Table 5.1 Climatological monthly total column O_3

	1	2	3	4	5	6	7	8	9	10	11	12
85°S	276	267	*307*	*289*	*264*	*240*	*265*	*265*	*170*	165	201	271
75°S	281	274	277	*274*	*266*	*257*	*248*	*241*	186	209	251	287
65°S	297	291	289	290	*286*	*287*	*287*	285	254	290	303	307
55°S	301	291	287	291	298	306	320	328	341	349	335	314
45°S	286	278	274	278	289	307	324	339	348	346	327	301
35°S	274	270	268	268	275	289	303	316	323	322	308	288
25°S	266	261	260	260	261	266	273	281	292	294	289	276
15°S	259	256	256	255	253	254	258	263	272	274	271	265
5°S	254	254	257	257	256	257	260	265	269	266	261	256
5°N	248	250	257	262	264	266	269	272	272	264	255	249
15°N	248	252	263	275	279	280	280	279	275	266	256	248
25°N	261	268	283	297	299	295	290	286	279	270	262	258
35°N	303	313	325	331	327	316	304	297	288	278	278	289
45°N	356	373	379	372	361	343	323	311	300	294	306	332
55°N	376	401	408	398	380	357	336	319	307	309	322	349
65°N	373	406	416	412	386	352	325	309	301	306	318	332
75°N	*389*	*411*	412	420	391	349	317	295	289	294	*327*	*354*
85°N	*385*	*401*	*405*	420	392	348	312	286	*296*	*302*	*323*	*360*

Note: Total column O_3 is in Dobson units, or 2.687×10^{16} molecules cm^{-3}. Central latitudes of 10° zones are listed, and numbers in bold italic denote estimated data with relatively large uncertainties.

latitudes, total column O_3 peaked in austral summer (290 DU) and showed minimum in austral spring (165 DU). Thus, there is a strong hemispheric asymmetry for total column O_3 in the polar stratosphere from the perspective of seasonal variations.

5.2 Mechanism of formation of the ozone hole

When the O_3 hole was first recognized in 1985, existing gas-phase photochemical mechanisms and atmospheric transport processes could not explain the rapid depletion of stratospheric O_3 over Antarctica. After decadal theoretical and experimental studies were completed, scientists found that reactive chlorine and bromine catalytically destroyed O_3 in the polar vortex via heterogeneous processes under 190 K. The details are elaborated below.

5.2.1 Gas-phase photochemical mechanism in the stratosphere

Photochemical reactions related to O_3 formation in the gas phase were relatively well understood when the Antarctic O_3 hole was first recognized. In the stratosphere, O_3 is believed to be formed only by the combination of O_2 and O, and the latter arises mainly from the photolysis of O_2 in the upper stratosphere, where short UV is sufficiently abundant. As the air stays in the stratosphere for a considerable time, O_3 is assumed to be at steady state. As O_2 concentration decreases with altitude and O concentration increases with altitude, O_3 concentration is maximal in the middle of the stratosphere at steady state. Without the involvement of odd hydrogen, odd nitrogen, and reactive halogen species, the steady-state concentration of O_3 is determined by the Chapman cycle, which usually overestimates O_3 concentration on a long-term average basis over a sufficiently large area in the stratosphere except in polar areas, where transport processes are important.

As water is ubiquitous in the stratosphere, odd hydrogen is naturally involved in reactions related to O_3. The addition of water-originated odd hydrogen to the Chapman system results in a lower concentration of O_3 at steady state. Besides odd hydrogen, the addition of odd nitrogen to the Chapman system is also believed to reduce O_3 concentration at steady state. Further reduction of O_3 concentration at steady state is caused by reactive halogen, and it is believed that each reactive halogen species may destroy a considerable amount of O_3 in the stratosphere. The gas-phase photochemical mechanism that includes the Chapman cycle and the destruction of O_3 by odd hydrogen, odd nitrogen, and reactive halogen may explain the steady-state O_3 concentration in the stratosphere. Diurnal variation of the steady-state concentration of O_3 is expected to be small in the stratosphere. At night, O_3 concentration may decrease slightly due to the absence of the production term and the presence of some loss terms.

However, the O_3 hole captured in the polar vortex over Antarctica occurred after a long polar night. In the first few weeks of polar dawn over Antarctica, stratospheric O_3 that survived from dark and cold winter was found to be rapidly depleted.

The early spring O_3 hole cannot be explained by a gas-phase photochemical mechanism. During winter time, the gas-phase photochemical engine stops. Thus, no significant amount of odd hydrogen, odd nitrogen, or reactive halogen is thought to be present in winter. When weak sunlight arrives in early spring, stratospheric O_3 is expected to gradually increase instead of rapidly decrease, based on a gas-phase photochemical mechanism related to O_3 in the stratosphere.

5.2.2   Unique environment of the O_3 hole

5.2.2.1   Dynamical features

Atmospheric transport is important for adjusting the radiative imbalance of different parts of the Earth. In order to maintain thermal balance in winter, tropical heat is transported to polar areas via thermal winds. Figure 5.8 depicts a theoretical circulation (Brewer–Dobson circulation) near the tropopause, which was inferred from horizontal fields of temperature, trends in O_3, tracers, and estimated age of air. It is seen that, as tropical air enters the lower stratosphere from the upper troposphere, it is transported to mid-latitudes and polar areas. In mid-latitudes, air mass from the lower stratosphere is transported to the upper troposphere. In polar areas, air mass descends to the bottom of the stratosphere. A polar vortex forms when zonally averaged monthly mean wind speed in high latitudes exceeds 15 m s^{-1}, which effectively blocks the air flow from mid-latitudes to polar areas.

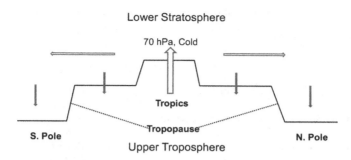

Figure 5.8 Illustration of theoretical Brewer–Dobson circulation near the tropopause.

Compared with the Arctic zone, Antarctica had unique features when the O_3 hole occurred. Firstly, the polar vortex over Antarctica was stronger and extended longer than over the Arctic zone due to geographical differences: while the surface of the Arctic zone is ocean, the surface of Antarctica is land; also, surrounding the Antarctic continent there are fairly uniform oceanic zones that are absent around the Arctic. During winter and early spring, a strong circular wind in the lower stratosphere isolated Antarctica for several months during the polar vortex period, and a strengthening westerly wind in high latitudes is usually associated with O_3 depletion in early spring. The polar vortex over the Arctic zone is usually weaker. Secondly, there are polar stratospheric clouds in the vortex over Antarctica, and the temperature within the cloud was often below 190 K. At this extremely low temperature, solid particles may be formed from H_2O, H_2SO_4, HNO_3, HCl, and HBr in the gas phase of the vortex.

5.2.2.2   Chemical features

Anthropogenic perturbation must be the driving force for the formation of the O_3 hole in the polar vortex over Antarctica. Since the wide use of stable halogen compounds for refrigeration, air conditioning, aerosol spray agents, foams, fire extinguishers, delicate electronic cleaning, etc., in the second half of the last century, anthropogenic halogen compounds have dominated over natural CH_3Cl and sea salt in the upper troposphere and stratosphere. So far, 50 non-radical stable halogen species, listed in Table 5.2, have been measured for their reactivities in stratospheric conditions. Via sunlight, these compounds are photolyzed to form reactive halogen radicals in the stratosphere.

5.2.3   Heterogeneous processes in polar stratospheric clouds

Laboratory experiments were conducted to study unique heterogeneous processes in the polar vortex over Antarctica. Within the range of measured concentrations, besides regular sulfuric acid aerosols, a fine aerosol haze composed of nitric acid and water in a molar ratio of about 1:3 started to form due to the low temperature in the early part of the Antarctic winter. In the later part of the Antarctic winter, larger water

Table 5.2 The 50 non-radical stable halogen species with photochemical data measured for stratospheric conditions

$1,2\text{-}(CF_3)_2\text{-}c\text{-}C_4F_6$	$CClF_3$	$CHCl_2CF_3$
	(CFC-13)	**(HCFC-123)**
C_4F_{10}	CF_3CF_3	$CHCl_2F$
	(PFC-116)	**(HCFC-21)**
C_5F_{12}	$CF_3CHFCHFCF_2CF_3$	$CHClF_2$
(PFC-41-12)	**(HFC-43-10mee)**	**(HCFC-22)**
C_6F_{14}	CF_4	$CHClFCF_3$
(PFC-51-14)	**(PFC-14)**	**(HCFC-124)**
CBr_2F_2	CH_2Br_2	CHF_2Br
(Halon-1202)		
$CBrF_2CBrF_2$	$CH_2ClCClF_2$	$CHF_2CF_2CF_2CHF_2$
(Halon-2402)	**(HCFC-132b)**	**(HFC-338pcc)**
$CBrF_3$	CH_2ClCF_3	CHF_2CF_3
(Halon-1301)	**(HCFC-133a)**	**(HFC-125)**
$c\text{-}C_4F_8$	CH_2F_2	CHF_3
	(HFC-32)	**(HFC-23)**
CCl_2F_2	CH_2FCF_3	Cl_2
(CFC-12)	**(HFC-134a)**	
CCl_2FCClF_2	CH_3Br	HBr
(CFC-113)		
CCl_2FCF_3	CH_3CCl_2F	HCl
(CFC-114a)	**(HCFC-141b)**	
CCl_3CF_3	CH_3CClF_2	HF
(CFC-113a)	**(HCFC-142b)**	
CCl_3F	CH_3CF_3	NF_3
(CFC-11)	**(HFC-143a)**	
CCl_4	CH_3CH_2F	SF_5CF_3
(CFC-10)	**(HFC-161)**	
$CClBrF_2$	CH_3CHF_2	SF_6
(Halon-1211)	**(HFC-152a)**	
$CClF_2CClF_2$	CH_3F	SO_2F_2
(CFC-114)	**(HFC-41)**	
$CClF_2CF_3$	$CHBr_3$	
(CFC-115)		

ice crystals started to form due to the extremely low temperature. The formation of these solid aerosol particles extended to early spring over Antarctica. As solid aerosols precipitate back to the surface, the conversion of HNO_3 and H_2O from the gas phase into the solid phase reduced the atmospheric concentrations of total nitrogen and total hydrogen in the stratosphere. Thus, denitrification and dehydration occurred in the polar vortex over Antarctica.

The O_3 hole must be related to the presence of the solid aerosol particles, which are unique in the polar vortex over Antarctica. On the surface of solid aerosol particles, reservoir chemicals of nitrogen and halogens, i.e. $ClONO_2$, $BrONO_2$, HCl, and HBr, indeed react to release halogen back to Cl_2, Br_2, BrCl, ClOOCl, ClOOBr, and BrOOBr, which are readily available for photochemical reactions in the gas phase in early spring. When sunlight appeared in early spring, the air in the polar vortex over Antarctica had been denitrified and dehydrated. Meanwhile, relatively large amounts of photochemically reactive chlorine and bromine were present. These reactive halogen species catalytically destroy O_3 in the usual manner, except with higher efficiency in the polar vortex over Antarctica in early spring than at other times and places due to less competition from odd nitrogen and odd hydrogen species. Since the presence of a polar vortex effectively blocked the transport of O_3 from mid-latitudes, stratospheric O_3 was depleted in the polar vortex over Antarctica in several weeks due to photochemical loss terms dominated by chlorine and bromine radicals.

Table 5.3 lists the major photochemical reactions responsible for the production and loss of O_3 in the stratosphere, which have come from several decades of research conducted by international communities with special sponsorship by NASA, USA for the compilation of data (Sanders et al., 2011). In Table 5.3, reactants and products of 111 photolysis reactions, 11 heterogeneous reactions and 299 gas-phase kinetic reactions are listed, and effective wavelengths are noted for photolysis reactions. Up-to-date cross-sections, quantum yields, parameters for reactions on aerosols in the polar vortex over Antarctica, and gas-phase kinetic parameters may be found from the original literature or the latest compilation.

5.3   Simulating the ozone hole in the stratosphere

To amend the O_3 hole, budget analysis and sophisticated computer simulations have been useful, as discussed below.

5.3.1   Budget analysis

The global long-term average of O_3 in the stratosphere is regulated by the atmospheric photochemical reactions listed in Table 5.3. The controlling factors of stratospheric O_3 abundance may be understood by conducting a chemical budget analysis for O_x, defined as $O_3 + O$. From Table 5.3, the only production term of O_x is the photolysis of O_2. The destruction terms include mainly reactions 2, 15, 19, 24, 49, 81, 145, 233, 254, 277, 279, 283, and 294. Thus, the steady-state concentration of O_3 is given by:

$$[O_3]_{ss} \approx J[O_2]/\{k_2[O] + k_{15}[H] + k_{19}[OH] + k_{24}[HO_2] + fk_{49}[NO] + k_{81}[F]$$
$$+ k_{145}[Cl] + k_{233}[Br] + k_{254}[SO_2] + k_{277}[S] + k_{279}[SO] + k_{283}[SH]$$
$$+ k_{294}[CS]\}$$

$$(F5.3.1a)$$

where f corrects for the fraction of $[NO_2]$ converted back to NO by O in reaction 29 versus the fraction by photolysis. In the extreme case, if all other species are constant

Table 5.3 Reactions for stratospheric O_3

(I) Gas-phase kinetic reactions
(a) Odd O

1	$O + O_2 \xrightarrow{M} O_3$
2	$O + O_3 \rightarrow O_2 + O_2$
3	$O^1D + O_2 \rightarrow O + O_2$
4	$O^1D + N_2 \rightarrow O + N_2$

(b) + Odd H

5	$O^1D + H_2 \rightarrow OH + H$
6	$O^1D + H_2O \rightarrow OH + OH$
7	$O_1D + NH_3 \rightarrow OH + NH_2$
8	$O^1D + CH_4 \rightarrow CH_3 + OH$
9	$O^1D + CH_4 \rightarrow CH_3O + H$
10	$O^1D + CH_4 \rightarrow CH_2O + H_2$
11	$O + OH \rightarrow O_2 + H$
12	$O + HO_2 \rightarrow OH + O_2$
13	$O + H_2O_2 \rightarrow OH + HO_2$
14	$H + O_2 \xrightarrow{M} HO_2$
15	$H + O_3 \rightarrow OH + O_2$
16	$H + HO_2 \rightarrow 2OH$
17	$H + HO_2 \rightarrow O + H_2O$
18	$H + HO_2 \rightarrow H_2 + O_2$
19	$OH + O_3 \rightarrow HO_2 + O_2$
20	$OH + H_2 \rightarrow H_2O + H$
21	$OH + OH \rightarrow H_2O + O$
22	$OH + OH \xrightarrow{M} H_2O_2$
23	$OH + H_2O_2 \rightarrow H_2O + HO_2$
24	$HO_2 + O_3 \rightarrow OH + 2O_2$
25	$HO_2 + HO_2 \rightarrow H_2O_2 + O_2$
26	$HO_2 + HO_2 \xrightarrow{M} H_2O_2 + O_2$
27	$HO_2 + HO_2 \rightarrow H_2O_2 + O_2$

(c) + Odd N

28	$O + NO \xrightarrow{M} NO_2$
29	$O + NO_2 \rightarrow NO + O_2$
30	$O + NO_2 \xrightarrow{M} NO_3$
31	$O^1D + N_2O \rightarrow N_2 + O_2$
32	$O^1D + N_2O \rightarrow NO + NO$
33	$O + NO_3 \rightarrow O_2 + NO_2$
34	$O + HNO_3 \rightarrow OH + NO_3$
35	$H + NO_2 \rightarrow OH + NO$
36	$OH + NO \xrightarrow{M} HONO$
37	$OH + NO_2 \overset{M}{\leftrightarrow} HNO_3$
38	$OH + HONO \rightarrow H_2O + NO_2$
39	$OH + HNO_3 \rightarrow H_2O + NO_3$
40	$OH + NH_3 \rightarrow H_2O + NH_2$
41	$HO_2 + NO \rightarrow NO_2 + OH$

Table 5.3 Reactions for stratospheric O_3—*cont'd*

42		$H_2O + NO_2 \rightarrow OH + HONO$
43		$HO_2 + NO_2 \overset{M}{\leftrightarrow} HO_2NO_2$
44		$HO_2 + NO_2 \rightarrow HONO + O_2$
45		$N + O_2 \rightarrow NO + O$
46		$N + O_3 \rightarrow NO + O_2$
47		$N + NO \rightarrow N_2 + O$
48		$N + NO_2 \rightarrow N_2O + O$
49		$NO + O_3 \rightarrow NO_2 + O_2$
50		$NO + NO_3 \rightarrow 2NO_2$
51		$NO_2 + O_3 \rightarrow NO_3 + O_2$
52		$NO_2 + NO_3 \rightarrow NO + NO_2 + O_2$
53		$NO_2 + NO_3 \overset{M}{\leftrightarrow} N_2O_5$
54		$NO_3 + NO_3 \rightarrow 2NO_2 + O_2$
55		$O_3 + HNO_2 \rightarrow O_2 + HNO_3$
56		$N_2O_5 + H_2O \rightarrow 2HNO_3$
(d) + CH$_4$		
57		$OH + CH_4 \rightarrow CH_3 + H_2O$
58		$CH_3 + O_2 \overset{M}{\rightarrow} CH_3O_2$
59		$CH_3O_2 + NO \rightarrow CH_3O + NO_2$
60		$CH_3O_2 + NO_2 \overset{M}{\leftrightarrow} CH_3O_2NO_2$
61		$HO_2 + CH_3O_2 \rightarrow CH_3OOH + O_2$
62		$CH_3O + O_2 \rightarrow CH_2O + HO_2$
(e) + F		
63		$O + FO \rightarrow F + O_2$
64		$O + FO_2 \rightarrow FO + O_2$
65		$OH + CH_3F \rightarrow CH_2F + H_2O$
66		$OH + CH_2F_2 \rightarrow CHF_2 + H_2O$
67		$OH + CHF_3 \rightarrow CF_3 + H_2O$
68		$OH + CH_2FCH_2F \rightarrow CHFCH_2F + H_2O$
69		$OH + CH_3CF_3 \rightarrow CH_2CF_3 + H_2O$
70		$OH + CH_2FCF_3 \rightarrow CHFCF_3 + H_2O$
71		$OH + CHF_2CHF_2 \rightarrow CF_2CHF_2 + H_2O$
72		$OH + CHF_2CF_3 \rightarrow CF_2CF_3 + H_2O$
73		$OH + CH_2FCF_2CF_3 \rightarrow CHFCF_2CF_3 + H_2O$
74		$OH + CF_3CH_2CF_3 \rightarrow CF_3CHCF_3 + H_2O$
75		$OH + CF_3CHFCF_3 \rightarrow CF_3CFCF_3 + H_2O$
76		$OH + CF_3OH \rightarrow CF_3O + H_2O$
77		$OH + CH_3OCF_3 \rightarrow CH_2OCF_3 + H_2O$
78		$OH + CHF_2OCHF_2 \rightarrow CF_2OCHF_2 + H_2O$
79		$OH + CHF_2OCF_3 \rightarrow CF_2OCF_3 + H_2O$
80		$F + O_2 \overset{M}{\leftrightarrow} FO_2$
81		$F + O_3 \rightarrow FO + O_2$
82		$F + H_2 \rightarrow HF + H$
83		$F + H_2O \rightarrow HF + OH$
84		$F + NO \overset{M}{\rightarrow} FNO$
85		$F + NO_2 \overset{M}{\rightarrow} FNO_2$

(*Continued*)

Table 5.3 Reactions for stratospheric O_3—*cont'd*

86	$F + HNO_3 \rightarrow HF + NO_3$
87	$F + CH_4 \rightarrow HF + CH_3$
88	$FO + NO \rightarrow NO_2 + F$
89	$FO + NO_2 \overset{M}{\rightarrow} FONO_2$
90	$FO + FO \rightarrow 2F + O_2$
91	$FO_2 + NO \rightarrow FNO + O_2$
92	$CF_3 + O_2 \overset{M}{\rightarrow} CF_3O_2$
93	$CF_3O + M \rightarrow F + CF_2O + M$
94	$CF_3O + O_2 \rightarrow FO_2 + CF_2O$
95	$CF_3O + O_3 \rightarrow CF_3O_2 + O_2$
96	$CF_3O + H_2O \rightarrow OH + CF_3OH$
97	$CF_3O + NO \rightarrow CF_2O + FNO$
98	$CF_3O + NO_2 \overset{M}{\rightarrow} CF_3ONO_2$
99	$CF_3O + CO \overset{M}{\rightarrow} CF_3OCO$
100	$CF_3O + CH_4 \rightarrow CH_3 + CF_3OH$
101	$CF_3O_2 + O_3 \rightarrow CF_3O + 2O_2$
102	$CF_3O_2 + CO \rightarrow CF_3O + CO_2$
103	$CF_3O_2 + NO \rightarrow CF_3O + NO_2$
104	$CF_3O_2 + NO_2 \overset{M}{\rightarrow} CF_3O_2NO_2$
(f) + Cl	
105	$O + ClO \rightarrow Cl + O_2$
106	$O + OClO \rightarrow ClO + O_2$
107	$O + OClO \overset{M}{\rightarrow} ClO_3$
108	$O + Cl_2O \rightarrow ClO + ClO$
109	$O + HCl \rightarrow OH + Cl$
110	$O + HOCl \rightarrow OH + ClO$
111	$OH + Cl_2 \rightarrow HOCl + Cl$
112	$OH + ClO \rightarrow Cl + HO_2$
113	$OH + ClO \rightarrow HCl + O_2$
114	$OH + OClO \rightarrow HOCl + O_2$
115	$OH + Cl_2O \rightarrow HOCl + ClO$
116	$OH + Cl_2O_2 \rightarrow HOCl + ClOO$
117	$OH + HCl \rightarrow H_2O + Cl$
118	$OH + HOCl \rightarrow H_2O + ClO$
119	$OH + ClNO_2 \rightarrow HOCl + NO_2$
120	$OH + CH_3Cl \rightarrow CH_2Cl + H_2O$
121	$OH + CH_2Cl_2 \rightarrow CHCl_2 + H_2O$
122	$OH + CHCl_3 \rightarrow CCl_3 + H_2O$
123	$OH + CH_2FCl \rightarrow CHClF + H_2O$
124	$OH + CHFCl_2 \rightarrow CFCl_2 + H_2O$
125	$OH + CHF_2Cl \rightarrow CF_2Cl + H_2O$
126	$OH + CH_3CCl_3 \rightarrow CH_2CCl_3 + H_2O$
127	$OH + CH_3CFCl_2 \rightarrow CH_2CFCl_2 + H_2O$
128	$OH + CH_3CF_2Cl \rightarrow CH_2CF_2Cl + H_2O$
129	$OH + CH_2ClCF_2Cl \rightarrow CHClCF_2Cl + H_2O$
130	$OH + CH_2ClCF_3 \rightarrow CHClCF_3 + H_2O$
131	$OH + CHCl_2CF_2Cl \rightarrow CCl_2CF_2Cl + H_2O$

Table 5.3 Reactions for stratospheric O_3—*cont'd*

132	$OH + CHFClCFCl_2 \rightarrow CFClCFCl_2 + H_2O$
133	$OH + CHCl_2CF_3 \rightarrow CCl_2CF_3 + H_2O$
134	$OH + CHFClCF_2Cl \rightarrow CFClCF_2Cl + H_2O$
135	$OH + CHFClCF_3 \rightarrow CFClCF_3 + H_2O$
136	$OH + CF_3CH_2CFCl_2 \rightarrow CF_3CHCFCl_2 + H_2O$
137	$OH + CCl_3CHO \rightarrow H_2O + CCl_3CO$
138	$HO_2 + Cl \rightarrow HCl + O_2$
139	$HO_2 + Cl \rightarrow OH + ClO$
140	$HO_2 + ClO \rightarrow HOCl + O_2$
141	$NO + OClO \rightarrow NO_2 + ClO$
142	$NO_3 + OClO \xrightarrow{M} O_2ClONO_2$
143	$NO_3 + HCl \rightarrow HNO_3 + Cl$
144	$Cl + O_2 \overset{M}{\leftrightarrow} ClOO$
145	$Cl + O_3 \rightarrow ClO + O_2$
146	$Cl + H_2 \rightarrow HCl + H$
147	$Cl + H_2O_2 \rightarrow HCl + HO_2$
148	$Cl + NO \xrightarrow{M} NOCl$
149	$Cl + NO_2 \xrightarrow{M} ClONO$
150	$Cl + NO_3 \rightarrow ClO + NO_2$
151	$Cl + N_2O \rightarrow ClO + N_2$
152	$Cl + CO \xrightarrow{M} ClCO$
153	$Cl + CH_4 \rightarrow HCl + CH_3$
154	$Cl + H_2CO \rightarrow HCl + HCO$
155	$Cl + CH_3OH \rightarrow CH_2OH + HCl$
156	$Cl + OClO \rightarrow ClO + ClO$
157	$Cl + ClOO \rightarrow Cl_2 + O_2$
158	$Cl + ClOO \rightarrow ClO + ClO$
159	$Cl + Cl_2O \rightarrow Cl_2 + ClO$
160	$Cl + ClNO \rightarrow NO + Cl_2$
161	$Cl + CH_3Cl \rightarrow CH_2Cl + HCl$
162	$Cl + CH_2Cl_2 \rightarrow HCl + CHCl_2$
163	$Cl + CHCl_3 \rightarrow HCl + CCl_3$
164	$Cl + CH_3F \rightarrow HCl + CH_2F$
165	$Cl + CH_2F_2 \rightarrow HCl + CHF_2$
166	$Cl + CHF_3 \rightarrow HCl + CF_3$
167	$Cl + CH_2FCl \rightarrow HCl + CHFCl$
168	$Cl + CHFCl_2 \rightarrow HCl + CFCl_2$
169	$Cl + CHF_2Cl \rightarrow HCl + CF_2Cl$
170	$Cl + CH_3CCl_3 \rightarrow CH_2CCl_3 + HCl$
171	$Cl + CH_3CH_2F \rightarrow HCl + CH_3CHF$
172	$Cl + CH_3CH_2F \rightarrow HCl + CH_2CH_2F$
173	$Cl + CH_3CHF_2 \rightarrow HCl + CH_3CF_2$
174	$Cl + CH_3CHF_2 \rightarrow HCl + CH_2CHF_2$
175	$Cl + CH_2FCH_2F \rightarrow HCl + CHFCH_2F$
176	$Cl + CH_3CFCl_2 \rightarrow HCl + CH_2CFCl_2$
177	$Cl + CH_3CF_2Cl \rightarrow HCl + CH_2CF_2Cl$

(*Continued*)

Table 5.3 Reactions for stratospheric O_3—*cont'd*

178	$Cl + CH_3CF_3 \rightarrow HCl + CH_2CF_3$
179	$Cl + CH_2FCHF_2 \rightarrow HCl + CH_2FCF_2$
180	$Cl + CH_2FCHF_2 \rightarrow HCl + CHFCHF_2$
181	$Cl + CH_2ClCF_3 \rightarrow HCl + CHClCF_3$
182	$Cl + CH_2FCF_3 \rightarrow HCl + CHFCF_3$
183	$Cl + CHF_2CHF_2 \rightarrow HCl + CF_2CHF_2$
184	$Cl + CHCl_2CF_3 \rightarrow HCl + CCl_2CF_3$
185	$Cl + CHFClCF_3 \rightarrow HCl + CFClCF_3$
186	$Cl + CHF_2CF_3 \rightarrow HCl + CF_2CF_3$
187	$Cl + C_2Cl_4 \overset{M}{\rightarrow} C_2Cl_5$
188	$ClO + O_3 \rightarrow ClOO + O_2$
189	$ClO + O_3 \rightarrow OClO + O_2$
190	$ClO + NO \rightarrow NO_2 + Cl$
191	$ClO + NO_2 \overset{M}{\rightarrow} ClONO_2$
192	$ClO + NO_3 \rightarrow ClOO + NO_2$
193	$ClO + ClO \rightarrow Cl_2 + O_2$
194	$ClO + ClO \rightarrow ClOO + Cl$
195	$ClO + ClO \rightarrow OClO + Cl$
196	$ClO + ClO \overset{M}{\leftrightarrow} Cl_2O_2$
197	$ClO + OClO \overset{M}{\leftrightarrow} Cl_2O_3$
198	$CH_2Cl + O_2 \overset{M}{\leftrightarrow} CH_2ClO_2$
199	$CHCl_2 + O_2 \overset{M}{\leftrightarrow} CHCl_2O_2$
200	$CCl_3 + O_2 \overset{M}{\rightarrow} CCl_3O_2$
201	$CFCl_2 + O_2 \overset{M}{\rightarrow} CFCl_2O_2$
202	$CF_2Cl + O_2 \overset{M}{\rightarrow} CF_2ClO_2$
203	$CCl_3O_2 + NO_2 \overset{M}{\rightarrow} CCl_3O_2NO_2$
204	$CFCl_2O_2 + NO_2 \overset{M}{\rightarrow} CFCl_2O_2NO_2$
205	$CF_2ClO_2 + NO_2 \overset{M}{\rightarrow} CF_2ClO_2NO_2$
206	$CH_2ClO + O_2 \rightarrow CHClO + HO_2$
207	$CH_2ClO_2 + HO_2 \rightarrow CH_2ClO_2H + O_2$
208	$CH_2ClO_2 + NO \rightarrow CH_2ClO + NO_2$
209	$CCl_3O_2 + NO \rightarrow CCl_2O + NO_2 + Cl$
210	$CCl_2FO_2 + NO \rightarrow CClFO + NO_2 + Cl$
211	$CClF_2O_2 + NO \rightarrow CF_2O + NO_2 + Cl$
(g) + Br	
212	$O + BrO \rightarrow Br + O_2$
213	$O + HBr \rightarrow OH + Br$
214	$O + HOBr \rightarrow OH + BrO$
215	$O + BrONO_2 \rightarrow NO_3 + BrO$
216	$OH + Br_2 \rightarrow HOBr + Br$
217	$OH + HBr \rightarrow H_2O + Br$
218	$OH + CH_3Br \rightarrow CH_2Br + H_2O$
219	$OH + CH_2Br_2 \rightarrow CHBr_2 + H_2O$
220	$OH + CHBr_3 \rightarrow CBr_3 + H_2O$

Table 5.3 Reactions for stratospheric O_3—*cont'd*

221	$OH + CHF_2Br \rightarrow CF_2Br + H_2O$
222	$OH + CH_2ClBr \rightarrow CHClBr + H_2O$
223	$OH + CH_2BrCF_3 \rightarrow CHBrCF_3 + H_2O$
224	$OH + CHFBrCF_3 \rightarrow CFBrCF_3 + H_2O$
225	$OH + CHClBrCF_3 \rightarrow CClBrCF_3 + H_2O$
226	$OH + CHFClCF_2Br \rightarrow CFClCF_2Br + H_2O$
227	$HO_2 + Br \rightarrow HBr + O_2$
228	$NO_3 + HBr \rightarrow HNO_3 + Br$
229	$Cl + CH_3Br \rightarrow HCl + CH_2Br$
230	$Cl + CH_2Br_2 \rightarrow HCl + CHBr_2$
231	$Cl + CHBr_3 \rightarrow CBr_3 + HCl$
232	$Cl + CH_2ClBr \rightarrow HCl + CHClBr$
233	$Br + O_3 \rightarrow BrO + O_2$
234	$Br + H_2O_2 \rightarrow HBr + HO_2$
235	$Br + NO_2 \overset{M}{\rightarrow} BrNO_2$
236	$Br + NO_3 \rightarrow BrO + NO_2$
237	$Br + H_2CO \rightarrow HBr + HCO$
238	$Br + OClO \rightarrow BrO + ClO$
239	$Br + Cl_2O \rightarrow BrCl + ClO$
240	$BrO + NO \rightarrow NO_2 + Br$
241	$BrO + NO_2 \overset{M}{\rightarrow} BrONO_2$
242	$BrO + ClO \rightarrow Br + OClO$
243	$BrO + ClO \rightarrow Br + ClOO$
244	$BrO + ClO \rightarrow BrCl + O_2$
245	$CH_2BrO_2 + NO \rightarrow CH_2O + NO_2 + Br$
(h) + S	
246	$O + SH \rightarrow SO + H$
247	$O + CS \rightarrow CO + S$
248	$O + H_2S \rightarrow OH + SH$
249	$O + OCS \rightarrow CO + SO$
250	$O + CS_2 \rightarrow CS + SO$
251	$O + SO_2 \overset{M}{\rightarrow} SO_3$
252	$O + CH_3SCH_3 \rightarrow CH_3SO + CH_3$
253	$O + CH_3SSCH_3 \rightarrow CH_3SO + CH_3S$
254	$O_3 + SO_2 \rightarrow SO_3 + O_2$
255	$OH + H_2S \rightarrow SH + H_2O$
256	$OH + CS_2 \rightarrow SH + OCS$
257	$OH + CS_2 \rightarrow CS_2OH$
258	$OH + CH_3SH \rightarrow CH_3S + H_2O$
259	$OH + CH_3SCH_3 \rightarrow H_2O + CH_2SCH_3$
260	$OH + S \rightarrow H + SO$
261	$OH + SO \rightarrow H + SO_2$
262	$OH + SO_2 \overset{M}{\rightarrow} HOSO_2$
263	$NO_3 + CH_3SCH_3 \rightarrow CH_3SCH_2 + HNO_3$
264	$Cl + H_2S \rightarrow HCl + SH$

(*Continued*)

Table 5.3 Reactions for stratospheric O_3—*cont'd*

265	$Cl + CH_3SH \rightarrow CH_3S + HCl$
266	$Cl + CH_3SCH_3 \rightarrow CH_3SCH_2 + HCl$
267	$Cl + CH_3S(O)CH_3 \rightarrow CH_3S(O)CH_2 + HCl$
268	$Cl + CH_3S(O)CH_3 \xrightarrow{M} CH_3(Cl)S(O)CH_3$
269	$ClO + SO \rightarrow Cl + SO_2$
270	$ClO + SO_2 \rightarrow Cl + SO_3$
271	$Br + H_2S \rightarrow HBr + SH$
272	$Br + CH_3SH \rightarrow CH_3S + HBr$
273	$Br + CH_3SCH_3 \rightarrow CH_3SCH_2 + HBr$
274	$Br + CH_3SCH_3 \xrightarrow{M} (CH_3)_2SBr$
275	$BrO + SO \rightarrow Br + SO_2$
276	$S + O_2 \rightarrow SO + O$
277	$S + O_3 \rightarrow SO + O_2$
278	$SO + O_2 \rightarrow SO_2 + O$
279	$SO + O_3 \rightarrow SO_2 + O_2$
280	$SO + NO_2 \rightarrow SO_2 + NO$
281	$SO + OClO \rightarrow SO_2 + ClO$
282	$SH + O_2 \rightarrow OH + SO$
283	$SH + O_3 \rightarrow HSO + O_2$
284	$SH + NO \xrightarrow{M} HSNO$
285	$SH + NO_2 \rightarrow HSO + NO$
286	$SH + N_2O \rightarrow HSO + N_2$
287	$SH + Cl_2 \rightarrow ClSH + Cl$
288	$SH + Br_2 \rightarrow BrSH + Br$
289	$SH + F_2 \rightarrow FSH + F$
290	$HSO + NO_2 \rightarrow HSO_2 + NO$
291	$HSO_2 + O_2 \rightarrow HO_2 + SO_2$
292	$HOSO_2 + O_2 \rightarrow HO_2 + SO_3$
293	$CS + O_2 \rightarrow OCS + O$
294	$CS + O_3 \rightarrow OCS + O_2$
295	$CS + NO_2 \rightarrow OCS + NO$
296	$CH_3S + NO_2 \rightarrow CH_3SO + NO$
297	$CH_3SO + NO_2 \rightarrow CH_3SO_2 + NO$
298	$CH_3SCH_2 + O_2 \xrightarrow{M} CH_3SCH_2O_2$
299	$CH_3SCH_2O_2 + NO \rightarrow CH_3S + CH_2O + NO_2$

(II) Photolysis reactions

1	$O_2 + h\nu \ (205\text{--}245 \ nm) \rightarrow O + O$
2	$O_3 + h\nu \ (122\text{--}828 \ nm) \rightarrow O_2 + fO^1D + (1-f)O$
3	$HO_2 + h\nu \ (190\text{--}260 \ nm) \rightarrow OH + fO^1D + (1-f)O$
4	$H_2O + h\nu \ (121\text{--}198 \ nm) \rightarrow H + OH$
5	$H_2O_2 + h\nu \ (190\text{--}350 \ nm) \rightarrow OH + OH$
6	$NO_2 + h\nu \ (241\text{--}663 \ nm) \rightarrow NO + O$
7	$NO_3 + h\nu \ (403\text{--}691 \ nm) \rightarrow fNO_2 + fO + (1-f)NO + (1-f)O_2$
8	$N_2O + h\nu \ (160\text{--}240 \ nm) \rightarrow N_2 + O^1D$
9	$N_2O_4 + h\nu \ (252\text{--}390 \ nm) \rightarrow 2NO_2$
10	$N_2O_5 + h\nu \ (200\text{--}420 \ nm) \rightarrow NO_3 + NO_2$
11	$HONO + h\nu \ (184\text{--}396 \ nm) \rightarrow OH + NO$

Table 5.3 Reactions for stratospheric O_3—*cont'd*

12	$HNO_3 + h\nu$ (192–350 nm) $\rightarrow$ OH + NO_2
13	$HO_2NO_2 + h\nu$ (190–350 nm) $\rightarrow$ HO_2 + NO_2
14	$CH_2O + h\nu$ (226–375 nm) $\rightarrow$ CO + 2H
15	$CH_3OOH + h\nu$ (210–405 nm) $\rightarrow$ CH_3OH + O
16	$HOCH_2OOH + h\nu$ (205–360 nm) $\rightarrow$ $HOCH_2O$ + OH
17	$CH_3ONO + h\nu$ (190–440 nm) $\rightarrow$ CH_3O + NO
18	$CH_3ONO_2 + h\nu$ (190–344 nm) $\rightarrow$ CH_3O + NO_2
19	$CH_3O_2NO_2 + h\nu$ (200–290 nm) $\rightarrow$ CH_3O_2 + NO_2
20	$HC(O)OH + h\nu$ (195–250 nm) $\rightarrow$ CO + OH + H
21	$FO_2 + h\nu$ (186–266 nm) $\rightarrow$ F + O_2
22	$F_2O + h\nu$ (210–546 nm) $\rightarrow$ F + FO
23	$F_2O_2 + h\nu$ (200–600 nm) $\rightarrow$ FO + FO
24	$FNO + h\nu$ (180–350 nm) $\rightarrow$ F + NO
25	$COF_2 + h\nu$ (186–229 nm) $\rightarrow$ COF + F
26	$COHF + h\nu$ (200–266 nm) $\rightarrow$ CO + HF
27	$CF_3O_2 + h\nu$ (191–276 nm) $\rightarrow$ CF_3O + O
28	$CF_3OH + h\nu$ (185–300 nm) $\rightarrow$ CF_3 + OH
29	$CF_3OOCF_3 + h\nu$ (200–263 nm) $\rightarrow$ $2CF_3O$
30	$CF_3C(O)F + h\nu$ (200–280 nm) $\rightarrow$ CF_3 + FCO
31	$CF_3C(O)Cl + h\nu$ (200–325 nm) $\rightarrow$ CF_3 + ClCO
32	$CF_3OONO_2 + h\nu$ (185–340 nm) $\rightarrow$ CF_3O_2 + NO_2
33	$Cl_2 + h\nu$ (260–550 nm) $\rightarrow$ Cl + Cl
34	$ClO + h\nu$ (245–316 nm) $\rightarrow$ Cl + O
35	$ClOO + h\nu$ (220–280 nm) $\rightarrow$ Cl + O_2
36	$OClO + h\nu$ (247–472 nm) $\rightarrow$ O + ClO
37	$ClOOCl + h\nu$ (200–420+ nm) $\rightarrow$ 2ClO
38	$HCl + h\nu$ (135–230 nm) $\rightarrow$ H + Cl
39	$HOCl + h\nu$ (200–420 nm) $\rightarrow$ Cl + OH
40	$ClNO + h\nu$ (190–500 nm) $\rightarrow$ Cl + NO
41	$ClNO_2 + h\nu$ (195–400 nm) $\rightarrow$ Cl + NO_2
42	$ClONO + h\nu$ (235–400 nm) $\rightarrow$ Cl + NO_2
43	$ClONO_2 + h\nu$ (196–432 nm) $\rightarrow$ ClO + NO_2
44	$CCl_4 + h\nu$ (174–275 nm) $\rightarrow$ Cl + CCl_3
45	$CH_3OCl + h\nu$ (230–374 nm) $\rightarrow$ CH_3O + Cl
46	$CHCl_3 + h\nu$ (180–256 nm) $\rightarrow$ $f\,CHCl_2 + f\,Cl + (1-f)CCl_3 + (1-f)H$
47	$CH_2Cl_2 + h\nu$ (176–256 nm) $\rightarrow$ CH_2Cl + Cl
48	$CH_3Cl + h\nu$ (174–236 nm) $\rightarrow$ CH_3 + Cl
49	$ClCHO + h\nu$ (240–307 nm) $\rightarrow$ CO + Cl + H
50	$ClCFO + h\nu$ (186–261 nm) $\rightarrow$ CO + F + Cl
51	$CFCl_3$ (CFC-11) $+ h\nu$ (174–260 nm) $\rightarrow$ Cl + $CFCl_2$
52	CF_2Cl_2 (CFC-12) $+ h\nu$ (170–240 nm) $\rightarrow$ Cl + CF_2Cl
53	CF_3Cl (CFC-13) $+ h\nu$ (172–220 nm) $\rightarrow$ Cl + CF_3
54	$CF_2ClCFCl_2$ (CFC-113) $+ h\nu$ (175–250 nm) $\rightarrow$ $Cl + f\,CF_2ClCFCl + (1-f)CF_2CFCl_2$
55	CF_2ClCF_2Cl (CFC-114) $+ h\nu$ (172–235 nm) $\rightarrow$ Cl + CF_2ClCF_2
56	CF_3CF_2Cl (CFC-115) $+ h\nu$ (172–230 nm) $\rightarrow$ Cl + CF_3CF_2

(Continued)

Table 5.3 Reactions for stratospheric O_3—*cont'd*

57	$CHFCl_2$ (HCFC-21) + $h\nu$ (174–236 nm) → $Cl + CHFCl$
58	CHF_2Cl (HCFC-22) + $h\nu$ (170–220 nm) → $Cl + CHF_2$
59	CH_2FCl (HCFC-31) + $h\nu$ (160–230 nm) → $Cl + CH_2F$
60	CF_3CHCl_2 (HCFC-123) + $h\nu$ (170–250 nm) → $Cl + CF_3CHCl$
61	CF_3CHFCl (HCFC-124) + $h\nu$ (170–230 nm) → $Cl + CF_3CHF$
62	CF_3CH_2Cl (HCFC-133) + $h\nu$ (160–245 nm) → $Cl + CF_3CH_2$
63	CH_3CFCl_2 (HCFC-141b) + $h\nu$ (170–240 nm) → $Cl + CH_3CFCl$
64	CH_3CF_2Cl (HCFC-142b) + $h\nu$ (170–230 nm) → $Cl + CH_3CF_2$
65	CH_2ClCHO + $h\nu$ (240–357 nm) → $CH_2Cl + CO + H$
66	$CHCl_2CHO$ + $h\nu$ (256–353 nm) → $CHCl_2 + CO + H$
67	CF_2ClCHO + $h\nu$ (235–370 nm) → $CF_2Cl + CO + H$
68	$CFCl_2CHO$ + $h\nu$ (235–370 nm) → $CFCl_2 + CO + H$
69	CCl_3CHO + $h\nu$ (200–344 nm) → $CCl_3 + CO + H$
70	$CH_3C(O)Cl$ + $h\nu$ (191–302 nm) → $CH_3 + CO + Cl$
71	$CH_2ClC(O)Cl$ + $h\nu$ (235–316 nm) → $CH_2Cl + CO + Cl$
72	$CHCl_2C(O)Cl$ + $h\nu$ (220–300 nm) → $CHCl_2 + CO + Cl$
73	$CCl_3C(O)Cl$ + $h\nu$ (167–338 nm) → $CCl_3 + CO + Cl$
74	$CF_3CF_2CHCl_2$ (HCFC-225ca) + $h\nu$ (160–239 nm) → $Cl + CF_3CF_2CHCl$
75	CF_2ClCF_2CHFCl (HCFC-225cb) + $h\nu$ (160–210 nm) → $Cl + CF_2CF_2CHFCl$
76	$CH_3C(O)CH_2Cl$ + $h\nu$ (210–360 nm) → $CO + CH_3 + CH_2Cl$
77	Br_2 + $h\nu$ (200–650 nm) → $2Br$
78	HBr + $h\nu$ (152–230 nm) → $H + Br$
79	BrO + $h\nu$ (287–385 nm) → $Br + O$
80	$OBrO$ + $h\nu$ (401–568 nm) → $O + BrO$
81	Br_2O + $h\nu$ (196–400 nm) → $BrO + Br$
82	$HOBr$ + $h\nu$ (250–550 nm) → $OH + Br$
83	$BrNO$ + $h\nu$ (189–708 nm) → $Br + NO$
84	$BrONO$ + $h\nu$ (200–364 nm) → $BrO + NO$
85	$BrNO_2$ + $h\nu$ (185–580 nm) → $Br + NO_2$
86	$BrONO_2$ + $h\nu$ (200–500 nm) → $BrO + NO_2$
87	$BrCl$ + $h\nu$ (200–600 nm) → $Br + Cl$
88	$BrOCl$ + $h\nu$ (230–390 nm) → $Br + ClO$
89	CH_3Br + $h\nu$ (174–290 nm) → $Br + CH_3$
90	CH_2Br_2 + $h\nu$ (174–300 nm) → $CH_2Br + Br$
91	$CHBr_3$ + $h\nu$ (170–362 nm) → $CHBr_2 + Br$
92	$BrCBrO$ + $h\nu$ (240–331 nm) → $CO + 2Br$
93	$BrCHO$ + $h\nu$ (240–324 nm) → $CO + Br + H$
94	CH_2ClBr (Halon-1011) + $h\nu$ (187–290 nm) → $CH_2Cl + Br$
95	$CHClBr_2$ (Halon-1012) + $h\nu$ (200–310 nm) → $CHClBr + Br$
96	$CHCl_2Br$ (Halon-1021) + $h\nu$ (200–320 nm) → $CHCl_2 + Br$
97	CCl_3Br (Halon-1031) + $h\nu$ (235–365 nm) → $CCl_3 + Br$
98	CHF_2Br (Halon-1201) + $h\nu$ (168–280 nm) → $CHF_2 + Br$
99	CF_2Br_2 (Halon-1202) + $h\nu$ (170–340 nm) → $CF_2Br + Br$
100	CF_2ClBr (Halon-1211) + $h\nu$ (170–320 nm) → $CF_2Cl + Br$
101	CF_3Br (Halon-1301) + $h\nu$ (168–300 nm) → $CF_3 + Br$

Table 5.3 Reactions for stratospheric O_3—cont'd

102	CF_3CH_2Br (Halon-2301) + $h\nu$ (190–294 nm) → CF_3CH_2 + Br
103	$CF_3CHClBr$ (Halon-2311) + $h\nu$ (170–310 nm) → CF_3CHCl + Br
104	CF_3CHFBr (Halon-2401) + $h\nu$ (190–280 nm) → CF_3CHF + Br
105	CF_2BrCF_2Br (Halon-2402) + $h\nu$ (170–320 nm) → CF_2BrCF_2 + Br
106	CF_3CF_2Br (Halon-2501) + $h\nu$ (190–300 nm) → CF_3CF_2 + Br
107	SO_2 + $h\nu$ (180–235 nm) → SO + O
108	SO_3 + $h\nu$ (180–330 nm) → SO_2 + O
109	H_2S + $h\nu$ (180–250 nm) → HS + H
110	CS_2+ $h\nu$ (275–370 nm) → CS + S
111	OCS + $h\nu$ (186–296 nm) → CO + S
(III) Aerosol reactions	
1	N_2O_5 + H_2O (s) → 2 HNO_3 (s)
2	$ClONO_2$ + HCl (s) → Cl_2 + HNO_3 (s)
3	$BrONO_2$ + HBr (s) → Br_2 + HNO_3 (s)
4	$ClONO_2$ + HBr (s) → BrCl + HNO_3 (s)
5	$BrONO_2$ + HCl (s) → BrCl + HNO_3 (s)
6	HOCl + HCl (s) → Cl_2 + H_2O (s)
7	HOBr + HBr (s) → Br_2 + H_2O (s)
8	HOCl + HBr (s) → BrCl + H_2O (s)
9	HOBr + HCl (s) → BrCl + H_2O (s)
10	$ClONO_2$ + H_2O (s) → HNO_3 (s) + HOCl
11	$BrONO_2$ + H_2O (s) → HNO_3 (s) + HOBr

and halogen radicals are denoted by equivalent chlorine (Cle), then formula (F5.3.1a) may be simplified to:

$$[O_3]_{ss}(\text{ppmv}) \approx 0.21 \times 10^6 / \{C + k[\text{Cle}]\} \qquad \text{(F5.3.1b)}$$

From the above formula, stratospheric O_3 would be inversely proportional to the equivalent chlorine concentration if the latter is high enough, so that the total rate of reactions 81, 145, and 233 dominates that of other reactions in the denominator of formula (F5.3.1a).

5.3.2 Computer simulations

To aid the mitigation of the O_3 hole, a computer simulation method has been applied to investigate the processes responsible for the O_3 hole and to predict O_3 recovery rates due to various control scenarios.

5.3.2.1 One-dimensional microphysical modeling of the O_3 hole

As the air in the polar vortex over Antarctica was effectively isolated from the rest of the atmosphere, one-dimensional microphysical modeling was used to simulate the formation of the observed O_3 hole when computational resources were limited in the early 1990s (Cicerone et al., 1991). In the simulation, seasonal concentrations of nitrogen and chlorine compounds estimated from two-dimensional models were used to initialize photochemical calculations in mid-winter due to the lack of measurements. During winter, fine particles of polar stratospheric clouds formed initially from sulfuric acid, nitric acid and water vapor in the gas phase. Coarse particles formed a little later. When particles were present, HCl (s) and $ClONO_2$ were converted to Cl_2 (g), $ClNO_2$ (g), HOCl (g), and HNO_3 (s) via heterogeneous reactions while N_2O_5 was converted to HNO_3 (s). The formation of coarse particles and heterogeneous reactions resulted in denitrification and a simultaneous increase in the ratio of reactive chlorine to reactive nitrogen. The simulation indicated that 90% of O_3 in the polar vortex could be destroyed by catalytic reactions of reactive chlorine in a month in early austral spring, when extra HCl was available to react with $ClONO_2$ during the sunlit period to reduce reactive nitrogen and to increase reactive chlorine. Inclusion of reactive hydrocarbons in the polar vortex was shown to increase O_3 in most cases.

5.3.2.2 Chemistry–climate modeling of the O_3 hole

As computational resources have increased dramatically in the past two decades, a number of three-dimensional atmospheric models have become available for simulating atmospheric events such as the O_3 hole, which is defined as when total column O_3 is below 220 DU in this section.

A recent study applied 10 computer models, which include chemistry–climate interactions, to investigate the recovering rate of the Antarctic O_3 hole from 2000 to 2100. These models simulated atmospheric parameters together with stratospheric O_3 chemistry in three-dimensional grid boxes, and included the affects of greenhouse gases. In particular, the 1960–2100 time series of surface halogen compounds was simulated from prescribed halogen emissions and surface concentrations of greenhouse gases, with fixed solar activity and without volcanic emissions. Observations since 1960 have not been used to correct the models. The quasi-biennial oscillation signal was simulated by a few models participating in the inter-comparison. Sea-surface temperature and sea ice cover was simulated by one participating model, while others used results from ocean model simulations. Chemistry–climate models forecasted O_3 mixing ratios, temperature, and winds at various pressure levels ranging from 20 to 150 hPa, corresponding to most of the O_3 situated over Antarctica.

Monthly mean total column O_3 zonally averaged from 60°S to 90°S was extracted for the months of September to December from 10 participating models for data analysis (Siddaway et al., 2012). As the equivalent chlorine started to increase rapidly before 1985, the average levels of 1970–1979, instead of the 1980 value, were used as the baseline value despite small perturbations to stratospheric O_3 by anthropogenic halogen emissions prior to this period. Simulated total column O_3 time series during 1960–2100 from the 10 participating models all showed a "V" shape, with a steep decrease from 1970 to the early 2000s and a gradual increase until 2100 for all four months. The minimum monthly total column O_3 was simulated to occur in October, followed by September and November. The simulated monthly O_3 change in December from 1960 to early 2000 was about half of that in October. Simulated vertical profiles of O_3 mixing ratio in the southern hemisphere were featured with the maximum loss at 50 hPa in September, at 70 hPa in October, and at 100 hPa in November and December. Meanwhile, the simulated O_3 mixing ratio at 150 hPa was closer to the baseline value than higher above. The simulated O_3 deviation from values on the baseline was closely correlated with simulated temperature deviation in the polar vortex. Between 1975 and 2000, simulated temperature at 100 hPa decreased by ~1 K per decade in September and October, which is close to observations by satellite, radiosonde, and lidar. Simulated temperature decrease was ~2 K per decade in November and December, which is larger than the observations of 1–1.5 K per decade. Simulated winds suggested the persistence of the polar vortex over Antarctica throughout the austral spring. Before 2000, simulated polar westerlies increased during November and December; after 2000, simulated polar westerlies decreased during November and December. During the months of September and October, no change was simulated for polar westerlies. In addition, the response of simulated wind change to changes in the O_3 mixing ratio in the polar vortex lagged for about a month.

Simulated total column O_3 from 1979 to 2011 was compared to satellite observations, and the difference between model and observations in September to November was larger than in December, which is consistent with the current status of our understanding of the O_3 hole. Despite the relatively large difference between individual models and satellite observations, the ensemble of 10 models captured the magnitudes of the observed monthly total column O_3 over Antarctica. Simulated vertical profiles of O_3 mixing ratios were also compared with ozonesonde observations, and model results were consistent with observations.

According to model simulations, the Antarctic O_3 hole centered at ~50 hPa during September and October is expected to gradually recover to the average levels of 1970–1979, around 2070 (±5). The recovery is expected to be sooner at 150 hPa, while no full recovery was predicted for O_3 mixing ratios at levels of 70–100 hPa before 2100.

5.4 Chronology of the ozone hole and remaining issues

The "O_3 hole" represents a success story resulting from international cooperation between research, industry, and government agencies. A brief recall of important events regarding the O_3 hole and remaining issues are discussed below.

5.4.1 Chronology of the O_3 hole

Since the discovery of the O_3 hole over Antarctica in 1985, significant efforts from global research, industry, and government communities have been conducted to prevent O_3 depletion from getting worse and to amend the O_3 hole. Table 5.4 lists the important events related to these efforts. Briefly, Farman et al. (1985) identified significant O_3 depletion in the polar vortex over Antarctica from ozonesonde observations during early austral spring. Careful review of satellite records confirmed that O_3 depletion was

Table 5.4 Chronology of important events related to stratospheric O_3 depletion

Year	Scientific assessment and international treaties
1981	The Stratosphere 1981: Theory and Measurements (WMO No. 11)
1985	**Discovery of ozone hole over Antarctica** (Farman et al., 1985)
	Atmospheric Ozone 1985 (WMO No. 16)
1987	*Montreal Protocol* (peak effective chlorine >15 ppb before 2100)
1988	International Ozone Trends Panel Report 1988 (WMO No. 18)
1989	Scientific Assessment of Stratospheric Ozone: 1989 (WMO No. 20)
1990	*London Adjustment and Amendment* (peak effective chlorine $= 10$ ppb in 2100 and continued to increase after that)
1991	Scientific Assessment of Ozone Depletion: 1991 (WMO No. 25)
1992	Methyl Bromide: Its Atmospheric Science, Technology, and Economics (UNEP, 1992)
1992	*Copenhagen Adjustment and Amendment* (peak effective chlorine < 3 ppb)
1994	Scientific Assessment of Ozone Depletion: 1994 (WMO No. 37)
1995	**Nobel Prize in Chemistry awarded for ozone research.** *Vienna Adjustment*
1997	Montreal Adjustment and Amendment
1998	Scientific Assessment of Ozone Depletion: 1998 (WMO No. 44)
1999	*Beijing Adjustment and Amendment* (peak effective chlorine < 3 ppb)
2002	Scientific Assessment of Ozone Depletion: 2002 (WMO No. 47)
2006	Scientific Assessment of Ozone Depletion: 2006 (WMO No. 50)
2007	Montreal Adjustment
2010	Scientific Assessment of Ozone Depletion: 2010 (WMO No.52)

occurring on a larger scale. In 1987, the first international treaty was established in Montreal, Canada, to reduce anthropogenic emission of organic halogen compounds (CFCs, CCl_4, CH_3CCl_3, halons, CH_3Br, and HCFCs), which are stable in the troposphere but photolyze in the stratosphere to destroy O_3. With the Montreal Protocol, the equivalent chlorine, defined as total halogen weighted with corresponding O_3 depletion potentials scaled to chlorine, would be halved by 2040. However, the equivalent chlorine would continue to increase in the stratosphere up to 15 ppb before 2080. In 1990, an amendment was made in London to further reduce anthropogenic emissions. With the amendment in London, the equivalent chlorine would be 60% lower than the Montreal Protocol in 2080. However, the equivalent chlorine would continue to rise after 2080; by 2100, it would reach 10 ppb. In 1992, another amendment was made in Copenhagen to thoroughly control anthropogenic emissions of O_3-depleting substances that contain chlorine and bromine. With the Copenhagen amendment, the equivalent chlorine would stop increasing before 2000, and the peak value in the stratosphere would be less than 3 ppb. In 1995, the Nobel Prize in Chemistry was awarded to three atmospheric chemists for their work on the formation and decomposition of O_3 in the stratosphere. Further amendments were made in Beijing and Montreal to eliminate anthropogenic emissions of O_3-depleing substances that contain chlorine and bromine.

5.4.2   Benefits and remaining issues

The global atmospheric burden of the equivalent chlorine started to decrease after anthropogenic emissions were abated. Initially, the decrease was mainly due to short-lived chemicals such as methyl chloride and methyl bromide. For example, during the decade 1993–2005, the equivalent chlorine in the troposphere decreased by 8%. As O_3-depleting substances that contain chlorine and bromine have typical lifetimes of several decades to several centuries, trends of the effective chlorine in the stratosphere need to be evaluated over a longer time. For example, if the equivalent chlorine in the stratosphere showed a uniform loss rate of 1% per year, the annual transport from the troposphere to the stratosphere must be less than 1% for the equivalent chlorine to decrease in the stratosphere.

The benefit to stratospheric O_3 recovering from worldwide efforts on reducing anthropogenic emissions of O_3-depleting substances that contain chlorine and bromine may be reflected by the global column O_3 listed in Table 5.1. In a recent decade, monthly column O_3 was within several percent of levels in 1960s over tropical, subtropical areas and mid-latitudes, while global thinning in column O_3 was a major concern in the 1990s. The O_3 hole in the polar vortex over Antarctica was still present in the last decade, probably because the threshold of effective chlorine for the formation of the O_3 hole was still exceeded. As the exact temporal and spatial extents of the formation of the O_3 hole is still unpredictable due to the nature of the long-term timescale and the complex natural and anthropogenic processes involved, continuing investigations are warranted for the benefits of residents and tourists in southern latitudes before Antarctic O_3 reaches safe levels.

As refrigeration, air conditioning, aerosol spray, foams, fire extinguishers, and delicate electronic cleaning are indispensable in civilized communities, halogen compounds that are either more stable or more reactive than O_3-depleting substances are employed as substitutes. For example, the HFCs and organic fluorides, listed in Table 5.2, are widely used to substitute for CFCs. While most HFCs decompose in the troposphere, some organic fluorides have lifetimes of thousands of years and may only decompose in the mesosphere, where the O_3 shield is not an issue at all. On one hand, emissions of HFCs instead of CFCs are expected to enhance the oxidizing power of the troposphere, and contribute to local air pollution. On the other hand, emissions of super-stable organic fluorides may impose significant perturbation to the mesosphere; the effect on human exposure is expected to be benign in the short term, while the effect on other atmospheric phenomena needs further investigation.

Summary

In this chapter, the discovery of the alarming springtime O_3 hole over Antarctica is considered by first describing methods of measurement of atmospheric O_3. Then, the identification of the O_3 hole over the South Pole is described, followed by an overview of the evidence of global O_3 thinning. The updated climatological column O_3 in a recent decade is tabulated for atmospheric chemical modeling. The mechanism of formation of the O_3 hole is elucidated by describing the stratospheric gas-phase mechanism, the unique dynamic and super-cold environment, and heterogeneous chemical reactions on polar stratospheric cloud particles. Computer simulations of the O_3 hole are analyzed from the perspectives of budget analysis, one-dimensional microphysical modeling, and three-dimensional chemistry–climate modeling. Finally, the chronology of important events regarding the O_3 hole, international control efforts and benefits, and remaining concerns are discussed.

Exercises

1. Use data from Table E5.1 to estimate the cumulative emissions for halogen compounds with a lifetime longer than 50 years, by assuming (1) the concentration is at steady state, (2) the emission started in 1970, and (3) the troposphere has been a closed system. Comment on the additional assumptions necessary for the estimation.

2. Conduct the same estimates as above for compounds with lifetimes shorter than 30 years by assuming the emission started only 10 years ago.

3. If all emissions are banned from 2011, estimate (1) when the total chlorine from the five compounds will be reduced by 50% from the level in 2010, and (2) how much total chlorine will be left in 2100. Assume that products from the dissociations of the five compounds precipitate to the surface instantly.

4. Assuming (1) total reactive chlorine was 5 ppb in the polar vortex over Antarctica in 2010, (2) the total reactive chlorine is proportional to the total chlorine from the five compounds listed in the table, and (3) the minimum column O_3 is inversely proportional to total reactive chlorine, estimate when the O_3 hole will recover if the O_3 depletion in 2010 was 90% of the reference level.

5. In the real atmosphere, reactive chlorine and bromine contributed about two-thirds and one-third of the O_3 depletion in the polar vortex over Antarctica respectively. Conduct the same analysis as in Exercise 4, by assuming (1) that the lifetimes in the polar vortex were the same as in the troposphere, (2) that total bromine had a lifetime of 65 years, and (3) that reactive bromine contributed one-third of the O_3 depletion in 2010.

6. A genius educator Steve designed a four-box model, shown in Figure E5.1, to introduce numerical modeling techniques to graduate students for explaining why the stratospheric O_3 mixing ratio could be higher over the polar area than over the tropics.

   In the above diagram, each box has the same amount of air mass and, over a season, the amount of air mass is about constant. In-situ measurements by a brave graduate student onboard a modern hot-air balloon found that the meridian wind speed (V) is

Table E5.1 Global surface concentrations, emissions, and lifetimes of selected halogen compounds in 2010

	Concentration (ppt)	Emission (Gg year^{-1})	Lifetime (years)
CFC-11	235	50	45
CFC-12	525	50	100
CFC-113	75	0	85
CCl_4	87	60	26
HCFC-22	200	350	12

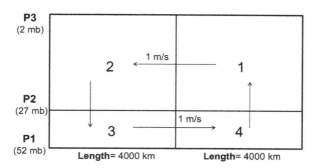

Figure E5.1 A four-box model.

about 1 m s^{-1} from box 1 to box 2 and from box 3 to box 4, while the length of each box, L, is 4000 km. Measurements of trace chemicals and radiation fields indicate that the photochemical steady-state mixing ratio of O_3, $P(i)$, is about 8, 3, 1, and 3.5 ppm for boxes 1–4 respectively, and the corresponding pseudo-first-order loss rate coefficient, $k(i)$, is about 70, 20, 2, and 2.5 × 10^{-8} s^{-1} respectively.

After the graduate student reported the above data to Steve, he worked until midnight to derive a formula that calculates the bulk O_3 mixing ratio for each box, $O_3(i)$, from $P(i)$, V/L, and $k(i)$.

Steve's equation was:

$$O_3(i) \cdot Y(i) = P(i)k(i)/R(i) + P(i-1)k(i-1)R/R(i, i-1)$$
$$+ P(i-2)k(i-2)R^2/R(i, i-1, i-2)$$
$$+ P(i-3) \cdot k(i-3) \cdot R^3/R(1, 2, 3, 4) \quad \text{(F6e)}$$

where:

$$R = V/L$$

$$R(i) = R + k(i)$$

$$R(i, i-1) = [R+k(i)] \cdot [R+k(i-1)]$$

$$R(i, i-1, i-2) = [R+k(i)] \cdot [R+k(i-1)] \cdot [R+k(i-2)]$$

$$R(1, 2, 3, 4) = [R+k(1)] \cdot [R+k(2)] \cdot [R+k(3)] \cdot [R+k(4)]$$

and

$$Y(i) = 1 - R^4/R(1, 2, 3, 4)$$

Index $i = (1, 2, 3, 4)$ while the corresponding index $i - 1 = (4, 1, 2, 3)$, which reflects that box $i - 1$ is upstream of box i for mass advection.

(a) Using Steve's formula (F6e), calculate the bulk O_3 mixing ratio for each box in the diagram. Survey the literature to assess if your results are reasonable.

(b) Prove Steve's formula by assuming that mass advection exactly balanced photochemical change for O_3 in each box and that the air mass was in hydrostatic balance.

6 Acid rain

Chapter Outline

Acid rain refers to rain, snow, and fog with a pH value below 5. In the modern atmosphere, unpolluted rain has a pH value of 5.6 due to CO_2. Anthropogenic activities, such as combustion of fossil fuels and biomass, emit large amounts of SO_2 and NO_x into the air. In the air, SO_2 is oxidized into sulfuric acid and NO_x is oxidized into inorganic and organic nitrates. Due to their solubility, SO_2, H_2SO_74, HNO_3, NH_3 and other soluble acids may be dissolved in atmospheric waters to form acid rain. As acid rain is deposited on the surface, ecosystems are disturbed. Since the 1980s, worldwide campaigns have been conducted to mitigate adverse effects from acid deposition. The details are elaborated below.

6.1 Observations of acid rain

6.1.1 Adverse effects of acid deposition

Acid deposition started to draw worldwide attention in the 1980s, when scientists from Europe, North America, and East Asia agreed that atmospheric sulfur deposition was linked to widespread damage to forests, soils and waters, and to materials including historic monuments. Under certain conditions, acid deposition may have harmful effects on human health.

Chemical Modeling for Air Resources. http://dx.doi.org/10.1016/B978-0-12-408135-2.00006-9

Atmospheric nitrogen deposition in the eastern USA in recent years accounted for 60–80%, 20–50%, and 15–25% of total nitrogen inputs in some coastal river estuaries in Chesapeake Bay and Tampa Bay (in northern New England), and in the Mississippi Delta region of the Gulf of Mexico respectively. This significant contribution is partly responsible for widespread eutrophication and associated hypoxia in those estuaries, and may have affected the growth of marine plankton in the ocean.

Acid deposition has adverse effects on soils. It accelerates the leaching of basic cations such as aluminum, calcium and magnesium, which help counteract acidification, from soils in acid-sensitive areas. Forest soils may accumulate sulfate and nitrate deposited from the atmosphere, and soil water may contain elevated aluminum due to acid precipitation. At high concentrations, aluminum may hinder uptake of water and nutrients by plant roots. In parts of the northeastern USA, acid precipitation has reduced available calcium in soils by 50% (Driscoll et al., 2001).

In developed countries, such as the USA and the European Union, sulfur emission has been reduced significantly in the past 30 years. It may be tempting to think that acid rain pollution has disappeared. However, scientists have warned that this is not the case for two reasons. Firstly, sensitive ecosystems have been depleted of acid neutralization capacity, and are thus more sensitive to acidification than before. Secondly, though SO_2 emission has decreased greatly, NO_x emission has increased in some areas. Though NH_4NO_3 is a nutrient for plants, nitrate is an important component of acid rain (Driscoll et al., 2001).

6.1.2 Transboundary acidification

The acidification of lakes was first discovered in Scandinavia in the 1960s. Continental Europe was believed to be partly responsible for the ecological damage due to large quantities of anthropogenic sulfur emissions that could be transported hundreds of kilometers due to the atmospheric lifetime of SO_2. In 1972, the United Nations Conference on the Human Environment in Stockholm initiated international cooperation to combat acidification. Member countries of the United Nations Economic Commission for Europe adopted the Convention on Long-range Transboundary Air Pollution at Geneva in November 1979, which has been effective since 1983, with 49 parties including North America due to the sharing of polar region and the fact that lakes in northeastern USA and eastern Canada are also acidified by anthropogenic acid deposition. Eight specific protocols have resulted from the convention, to monitor and reduce SO_x, NO_x, NH_3, VOCs, heavy metals, persistent organic pollutants, acidification, eutrophication, and ground-level O_3. Since the early 1980s, acid rain has been monitored in many developing countries, such as China, as well as in the European Union and North America.

6.1.3 Acid deposition in the USA

The USA monitors acid deposition at a number of ground stations representing urban, rural, and background conditions, under the umbrella of the National Acid Precipitation Assessment Program (NAPAP). To assess the effectiveness of the acid rain program that aims to reduce emissions of SO_2 and NO_x from electricity generating units, the US EPA and National Park Service established the Clean Air Status and Trends Network (CASTNET) to track real-world results as emission reductions became effective. Since 1987, dry deposition of sulfate and ammonium nitrate have been measured at the CASTNET network, which consisted of 84 monitoring stations throughout contiguous North America in 2010.

Figure 6.1 illustrates the annual mean SO_2 and sulfate concentrations ($\mu g\ m^{-3}$) at surface stations over the USA in 2010. It is shown that SO_2 concentration peaked in upper midwest areas with a value of 7.3. In the western USA, SO_2 concentration was typically below 1. The sulfate concentration ranged from 2.4 to 3.6 over about one-third of the USA, with a peak value in the upper midwest. In the western half of the USA, sulfate ranged from 0.4 to 1.1. Figure 6.2 shows annual mean concentrations of total nitrate ($NO_3^- + HNO_3$) and ammonium ($\mu g\ m^{-3}$) at ground stations over the USA in 2010. It is seen that the total nitrate concentration was relatively high in the upper midwest, with values of 3–4; in the western USA, total nitrate was below 2 except in southern California. Ammonium concentration ranged from 1 to 1.6 in the midwest and surrounding areas, and was below 0.6 in the rest of the contiguous USA.

Dry deposition was calculated from atmospheric concentration and modeled dry deposition velocities. Dry deposition of sulfur via SO_2 and sulfate had a similar pattern, not shown, to their atmospheric concentrations, with values of 2–5.5 in the midwest and surrounding areas; in the western USA, the value was below 1. Total sulfur deposition was 4–9.8 kg ha^{-1} year^{-1} in the upper midwest and surrounding areas, and was below 1.5 in the western USA. Total nitrogen deposition was 5–7 kg ha^{-1} year^{-1} at most ground stations in the upper midwest and surrounding areas, and was 1–4 kg ha^{-1} year^{-1} in the western USA.

Table 6.1 lists observed wet, dry, and total depositions of sulfur and nitrogen from the atmosphere to various regions of the eastern USA in 1989–1991 and 2007–2009 respectively on annual average.

It is seen that total sulfur deposition in the eastern regions of the USA was 8–16 kg ha^{-1} in 1989–1991 and 4.6–8.0 kg ha^{-1} in 2007–2009. Meantime, total inorganic nitrogen deposition was 6–9 kg ha^{-1} in 1989–1991 and 4.4–6.9 kg ha^{-1} in 2007–2009. It is also seen that wet deposition dominated over dry deposition in the east coast for sulfur, while the two pathways were comparable in the midwest; total sulfur deposition halved due to extraordinary efforts to reduce emissions of acid rain precursors during the 14 years. For ammonium nitrate, observations showed that wet deposition dominated over dry deposition in all eastern regions; total deposition of nitrogen decreased by 20–30% in 14 years with a little larger reduction in dry deposition than wet deposition. The relatively modest reduction in nitrogen deposition compared to sulfur deposition in the same period reflects the fact that emission of power plants predominated SO_2 emission while other sectors were more important for NO_x emissions. For example, the power sector accounted for ~70% of SO_2 emission, but only ~20% of NO_x emission. In addition, the major source of NH_3 is agricultural emission, which was not regulated by the Clean Air Act during the period. Moreover, unmeasured nitrogen compounds might contribute significantly to deposition fluxes of total nitrogen besides ammonium nitrate.

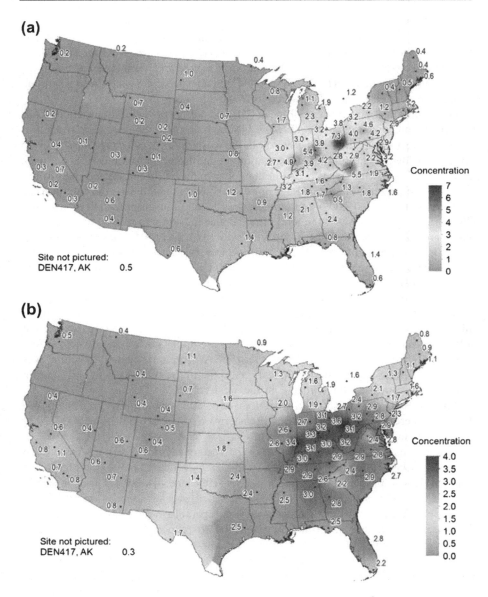

Figure 6.1 Annual mean SO_2 (a) and sulfate (b) concentrations ($\mu g\ m^{-3}$) at surface stations over the USA in 2010.
Obtained from USA Clean Air Status and Trends Network (2012).

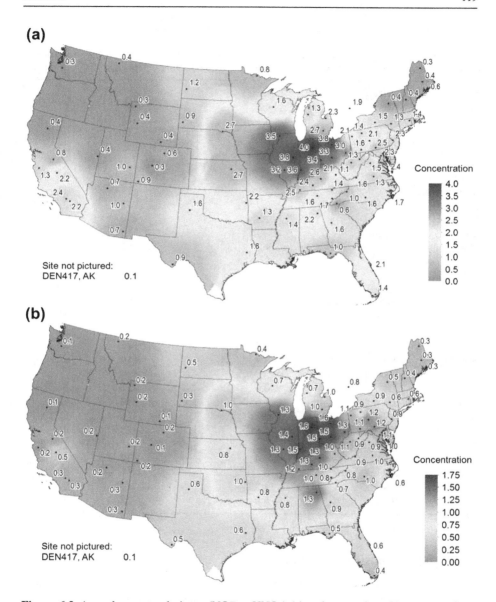

Figure 6.2 Annual mean total nitrate (NO_3^- + HNO_3) (a) and ammonium (b) concentrations ($\mu g\ m^{-3}$) in the USA in 2010.
Obtained from USA Clean Air Status and Trends Network (2012).

Table 6.1 Sulfur and nitrogen deposition in eastern USA (NAPAP, 2011)

Region	Average deposition (kg ha^{-1}), 1989–1991	Average deposition (kg ha^{-1}), 2007–2009	Percentage change	Number of sites
Wet	**Sulfate**			
Northeast	7.5	4.3	−43	17
Mid-Atlantic	9.2	5.3	−42	11
Southeast	6.1	3.5	−43	23
Midwest	7.1	4	−44	27
Dry	**Sulfur**			
Northeast	2.9	1	−66	3
Mid-Atlantic	6.7	2.9	−57	11
Southeast	1.2	0.7	−42	2
Midwest	6.5	2.8	−57	10
Total	**Sulfur**			
Northeast	9.8	4.7	−52	3
Mid-Atlantic	16	8	−50	11
Southeast	8	4.6	−43	2
Midwest	15	7	−53	10
Wet	**Inorganic nitrogen**			
Northeast	5.6	4.1	−27	17
Mid-Atlantic	6.2	4.5	−27	11
Southeast	4.4	3.4	−23	23
Midwest	5.8	4.9	−16	27
Dry	**Inorganic nitrogen**			
Northeast	1.4	0.6	−57	3
Mid-Atlantic	2.5	1.5	−40	11
Southeast	0.9	0.8	−11	2
Midwest	2.5	1.8	−28	10
Total	**Inorganic nitrogen**			
Northeast	6.5	4.4	−32	3
Mid-Atlantic	8.7	6	−31	11
Southeast	5.9	4.8	−19	2
Midwest	9	6.9	−23	10

6.1.4   Acid deposition in China

China has monitored acid rain since the 1980s, when the pH value of fog water collected at a scenery mountain station at an elevation of 3 km ASL was measured to be ~3 and hundreds of organic acids besides sulfate, nitrate, chloride, sodium, and potassium ions were detected in rainwater. As economic development took off rapidly in the last 30 years, total energy consumption and associated emissions of air pollutants have increased significantly, partly due to the growth of the population and partly due to the growth of energy consumption on a per-capita basis. By 2011, there were 468 ground stations at county level. It was found that the frequency of acid rain exceeded 75% at 44 stations, and the pH value was below 4.5 at 30 stations in 2011.

Severe acid rain mainly occurred in southern provinces in 2011, as shown in Figure 6.3. In mainland China, major ions detected in precipitation were SO_4^{2-}, Ca^{2+}, NH_4^+, Cl^-, NO_3^-, Na^+, Mg^{2+}, F^-, K^+, and H^+ on an equivalent basis in 2011, with $CaSO_4$ contributing over 50% of the total equivalence. Dissolved nitrogen-containing organic compounds could account for 25–30% of total dissolved nitrogen in rainwater over China from 2005 to 2009 (Zhang et al., 2008a, 2012), but their contributions to acidity remain to be determined. In 2011, national emissions amounted to 22.2 million tons for SO_2 and 24 million tons for NO_x in China; industrial emissions were dominant for both SO_2 and NO_x, though vehicle contributions have become more and more important in metropolitan areas (Cai and Xie, 2007; Guo et al., 2007; Zhang et al., 2008a). Atmospheric concentrations of SO_2 and NO_x at ground stations in 2011 are shown in Figure 6.4. It is seen that SO_2 concentration was 20–60 µg m^{-3} at most stations with a peak value of 100 µg m^{-3}. NO_x concentration was below 40 µg m^{-3} at most stations and 40–80 µg m^{-3} at a significant number of stations.

Since April 1990, the Taiwan Environmental Protection Agency has started to monitor acid rain pollution. The first monitoring station was operated by Taipei Weather Station that has experienced frequent acid rain with a minimum pH value

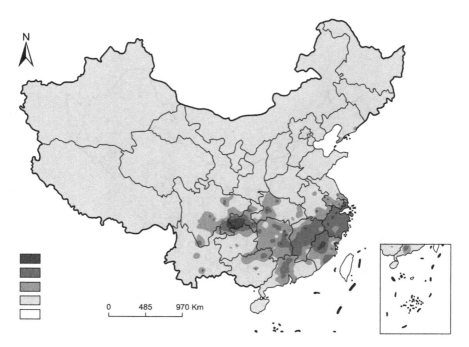

Figure 6.3 Smoothed distribution of annual average pH values of precipitation at 468 observation stations in China in 2011. Key from top to bottom: pH < 4.5; pH 4.5–5.0; pH 5.0–5.6; pH > 5.6; and white area: Taiwan; see text for more information. Monitoring stations were at county level.
Obtained from China Ministry of Environmental Protection (2012).

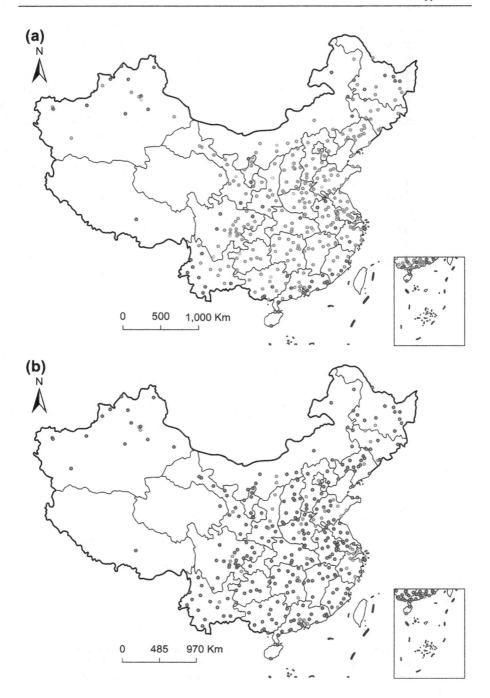

of 3.8 so far. On average, 57% of rain in Taiwan had a pH value less than 5. In Taiwan, acid rain mainly occurred in winter months, when the frequency of acid rain exceeded 70% with an average pH value of 4.8. From May to September, the pH value of rain exceeded 5 in most cases. In particular, northern Taiwan experienced severe acid rain pollution with a frequency of 80% and a minimum pH value of 3.0. For comparison, tomato juice has a pH value of 4.2 and orange juice has a pH value of 3.5, while vinegar has a pH value of 2.9. During a typhoon event, a mountain station (Yangmingshan) experienced 9 mm of rain with a pH value of 3.2; as the air stream swept across part of mainland China before reaching the observational station, it was speculated that about half of the acidity originated from local pollutants while the rest was imported from marine vessels and continental areas including mainland China and other Asian countries which the air stream passed through. As acid rain damages rivers, lakes, forest ecosystems, agricultural harvests, architectural structures, and public health, it was estimated that the economic loss due to acid rain has exceeded NT$10 billion in Taiwan. To reduce acid rain, the Taiwan EPA has implemented aggressive measures to control regional anthropogenic emissions of acid rain precursors. The standards of SO_x content in liquid fuels and the emission standards of NO_x and PM from diesel vehicles in Taiwan followed the USA within 3 years over the period 1989–2012.

6.2 Mechanism of formation of acid rain

Decadal chemical analyses of rain, snow, and fog samples have revealed that sulfuric acid and nitric acid are responsible for most of the acidity in acid deposition. In the past 30 years, the mechanism of formation of sulfuric acid and nitric acids from their precursors SO_2 and NO_x in the atmosphere has been thoroughly investigated. An example photochemical mechanism for simulating acid rain (Liang and Jacobson, 1999) is elaborated below.

Table 6.2 lists a number of chemicals important for describing photochemical reactions of SO_2 and NO_x in the gas and aqueous phases of the global troposphere. It is seen that 12 forms of sulfur species may be represented by 7 model species and 12 oxidants may be represented by 11 model species, while 13 nitrogen species may be represented by 10 model species. It is also seen that, in the global troposphere, 25 organic species may be represented by 18 model species. Thus, it is observed that 46 simultaneous chemical ordinary differential equations may be used to describe temporal changes of 62 chemicals at a point in the global troposphere if the point is treated as a closed system.

Detailed gas–aqueous photochemical reactions and corresponding kinetic parameters of the 62 chemicals listed in Table 6.2 are included in Table 6.3. In the

Figure 6.4 Annual average concentrations of acid rain precursors in urban areas of China in 2011. (a) SO_2 – black: $\leq 20\ \mu g\ m^{-3}$; gray: 20–60 $\mu g\ m^{-3}$; pale gray: 60–100 $\mu g\ m^{-3}$. (b) NO_x – black: $\leq 40\ \mu g\ m^{-3}$; gray: 40–80 $\mu g\ m^{-3}$. Monitoring stations were at district level. Obtained from China Ministry of Environmental Protection (2012).

Table 6.2 Important chemicals for the formation of SO_4^{2-} and NO_3^-

Sulfur compounds	
Sulfur compounds	23. $NO_3 = NO_3$ (g)
1. $SO_2 = SO_2$(g)	24. $N_2O_5 = N_2O_5$ (g)
2. $S_{4,T} = SO_2$ (aq) $+ HSO_3^- + SO_3^{2-}$	25. $NH_3 = NH_3$ (g)
3. HMSA (aq)	26. $HNO_3 = HNO_3$ (g)
4. HSO_5^-	27. NH_4^+
5. SO_5^-	28. NO_3^-
6. SO_4^-	**Organic species**
7. $SO_{4,T}^{2-} = H_2SO_4$ (g) $+ H_2SO_4$ (aq) $+ HSO_4^- + SO_4^{2-}$	29. $CH_4 = CH_4$ (g)
Oxidants	30. $C_2H_6 = C_2H_6$ (g)
8. $O_3 = O_3$ (g)	31. $HCHO_T = HCHO$ (g) $+ HCHO$ (aq) $+ H_2C(OH)_2$
9. $O = O$ (g)	32. $CH_3CHO = CH_3CHO$ (g)
10. $O^1D = O^1D$ (g)	33. $CH_3O_2 = CH_3O_2$ (g)
11. $OH = OH$ (g)	34. CH_3O_2 (aq)
12. $HO_2 = HO_2$ (g)	35. $C_2H_5O_2 = C_2H_5O_2$ (g)
13. $H = H$ (g)	36. $CH_3OOH = CH_3OOH$ (g)
14. $H_2O_2 = H_2O_2$ (g)	37. CH_3OOH (aq)
15. O_3 (aq)	38. $C_2H_5OOH = C_2H_5OOH$ (g)
16. OH (aq)	39. $HCOOHA_T = HCOOH$ (aq) $+ HCOO^-$
17. H_2O_2 (aq)	40. $HCOOH = HCOOH$ (g)
18. HO_2 (aq, T) $= HO_2$ (aq) $+ O_2^-$	41. $CH_3COOH_T = CH_3COOH$ (g) $+ CH_3COOH$ (aq) $+ CH_3COO^-$ (aq)
Nitrogen compounds	42. $C_2H_5OH = C_2H_5OH$ (g)
19. $NO_{2,T} = NO_2$ (g) $+ NO_2$ (aq)	43. $CH_3OH_T = CH_3OH$ (g) $+ CH_3OH$ (aq)
20. $NO_T = NO$ (g) $+ NO$ (aq)	44. $CH_3CO_3 = CH_3CO_3$ (g)
21. $HNO_{2,T} = HNO_2$ (g) $+ HNO_2$ (aq)	45. $CH_3C(O)OOH_T = CH_3C(O)OOH$ (g) $+ CH_3C(O)OOH$ (aq)
22. $HNO_4 = HNO_4$ (g)	46. PAN $= CH_3C(O)O_2NO_2$ (g)

light of the potential importance of chlorine in the global troposphere due to large oceanic flux and its significant presence in rainwater along coastal areas, Table 6.3 also lists 18 aqueous reactions involving three chlorine species, namely, Cl^-, Cl (aq), and Cl_2^-. In Table 6.3, O_2, H_2O, and CO_2 are often omitted from the product lists, as their concentrations in the global troposphere are usually calculated independently due to negligible perturbations from the formation of sulfate and nitrate.

6.3 Simulating acid rain

Simulation of acid deposition has been attempted since the 1980s (Chang et al., 1987; Arndt et al., 1997; Wang et al., 2000; Carmichael et al., 2002a, b). In the last two decades, three-dimensional models have been developed to describe concurrent transformation and transport processes of SO_2 and NO_x in the atmosphere.

Table 6.3 Photochemical mechanisms for simulating the formation of SO_4^{2-} and NO_3^-

Reaction	Rate constant	k_1	k_2
Gas phase			
$O_3 + hv \rightarrow O^1D + O_2$	a		
$O_3 + hv \rightarrow O + O_2$	a		
$O^1D + N_2 \rightarrow O + N_2$	b	1.8×10^{-11}	-107
$O^1D + O_2 \rightarrow O + O_2$	b	3.2×10^{-11}	-67
$O^1D + H_2O \rightarrow OH + OH$		2.2×10^{-10}	
$O + O_2 + M \rightarrow O_3 + M$	c	5.7×10^{-34}	2.6
	$k_3 = 1, k_4 = 0$	$k_5 = 0, k_6 = 1$	$k_7 = 0, k_8 = 0$
$NO_2 + hv \rightarrow NO + O$	a		
$O_3 + NO \rightarrow NO_2 + O_2$	b	1.4×10^{-12}	1310
$O_3 + OH \rightarrow HO_2 + O_2$	b	1.7×10^{-12}	940
$O_3 + HO_2 \rightarrow OH + O_2 + O_2$	b	1.4×10^{-14}	600
$O_3 + NO_2 \rightarrow O_2 + NO_3$	b	1.4×10^{-13}	2470
$H_2O_2 + hv \rightarrow OH + OH$	a		
$H + O_2 + M \rightarrow HO_2 + M$	c	4.3×10^{-32}	1.8
	$k_3 = 7.5 \times 10^{-11},$ $k_4 = 0$	$k_5 = -0.5978,$ $k_6 = 1$	$k_7 = 0, k_8 = 0$
$OH + HO_2 \rightarrow H_2O + O_2$	b	4.8×10^{-11}	-250
$HO_2 + NO \rightarrow OH + NO_2$	b	3.6×10^{-12}	-270
$HO_2 + HO_2 \rightarrow H_2O_2$	b	2.2×10^{-13}	-600
$HO_2 + HO_2 + M \rightarrow H_2O_2$	d	1.5×10^{-14}	-980
$HO_2 + HO_2 + H_2O \rightarrow H_2O_2$	b	3.1×10^{-34}	-2800
$HO_2 + HO_2 + H_2O + M \rightarrow H_2O_2$	d	2.1×10^{-35}	-3180
$OH + H_2 \rightarrow H_2O + H$	b	7.7×10^{-12}	2100
$CO + OH \rightarrow H + CO_2$	e	1.3×10^{-13}	0.6
$OH + CH_4 \rightarrow CH_3O_2 + H_2O$	b	1.85×10^{-12}	1690

(Continued)

Table 6.3 Photochemical mechanisms for simulating the formation of SO_4^{2-} and NO_3^-—*cont'd*

Reaction	Rate constant	k_1	k_2
$CH_3O_2 + NO \rightarrow HCHO + HO_2 + NO_2$	b	2.3×10^{-12}	-360
$CH_3O_2 + HO_2 \rightarrow CH_3OOH + O_2$	b	3.8×10^{-13}	-780
$CH_3O_2 + CH_3O_2 \rightarrow CH_3OH + HCHO$	f_1, $k_3 = 5.94 \times 10^{-13}$, $k_4 = 505$	1.1×10^{-13}	-365
$CH_3O_2 + CH_3O_2 \rightarrow 2HCHO + 2HO_2$	b	5.9×10^{-13}	505
$CH_3OOH + hv \rightarrow HCHO + H + OH$	a		
$CH_3OOH + OH \rightarrow CH_3O_2 + H_2O$	b	1.9×10^{-12}	-190
$CH_3OOH + OH \rightarrow HCHO + OH + H_2O$	b	1.0×10^{-12}	-190
$HCHO + OH \rightarrow CO + HO_2 + H_2O$	b	5.4×10^{-12}	-135
$HCHO + hv \rightarrow 2H + CO$	a		
$HCHO + hv \rightarrow H_2 + CO$	a		
$NO_2 + OH + M \rightarrow HNO_3 + M$	c, $k_3 = 4.1 \times 10^{-11}$, $k_4 = 0$	2.6×10^{-30}, $k_5 = -0.9163$, $k_6 = 1.255$	3.0, $k_7 = 0$, $k_8 = 0$
$HNO_3 + hv \rightarrow OH + NO_2$	a		
$HNO_3 + OH \rightarrow H_2O + NO_3$	f_2, $k_3 = 4.1 \times 10^{-16}$, $k_4 = 1440$	7.2×10^{-15}, $k_5 = 1.9 \times 10^{-33}$, $k_6 = 725$	785
$NO + OH + M \rightarrow HNO_2 + M$	c, $k_3 = 4.5 \times 10^{-11}$, $k_4 = 0$	5.8×10^{-31}, $k_5 = -0.1054$, $k_6 = 1$	2.4, $k_7 = 0$, $k_8 = 0$
$HNO_2 + hv \rightarrow OH + NO$	a		
$HNO_2 + OH \rightarrow H_2O + NO_2$	b	2.5×10^{-12}	-260
$HO_2 + NO_2 + M \rightarrow HNO_4 + M$	c, $k_3 = 4.7 \times 10^{-12}$, $k_4 = 0$	1.4×10^{-31}, $k_5 = -0.5108$, $k_6 = 1$	3.2, $k_7 = 0$, $k_8 = 0$

Reaction				
$HNO_4 + M \rightarrow HO_2 + NO_2 + M$	c	$k_3 = 2.6 \times 10^{15},$ $k_4 = 0$	4.0×10^{-6} $k_5 = -0.5108,$ $k_6 = 1$	0 $k_7 = 10{,}000,$ $k_8 = 10{,}900$
$HNO_4 + h\nu \rightarrow OH + NO_3$	a			
$HNO_4 + h\nu \rightarrow HO_2 + NO_2$	a			
$HNO_4 + OH \rightarrow 0.5\ NO_2 + 0.5H_2O_2 + 0.5NO_3$	b		1.9×10^{-12}	-270
$NO_3 + h\nu \rightarrow NO_2 + O$	a			
$NO_3 + h\nu \rightarrow NO + O_2$	a			
$NO_3 + NO \rightarrow 2NO_2$	b		1.8×10^{-11}	-110
$NO_3 + HCHO \rightarrow HNO_3 + HO_2 + CO$		5.8×10^{-16}		
$NO_2 + NO_3 + M \rightarrow N_2O_5 + M$	c	$k_3 = 2 \times 10^{-12},$ $k_4 = -0.2$	2.1×10^{-30} $k_5 = -1.11,$ $k_6 = 1$	3.4 $k_7 = 0, k_8 = 0$
$N_2O_5 + M \rightarrow NO_2 + NO_3 + M$	c	$k_3 = 9.7 \times 10^{14},$ $k_4 = -0.1$	7.9×10^{-4} $k_5 = -1.11,$ $k_6 = 1$	3.5 $k_7 = 11{,}000,$ $k_8 = 11{,}080$
$N_2O_5 + h\nu \rightarrow NO_3 + NO_2$	a			
$N_2O_5 + h\nu \rightarrow NO_3 + NO + O$	a			
$SO_2 + OH \rightarrow H_2SO_4 + HO_2$	c	$k_3 = 2.0 \times 10^{-12},$ $k_4 = 0$	4.0×10^{-31} $k_5 = -0.8, k_6 = 1$	3.3 $k_7 = 0, k_8 = 0$
$SO_2 + O \rightarrow H_2SO_4 + M$	b		3.2×10^{-32}	1000
$CH_3CHO + h\nu \rightarrow CH_3O_2 + HO_2 + CO$	a			
$CH_3CHO + h\nu \rightarrow CH_4 + CO$	a			
$C_2H_5OOH + h\nu \rightarrow OH + HO_2 + CH_3CHO$	a			
$CH_3CHO + OH \rightarrow CH_3CO_3 + H_2O$	b		4.4×10^{-12}	-365
$CH_3CO_3 + NO_2 + M \rightarrow PAN + M$	c	$k_3 = 1.2 \times 10^{-11},$ $k_4 = 0.9$	2.1×10^{-28} $k_5 = -1.2,$ $k_6 = 1$	7.1 $k_7 = 0, k_8 = 0$

(Continued)

Table 6.3 Photochemical mechanisms for simulating the formation of SO_4^{2-} and NO_3^-—cont'd

Reaction	Rate constant	k_1	k_2
PAN → CH_3CO_3 + NO_2	c, $k_3 = 5.4 \times 10^{16}$, $k_4 = 0$	3.9×10^{-3}, $k_5 = -1.2$, $k_6 = 1$	0, $k_7 = 12{,}100$, $k_8 = 13{,}830$
OH + PAN → CH_3CO_3 + HNO_3	b	9.5×10^{-13}	-650
CH_3CO_3 + NO → CH_3O_2 + NO_2 + CO_2	b	7.5×10^{-12}	-290
C_2H_6 + OH → $C_2H_5O_2$ + H_2O	b	6.9×10^{-12}	1000
$C_2H_5O_2$ + NO → CH_3CHO + NO_2 + HO_2	b	2.6×10^{-12}	-380
C_2H_5OH + OH → HO_2 + CH_3CHO	b	4.1×10^{-12}	70
$C_2H_5O_2$ + HO_2 → C_2H_5OOH + O_2	b	3.8×10^{-13}	-900
HO_2 + CH_3CO_3 → CH_3COOH + O_3	b	8.5×10^{-14}	-1040
HO_2 + CH_3CO_3 → CH_3CO_3H	b	3.4×10^{-13}	-1040
CH_3CO_3 + CH_3CO_3 → $2CH_3O_2$	b	2.9×10^{-12}	-500
CH_3CO_3 + CH_3O_2 → HCHO + CH_3O_2 + HO_2	b	4.0×10^{-12}	-272
CH_3CO_3 + CH_3O_2 → CH_3COOH + HCHO	b	4.4×10^{-13}	-272
Gas–aqueous interaction			
N_2O_5 → $2HNO_3$	g	108	0
O_3 → O_3 (aq)	g	48	0.1
O_3 (aq) → O_3	h	1.1×10^{-2}	-4.76
HO_2 → HO_2 (aq)	g	33	0.1
HO_2 (aq) → HO_2	h	$2.0\,(\pm1) \times 10^{3}$	-13.2
OH → OH (aq)	g	17	0.1
OH (aq) → OH	h	20	-10.5
H_2O_2 → H_2O_2 (aq)	g	34	0.1
H_2O_2 (aq) → H_2O_2	h	$1.1\,(\pm0.36) \times 10^{5}$	-13.2
HCOOH → HCOOH (aq)	g	46	0.1
HCOOH (aq) → HCOOH	h	$5.6\,(\pm2) \times 10^{3}$	-11.4

Reaction			
$CH_3OOH \rightarrow CH_3OOH\,(aq)$	g	48	0.1
$CH_3OOH\,(aq) \rightarrow CH_3OOH$	h	220	-11.2
$SO_2 \rightarrow SO_2\,(aq)$	g	64	0.1
$SO_2\,(aq) \rightarrow SO_2$	h	1.2 ($\pm$0.03)	-6.27
$CH_3O_2 \rightarrow CH_3O_2\,(aq)$	g	47	0.1
$CH_3O_2\,(aq) \rightarrow CH_3O_2$	h	5.9	-11.2
Aqueous phase			
$OH\,(aq) + HO_2\,(aq) \rightarrow H_2O\,(aq) + O_2\,(aq)$	i	6.6 ($\pm$0.55) $\times 10^9$	3
$OH\,(aq) + O_2^- \rightarrow OH^- + O_2\,(aq)$	i	0.85 ($\pm$0.15) $\times 10^{10}$	3
$HO_2\,(aq) + O_2^- \rightarrow H_2O_2\,(aq) + O_2\,(aq) + OH^-$	i	0.97 ($\pm$0.06) $\times 10^8$	2.1
$HO_2\,(aq) + HO_2\,(aq) \rightarrow H_2O_2\,(aq) + O_2\,(aq)$	i	8.3 ($\pm$0.7) $\times 10^5$	5.4 ($\pm$0.5)
$OH\,(aq) + H_2O_2\,(aq) \rightarrow HO_2\,(aq) + H_2O\,(aq)$	i	2.7 $\times 10^7$	3.4
$O_2^- + O_3\,(aq) \rightarrow OH\,(aq) + OH^- + 2O_2\,(aq)$	i	1.5 ($\pm$0.05) $\times 10^9$	3
$H_2O_2\,(aq) + h\nu \rightarrow OH\,(aq) + OH\,(aq)$	a		
$OH\,(aq) + HSO_3^- \rightarrow SO_5^- + H_2O\,(aq)$	i	4.5 $\times 10^9$	3
$HSO_3^- + H^+ + H_2O_2\,(aq) \rightarrow SO_4^{2-} + 2H^+$	j	7.5 ($\pm$1.5) $\times 10^7$	9.45 ($\pm$0.05)
$H_2C(OH)_2 + OH\,(aq) \rightarrow HO_2\,(aq) + HCOOH\,(aq)$	i	7.7 ($\times$3.8) $\times 10^8$	2 (1 $\pm$ 0.34)
$OH\,(aq) + SO_3^{2-} \rightarrow SO_5^- + OH^-$	i	5.5 ($\pm$0.35) $\times 10^9$	3
$HCOO^- + OH\,(aq) \rightarrow CO_2 + HO_2\,(aq) + OH^-$	i	3.2 ($\times$3.8) $\times 10^9$	2.5 (1 $\pm$ 0.51)
$HSO_3^- + H^+ + CH_3OOH\,(aq) \rightarrow SO_4^{2-} + 2H^+ + CH_3OH\,(aq)$	k	1.9 $\times 10^7$	7.56
$SO_3^{2-} + SO_4^- \rightarrow SO_4^{2-} + SO_5^-$	i	7.5 $\times 10^8$	3
$HSO_3^- + SO_4^- \rightarrow SO_4^{2-} + SO_5^- + H^+$	i	7.5 ($\pm$0.6) $\times 10^8$	3
$HSO_3^- + H_2C(OH)_2 \rightarrow HOCH_2SO_3^-$	i	0.43 ($\pm$0.02)	6 ($\pm$0.2)
$SO_3^{2-} + H_2C(OH)_2 \rightarrow HOCH_2SO_3^- + OH^-$	i	1.4 ($\pm$0.03) $\times 10^4$	4.9 ($\pm$0.1)
$HSO_3^- + O_3\,(aq) \rightarrow SO_4^{2-} + H^+$	i	3.7 $\times 10^5$	11.06
$SO_3^{2-} + O_3\,(aq) \rightarrow SO_4^{2-}$	i	1.5 $\times 10^9$	10.56
$SO_4^- + OH^- \rightarrow SO_4^{2-} + OH\,(aq)$	i	8.0 $\times 10^7$	3
$SO_4^- + H_2O_2\,(aq) \rightarrow H^+ + SO_4^{2-} + HO_2\,(aq)$	i	1.2 ($\pm$0.1) $\times 10^7$	4

(Continued)

Table 6.3 Photochemical mechanisms for simulating the formation of SO_4^{2-} and NO_3^-—*cont'd*

Reaction	Rate constant	k_1	k_2
$SO_4^- + H_2O\ (aq) \rightarrow SO_4^{2-} + H^+ + OH\ (aq)$	l	$440\ (\pm 50)$	$3.7\ (\pm 0.1)$
$SO_4^- + HCOO^- \rightarrow SO_4^{2-} + CO_2\ (aq) + HO_2\ (aq)$	i	$1.1\ (\pm 0.05) \times 10^8$	3
$CH_3O_2\ (aq) + O_2^- \rightarrow CH_3OOH\ (aq) + OH^- + O_2\ (aq)$	i	5.0×10^7	2.1
$HCOOH\ (aq) + OH\ (aq) \rightarrow H_2O\ (aq) + CO_2\ (aq) + HO_2\ (aq)$	i	$1.1\ (\times 3.9) \times 10^8$	$2\ (1 \pm 0.41)$
$O_3\ (aq) + H_2O_2\ (aq) + OH^- \rightarrow OH\ (aq) + O_2^- + O_2\ (aq) + H_2O\ (aq)$	k	4.4×10^8	-8
$HOCH_2SO_3^- + OH^- \rightarrow SO_3^{2-} + H_2C(OH)_2$	i	$3.7\ (\pm 1) \times 10^3$	9
$CH_3OOH\ (aq) + OH\ (aq) \rightarrow H_2C(OH)_2 + OH\ (aq)$	i	1.9×10^7	3.7
$CH_3OOH\ (aq) + OH\ (aq) \rightarrow CH_3O_2\ (aq) + H_2O$	i	2.7×10^7	3.4
$HOCH_2SO_3^- + OH\ (aq) \rightarrow H_2C(OH)_2 + SO_5^- + H^+ + OH^-$	i	1.3×10^9	3
$SO_4^- + HO_2\ (aq) \rightarrow SO_4^{2-} + H^+ + O_2\ (aq)$	i	5.0×10^9	3
$SO_4^- + O_2^- \rightarrow SO_4^{2-} + O_2\ (aq)$	i	5.0×10^9	3
$HCOO^- + O_3\ (aq) \rightarrow CO_2\ (aq) + OH\ (aq) + O_2^-$	i	1.0×10^2	11
$SO_5^- + HCOO^- \rightarrow HSO_5^- + CO_2\ (aq) + O_2^-$	i	1.4×10^4	8
$SO_4^- + NO_2^- \rightarrow SO_4^{2-} + NO_2\ (aq)$	i	9.8×10^8	3
$SO_5^- + HSO_3^- \rightarrow HSO_5^- + SO_5^-$	i	2.5×10^4	7.7
$HSO_5^- + OH\ (aq) \rightarrow SO_5^- + H_2O\ (aq)$	i	1.7×10^7	3.8
$HSO_5^- + HSO_3^- + H^+ \rightarrow 2SO_4^{2-} + 3H^+$	k	$1.7\ (\pm 1) \times 10^7$	$4\ (\pm 2.2)$
$SO_5^- + HSO_3^- \rightarrow SO_4^- + SO_4^{2-} + H^+$	i	7.5×10^4	7
$O_2^- + SO_5^- \rightarrow O_2\ (aq) + HSO_5^- + OH^-$	i	1.0×10^8	2.1
$NO_2^- + OH\ (aq) \rightarrow NO_2\ (aq) + OH^-$	i	1.0×10^{10}	3
$H_2O\ (aq) \leftrightarrow OH^- + H^+$	m	1.0×10^{-14}	13.34
$HO_2\ (aq) \leftrightarrow O_2^- + H^+$	n	1.6×10^{-5}	
$HCOOH\ (aq) \leftrightarrow HCOO^- + H^+$	n	1.8×10^{-4}	0.3

Reaction			
HNO_2 (aq) $\leftrightarrow$ NO_2^- + H^+	n	5.1×10^{-4}	2.5
SO_2 (aq) $\leftrightarrow$ HSO_3^- + H^+	n	1.5×10^{-2}	-4
HSO_3^- $\leftrightarrow$ SO_3^{2-} + H^+	n	6.3×10^{-8}	-2.99
HSO_4^- $\leftrightarrow$ SO_4^{2-} + H^+	n	0.01	-5.44
CH_3COOH (aq) + OH (aq) $\rightarrow$ HO_2 (aq) + $H_2C(OH)_2$	i	$1.8 \ (\times 4.1) \times 10^7$	2.6 (1 ± 0.26)
CH_3COO^- + OH (aq) $\rightarrow$ O_2^- + $H_2C(OH)_2$	i	$7.5 \ (\times 3.8) \times 10^7$	3.5 (1 ± 0.23)
CH_3COOH $\leftrightarrow$ CH_3COOH (aq)	o	8.8×10^3	-12.8
CH_3COOH (aq) $\leftrightarrow$ CH_3COO^- + H^+	n	1.7×10^{-5}	-0.1
$HCHO$ $\leftrightarrow$ $H_2C(OH)_2$	o	3.0×10^3	-14.3
HNO_2 $\leftrightarrow$ HNO_2 (aq)	o	49	-9.5
Chlorine reactions			
Cl^- + OH (aq) + H^+ $\rightarrow$ Cl (aq)	i	$1.4 \ (\pm 0.1) \times 10^{10}$	
Cl_2^- + HO_2 (aq) $\rightarrow$ $2Cl^-$ + H^+ + O_2 (aq)	i	$4.2 \ (\pm 0.3) \times 10^9$	
Cl_2^- + O_2^- $\rightarrow$ $2Cl^-$ + O_2 (aq)	i	1.0×10^9	
Cl_2^- + H_2O_2 (aq) $\rightarrow$ $2Cl^-$ + H^+ + HO_2 (aq)	i	1.4×10^5	
H_2O_2 (aq) + Cl (aq) $\rightarrow$ Cl^- + HO_2 (aq) + H^+	i	4.5×10^7	
NO_3 (aq) + Cl^- $\rightarrow$ NO_3^- + Cl (aq)	i	$1.0 \ (\pm 0.2) \times 10^7$	
Cl (aq) $\rightarrow$ Cl^- + H^+ + OH (aq)	l	1.0×10^5	8.6 (±1)
SO_4^- + Cl^- $\rightarrow$ SO_4^{2-} + Cl (aq)	i	2.0×10^8	
CH_3OH (aq) + Cl_2^- $\rightarrow$ $2Cl^-$ + $H_2C(OH)_2$ + H^+ + HO_2 (aq)	i	3.5×10^3	
$HCOOH$ (aq) + Cl_2^- $\rightarrow$ $2Cl^-$ + CO_2 (aq) + H^+ + HO_2 (aq)	i	$5 \ (\pm 1) \times 10^3$	
$HCOO^-$ + Cl_2^- $\rightarrow$ $2Cl^-$ + CO_2 (aq) + HO_2 (aq)	i	$1.6 \ (\pm 0.2) \times 10^6$	
Cl_2^- + OH^- $\rightarrow$ $2Cl^-$ + OH (aq)	i	4.0×10^6	4.3
Cl_2^- + HSO_3^- $\rightarrow$ $2Cl^-$ + SO_5^- + H^+	i	3.4×10^8	3
Cl_2^- + SO_3^{2-} $\rightarrow$ $2Cl^-$ + SO_5^-	i	1.6×10^8	3
NO_2^- + Cl_2^- $\rightarrow$ NO_2 (aq) + $2Cl^-$	i	$2.3 \ (\pm 0.2) \times 10^8$	

(Continued)

Table 6.3 Photochemical mechanisms for simulating the formation of SO_4^{2-} and NO_3^- —cont'd

Reaction	Rate constant	k_1	k_2
$Cl_2^- \leftrightarrow Cl\ (aq) + Cl^-$	l	5.3×10^{-6}	
$Cl\ (aq) + h\nu \rightarrow Cl^- + H^+ + OH\ (aq)$	a		
$Cl_2^- + h\nu \rightarrow 2Cl^- + H^+ + OH\ (aq)$	a		

Note: Gas-phase reaction parameters are from Atkinson et al. (2004, 2006) and Horowitz (1997), and aqueous-phase reaction parameters are from Liang and Jacobson (1999) and references therein.

a. Photolysis reactions.

b. Rate constant $k = k_1 \exp(-k_2/T)$. The units are s^{-1} for first-order reactions, molecule^{-1} cm^3 s^{-1} for second-order reactions, and molecule^{-2} cm^6 s^{-1} for third-order reactions. T is temperature in kelvin.

c. Rate constant $k = [AB/(A+B)] \cdot \exp\{k_5[1 + (\log_{10}(A/B)/k_6)^2]^{-1}\}$, where $A = k_1[M](300/T)^{k2} \cdot \exp(-k_7/T)$, $B = k_3(300/T)^{k4} \cdot \exp(-k_8/T)$, and [M] denotes the number concentration of air molecules in molecules cm^{-3}.

d. Rate constant $k = 10^{-19}$[M]$k_1\exp(-k_2/T)$, and [M] is as in footnote c.

e. Rate constant $k = k_1(1 + k_2P)$, where P is pressure in atmospheres.

f_1. Rate constant $k = k_1\exp(-k_2/T) - k_3\exp(-k_4/T)$.

f_2. Rate constant $k = k_1\exp(k_2/T) + R_2R_3/(R_2 + R_3)$, where $R_2 = k_2\exp(k_4/T)$, $R_3 = k_5[M]\exp(k_6/T)$, and [M] is as in footnote c.

g. Rate constant $k = \{3D_g \cdot L\}/\{(\Lambda^2)Dg = 9.45 \times 10^{17}[M] \cdot [T(0.03472 + 1/k_1)]^{0.5}\} \cdot L$ is the volume mixing ratio of liquid water. $\Lambda = 1 + (\lambda + 1.3(1/k_2 - 1))$, and $\lambda = (0.71 + 1.3\beta)/(1 + \beta)$, $\beta = 4.54 \times 10^{-15} \cdot (V_g^2 + V_{air}^2)^{0.5} \cdot V_g = [8RT/(\pi k_1)]^{0.5}$ and $V_{air} = [8RT/(28.8\pi)]^{0.5} \cdot R = 8.31 \times 10^7$ is the ideal gas constant multiplied by a factor to keep V_s in units of cm s^{-1}. r is the radius of an aerosol droplet in cm.

h. Rate constant $k = k_{n-1}/(0.082TLC)$, $C = k_1\exp(-500k_2(1/T - 1/298))$. k_{n-1} is the rate constant of its reverse reaction. L is as in g.

i. Rate constant $k = ClD^2 \cdot C$ is as in h. $D = 6.023 \times 10^{20}L$, and L is as in g.

j. Rate constant $k = Cl(D^2 + 13D[H^+])$. C and D are as in i, and $[H^+]$ is in the unit of the number of protons per cubic centimeter of air.

k. Rate constant $k = ClD^2 \cdot C$ and D are as in i.

l. Rate constant $k = C$, and C is as in h.

m. Rate constant $k = C \cdot D^2$. C and D are as in i.

n. Rate constant $k = C \cdot D$. C and D are as in i.

o. Rate constant $k = C \cdot E$. C is as in h. $E = 0.082TL$, and L is as in g.

Anthropogenic emissions of SO_2 and NO_x are calculated separately from detailed emission inventories, using source-specific emission profiles and activity data in combination with meteorological fields when evaporation and plume rises are important. Deposition velocities are parameterized from meteorological variables with observational constraints; a popular algorithm is the resistance-in-series method, which is widely used in chemical transport models such as the RADM model (Chang et al., 1987), the acid deposition modeling system developed for China (Wang et al., 2000), and CMAQ developed by the US EPA. Wind, temperature, pressure, and humidity fields are simulated by meteorological models that solve partial differential equations describing conservation of masses, momentum (vorticity), and energy in the atmosphere; high-order Runge–Kutta methods are usually used to advance time steps constrained by application conditions. Due to numerical limitations in meteorological simulations, for example, the solution field is sensitive to initial values, and observations are usually used to calibrate modeled meteorological variables as much as possible to prevent the solution from becoming chaotic. Release of chemical energy from photochemical reactions has been shown to slightly perturb the energy budget in the troposphere, and resulting perturbations to meteorological fields and climate systems are small (Jacobson, 1998b). In addition, atmospheric modeling with interactive chemistry and meteorology has been conducted recently to simulate gas–aerosol air pollution from urban through to global scales using a nearly explicit gas-phase photochemical mechanism (Jacobson and Ginnebaugh, 2010).

In typical atmospheric chemical modeling, the aqueous-phase reactions listed in Table 6.3 are often simplified to alleviate the computational burden. An example is given below to describe the simulation of the benefit of reducing sulfur emissions from marine vessels on winter sulfate in central California.

6.3.1 Case descriptions

Multiple-year observations in the California coastal area indicated that the background sulfate in the air over California ranged from 0.5 µg m^{-3} in central California to 1.5 µg m^{-3} in southern California. For comparison, PM sulfate was 4–5 µg m^{-3} in central California during a winter episode (Liang et al., 2006b). Ship emissions, besides biogenic emissions, in the Pacific Ocean, could be responsible for the background sulfate. To assess the impact of ship emissions on the PM sulfate in central California, a grid-based, three-dimensional, photochemical–microphysical transport model (US EPA CMAQ) was used. The model has been evaluated by simulating an extended PM episode (December 25, 2000 through to January 7, 2001) captured in central California (Livingstone et al., 2009).

6.3.2 Model domain and setup

The model domain, as shown in Figure 6.5, was designed to cover central and northern California and adjacent areas. The horizontal area is the same as for the winter modeling of the California Regional PM$_{10}$/PM$_{2.5}$ Air Quality Study

Figure 6.5 Horizontal domain for simulating benefits of reducing sulfur emissions from ships to PM sulfate over central California.

(Livingstone et al., 2009) and the Summer Central California Ozone Study (CCOS). The model has 185×185 horizontal grids with a resolution of 4 km $\times$ 4 km, and 15 expanding vertical layers of ~30 m thickness for the bottom layer extending to the upper troposphere. More details can be found in Livingstone et al. (2009) and references therein.

The Community Multiscale Air Quality Model (CMAQ) was employed with updates for California conditions in this study. The science and algorithm of CMAQ was described in US EPA (1999) with updates available at http://www.cmascenter.org/. The setup for chemical mechanisms includes SAPRC99 for the gas-phase mechanism with the addition of dimethylsulfide and methane, RADM for the aqueous-phase mechanism with additional constraints for catalytic concentrations and the calculation of activity coefficients, and bulk equilibria between the gas and liquid phases for secondary inorganic and organic aerosols respectively. Photolysis rates were calculated using climatological profiles, with cloud radiative effects included as a function of cloud optical depth and relative position. Relative humidity and column liquid water were constrained with observational diagnosis in California San Joaquin Valley, USA.

Boundary and initial concentrations of species except ozone were set to be small, so that they have negligible effects on simulated results in central California, USA. The boundary profile of ozone was derived from multiple-year ozonesonde observations measured at a coastal station in northern California, USA (Cooper et al., 2005).

Emissions of trace gases and PM mass were prepared at the California Air Resources Board (CARB), California, USA. Speciated, size-resolved PM emissions were prepared at the University of California at Davis, using the gridded PM emission inventory provided at CARB. Ship emissions were distributed along ship lanes and ports, using monthly averaged profiles prepared at CARB, while ship emission of sulfur species was based on a final report sponsored by the CARB, as described in Wang et al. (2008).

Meteorological parameters were extracted from outputs of the Fifth-Generation Mesoscale Meteorological Model (MM5) developed at NCAR and Pennsylvania State University. The MM5 simulations used for this study involved three runs with progressive one-way nested grids. Analysis nudging was applied in two parent grids (12 km × 12 km and 36 km × 36 km), but no FDDA was applied in the final grid (4 km × 4 km) used for CMAQ. Planetary boundary layer height was calculated using surface similarity theory. The simulation period was from December 17, 2000 to January 7, 2001, with the initial 8 days for model spin-up, and results in the last 2 weeks are discussed below.

6.3.3 Simulation results

Dozens of 22-day simulations were conducted to establish the base case for this study, when model results of gases and PM components reproduced observations within reasonable certainty. Detailed evaluations of inorganic and organic gas species and PM components in the model were reported in Livingstone et al. (2009).

To illustrate the benefit of reducing sulfur emission from ships during winter, an additional pair of simulations was conducted using a historical sulfur emission inventory from ships and a hypothetical 40% reduction in sulfur emissions from ships operating in the model domain respectively. Figure 6.6 shows horizontal distributions of daily average mixing ratio of SO_2 from the two simulations. It is seen that, in a historical condition, the SO_2 mixing ratio was outstanding along the ship lanes in the coastal Pacific and over the San Francisco Bay area during the Christmas–New Year holiday period; large areas along the coast experienced an SO_2 mixing ratio over 10 ppb, and the peak value of 40 ppb occurred near San Francisco. With a hypothetical 40% reduction in the historical sulfur emission from ships, areas with a high SO_2 mixing ratio decreased significantly and the peak value also decreased to 24 ppb. Thus, simulations suggested that controlling ship emissions could significantly reduce the SO_2 mixing ratio in coastal air.

Figure 6.7 shows the simulated daily average concentration of PM sulfate in the bottom layer using a historical inventory of sulfur emission from ships, and simulated reduction in the surface concentration of PM sulfate from a hypothetical 40% reduction in sulfur emission from ships in the model domain. It is seen that, on a historical winter day, the concentration of PM sulfate was 0.5–2 $\mu g\ m^{-3}$ in surface air over the coastal Pacific, San Francisco Bay area, and California Central Valley; the peak value reached 13.3 $\mu g\ m^{-3}$. With a hypothetical 40% reduction in sulfur emission from ships in the model domain, the concentration of PM sulfate decreased significantly. It is observed that the maximum decrease in daily average concentration

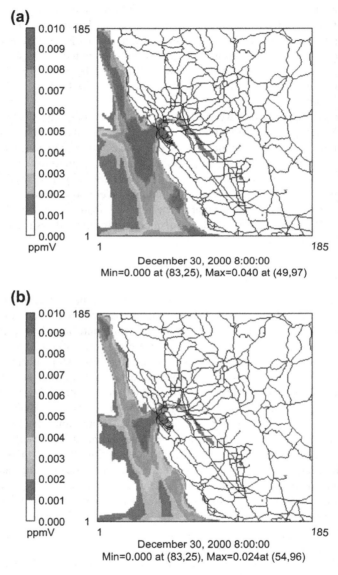

Figure 6.6 Simulated daily average mixing ratio of SO_2 in the bottom layer. (a) A historical sulfur emission inventory from ships was used. (b) A hypothetical 40% reduction in sulfur emission from ships was applied in the model domain.

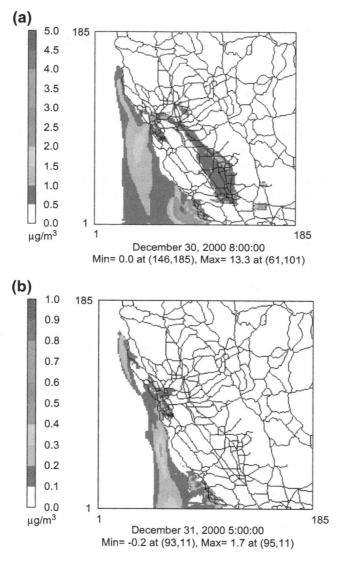

Figure 6.7 (a) Simulated daily average concentration of PM sulfate in the bottom layer using a historical inventory of sulfur emission from ships. (b) Simulated reduction in the surface concentration of PM sulfate from a hypothetical 40% reduction in sulfur emission from ships in the model domain.

of PM sulfate was 1.7 μg m^{-3}, and the benefit for PM sulfate concentration in several grids north of San Francisco exceeded 1 μg m^{-3}. As the formation of sulfate is sensitive to liquid water content, the exact location with the peak benefit depends on the weather.

6.3.4 Contribution from marine biota

Marine biota emit organic sulfur compounds, such as dimethylsulfide (DMS), to the marine boundary layer. In the northern California model domain during winter, the daily emission of DMS was estimated to be ~12 tons sulfur using an emission rate of 5×10^{-4} mol s^{-1} in each surface grid over the Pacific Ocean; for comparison, ship emission rate was 18–32 tons of sulfur per day. Over the land, total emission was 56 tons of sulfur per day; gasoline and diesel vehicles emitted ~6.5 tons of sulfur per day during weekdays, and diesel emissions were significantly lower during weekends in the model domain.

The photochemical oxidation of DMS into sulfate was implemented into CMAQ using the mechanism of Karl et al. (2007). Simulated mixing ratio of DMS in surface grids over the Pacific Ocean in northern California was within 90 ppt during winter, which is consistent with observations. The effect of oceanic emission of DMS on wet deposition of sulfate is shown in Figure 6.8, together with simulated column water content during daytime. It is seen that, during a wet day as indicated by large values of column water content in the figure, the coastal Pacific between San Francisco and Santa Barbara in northern California, USA could receive up to 1 g ha^{-1} hour^{-1} more sulfate via wet deposition due to oceanic emission of DMS during daytime in winter.

Oceanic emission of DMS varies with nutritional conditions in surface water as well as seasons (Guo et al., 2010). For example, some measurements in the coastal Pacific near San Diego did not detect any DMS. When DMS was absent, it is conceivable that reducing sulfur emission from ships was relatively more effective for reducing sulfur burden in coastal areas (Vutukuru and Dabdub, 2008).

6.4 Current issues on acid rain

6.4.1 Critical load

Substantial efforts have been conducted in developed countries to control transboundary acidification in the last 30 years. For example, in a campaign to control transboundary transport of acid rain pollution in Europe, 21 European countries reached a protocol in 1985 to reduce sulfur emission or transboundary transport by more than 30% from 1980 to 1993. By 1993, all parties had met the goal, and 11 parties had achieved more than 60% reduction while the total emission was halved using the baseline values in 1980. As a result, the annual emission of SO_2 in Europe decreased from 62 million tons in 1980 to 24 million tons in 1993. In 1994, a subsequent protocol was reached to further reduce sulfur emission. The USA and Canada also reached a bilateral agreement with the acid rain Annex in 1991. From

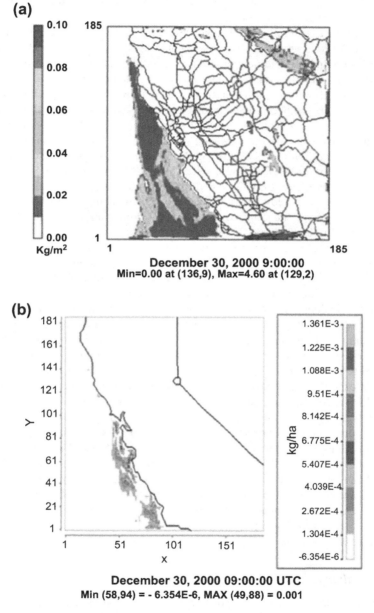

Figure 6.8 (a) Simulated column water content averaged over eight hours during daytime. (b) Effects of DMS emission from oceanic biota on wet deposition of sulfate. Units are kg ha^{-1} h^{-1}.

1970 to 2003, NO_x emission was reduced by 25% in the USA, despite rising vehicle miles driven and energy consumption.

In the USA, a federal interagency program, the National Acid Precipitation Assessment Program (NAPAP), has coordinated acid rain research and reported to Congress on findings of acid precipitation, especially on sensitive ecosystems, in the last two decades. Significant efforts have been conducted to reduce emissions of acid rain precursors. For example, emission-control technologies, especially flue-gas desulfurization (or scrubbers) have been increasingly applied, while the amount of fuel used at sources regulated by the Acid Rain Program has decreased. In addition, cleaner fuels have been used at regulated facilities. By 2009, emissions from power plants were 5.7 million tons for SO_2 and 2 million tons for NO_x; these values were 64% lower than 1990 level and 67% lower than 1980 emissions for SO_2 and 67% lower than the 1995 level for NO_x, aided by the cap-and-trade program. It was estimated that the emission reduction provided economic benefits of $170–430 billion in 2010 for human health due to improved air quality alone; emission reduction has also benefited visibility and ecological systems.

Despite significant efforts to reduce anthropogenic emissions of acid rain precursors over the past 30 years, there are remaining problems. For example, scientists have observed delays in ecosystem recovery in the eastern USA, and ammonium nitrate deposition has impacted western ecosystems.

Atmospheric deposition of sulfate, ammonium nitrate, mercury, and other toxics may still exceed "critical loads" for impacted ecosystems in changing ecological and environmental systems in the USA. For example, the critical load of sulfur and nitrogen for streams in the central Appalachian region of the eastern USA was calculated to be 370 meq m^{-2} $year^{-1}$ on average. From 1989 to 1991, atmospheric deposition exceeded the critical load of 40% of streams in the central Appalachian region; the excess was estimated to be 30% or 35,000 km of streams in the area from 2006 to 2008. It is noteworthy that values of critical load depend on conditions of receptors. For example, a recent study discovered that lakes in the Rocky Mountain National Park of the western USA experienced a shift in phytoplankton species in the 1960s when nitrogen deposition exceeded 1.5 kg ha^{-1} $year^{-1}$, the demonstrated critical load for the eutrophication of the lakes, according to records in lake sediment cores. Over wide areas of the USA, including the western USA, New England, and Florida, as shown in Figure 6.9, ecosystems are sensitive to acid deposition though atmospheric deposition has not yet exceeded critical loads in these areas.

6.4.2 Challenges in acid rain simulations

Modeling regional dry and wet depositions of sulfate, nitrate, and ammonium from the atmosphere may be conducted by US EPA CMAQ and similar air quality modeling systems, such as CAMx by Environ Inc., California, USA and WRF-Chem by the National Center for Atmospheric Research, USA. With these state-of-the-art regional modeling systems, major challenges still remain in simulating acid rain. Firstly, concentrations in models and observations may need to include organic

Figure 6.9 Areas designated by the US EPA as ecosystems sensitive to acid precipitation. Obtained from NAPAP (2011).

sulfates and organic nitrates in atmospheric boundary layers. A recent, 1-year field campaign in eastern China revealed a significant difference between inorganic sulfate and total sulfur in daily fine PM samples collected in the breathing zone of a subtropical metropolitan area (Sun et al., 2013). Corresponding emission inventories and chemical mechanisms need to be developed to elucidate the problem.

Secondly, atmospheric acid deposition may need to be integrated into an environmental system model to simulate its fate in soil, surface water, and terrestrial biota. In California, ammonium nitrate has become the dominant component of acid deposition in urban areas and farmlands due to the aggressive reduction in anthropogenic sulfur emission and the huge number of vehicle miles driven each year. In areas without enough ammonia near the surface, acid deposition could gradually deplete the neutralization capacity and eventually exceed the critical loads of surface ecosystems.

Summary

In this chapter, acid rain pollution has been discussed from the scientific and regulatory perspectives. Adverse effects of acid deposition and transboundary acidification is introduced first. Then, observational efforts and results are elaborated for acid rain precursors, components, and depositions in the USA and China. After that, a photochemical mechanism is presented to describe the formation of inorganic sulfate and nitrate in the global troposphere from anthropogenic emissions. Then, a real regulatory simulation case in support of the formation of the third sulfur emission control area, including coastal California, USA, is illustrated step by step. Finally, challenges in simulating acid rain are presented.

Exercises

1. As a final step to survey acid rain pollution status over many urban areas in a nation, the chief scientist condensed a year-long dataset from all ground stations into the information given in Table E6.1. Only bicarbonate ion was ignored from the dataset. Estimate: (a) the total concentration of all ions in units of μg solutes per liter rain, using constant CO_2 partial pressure of 38 Pa and constant air temperature of 290 K; (b) the pH value.

Table E6.1 Mean ionic composition of acid rain in 2011 (%)

SO_4^{2-}	28.1	Ca^{2+}	25.1
NO_3^-	7.4	Mg^{2+}	5.4
F^-	3.4	Na^+	5.7
Cl^-	8.3	K^+	2.3
NH_4^+	12.6	H^+	1.7

2. Assuming a fictitious ion $CH_3-SO_3^-$ accounted for 10% of total sulfur in all acid rain samples summarized in Table E6.1. Repeat the above calculations.

3. Dry deposition is an important atmospheric phenomenon for chemicals that react with the Earth's surface. Suppose a DMS molecule travels with an air parcel to a city with twenty-third century uniform architecture: all trees have the same unique characteristics for DMS to deposit; all buildings have the same unique characteristics for DMS to deposit, as do all urban streets. Write a formula for modeling dry deposition velocity of the DMS molecule in this future city, using the resistance-in-series algorithm.

4. The oxidation of DMS in the atmosphere has been described by a handful of photochemical mechanisms with the number of reactions ranging from a few to over 100. Read the journal article by Karl et al. (2007). Construct a 20-reaction photochemical mechanism for DMS degradation in the atmosphere.

5. Using the concept of critical load, discuss which area is more susceptible to acid rain: (a) an arid plateau with annual precipitation of 50 mm, or (b) a subtropical farmland with annual precipitation of 1.3 m, or (c) a tropical forest with annual rainfall of 3 m.

7 Climate change

Climate is the long-term average of weather on regional and global scales, and climate change refers to the change in temperature, precipitation, and wind parameters in a time range from months to millions years, as distinct from weather change. Global weather observations have been conducted for over 100 years and earlier climate parameters have been inferred geochemically. According to geochemical records, the climate system, defined as the atmosphere and its boundary with solid Earth, had changed periodically prior to human civilization. However, human activities are probably responsible for most observed changes in the climate system since the 1950s, due to emissions of greenhouse gases and deforestation. To investigate anthropogenic perturbation to the climate system, a number of chemistry–climate models have been developed in recent decades. The details are discussed below.

Chemical Modeling for Air Resources. http://dx.doi.org/10.1016/B978-0-12-408135-2.00007-0

7.1 Observed climate parameters

7.1.1 Temperature

Global observations of climate parameters have been conducted for over 150 years. So far, over 400 million temperature records are available from land stations, and over 140 million records for sea surface temperature along ship lanes and on over 1000 buoys. Figure 7.1 shows sea surface temperature reconstructed by NOAA, USA, for January 1854 and July 2012. It is seen that sea surface temperature was freezing in polar areas and fairly warm in tropical areas, where sea surface temperature was about 25°C. The peak sea surface temperature was 30°C in January 1854 and 33°C in July 2012. As the oceanic fraction is larger in the southern hemisphere than in the northern hemisphere, the difference in the two peak values may show the difference in heat capacity of surface water as well as climate change in the last 150 years.

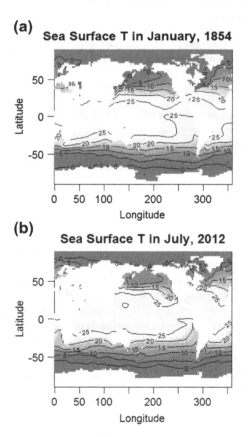

Figure 7.1 Sea surface temperature (°C) for January 1854 (a) and July 2012 (b). Data reconstructed by NOAA, USA.

7.1.2   Surface pressure

Surface pressure at average sea level is fairly constant, $\sim 1.013 \times 10^5$ Pa, and its variation due to temperature, water content, and trace species is usually below $\sim 1\%$. Figure 7.2(a) shows mean sea-level pressure on July 1, 2010, from NOAA, USA. The pressure reduced to mean sea level ranged from 0.943×10^5 to 1.037×10^5 Pa, with

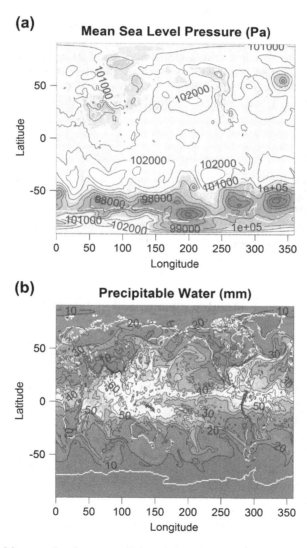

Figure 7.2 (a) Mean sea-level pressure (Pa) on July 1, 2010. (b) Precipitable water (mm) on July 1, 2010.
Data reconstructed by NOAA, USA.

an average value of 1.01×10^5 Pa. It is seen that there was a high-pressure area around Bermuda in the Atlantic Ocean, and there was a high-pressure area in the northeastern Pacific Ocean. There were also several high-pressure areas in mid-latitudes of the southern hemisphere.

7.1.3 Precipitable water

Precipitable water in the atmosphere is an important climate parameter, which is expected to increase with global mean sea surface temperature. Figure 7.2(b) shows precipitable water on July 1, 2010 from NOAA, USA. The precipitable water ranged from 0.1 to 78.1 mm with a global average of 21.6 mm, at a resolution of $1° \times 1°$. It is seen that precipitable water was abundant over the tropics, Indian Ocean, Gulf of Mexico, southeastern USA and coastal China. Over polar areas, winter hemisphere, Tibetan plateau and the Middle East, the precipitable water was below 20 mm.

7.1.4 Winds

Sea surface wind speed is an important climate parameter, as it affects water evaporation rates. Figure 7.3 shows monthly sea surface wind speed at 10 m ASL in July 2011, from NOAA, USA. The wind speed ranged from 0 to 27.6 m s^{-1}, with an average of 8.0 m s^{-1} over global oceans. It is seen that the wind speed was higher over open oceans than coastal areas, in general. The wind speed was ~6 m s^{-1} in coastal oceans and ~10 m s^{-1} in winter high latitudes.

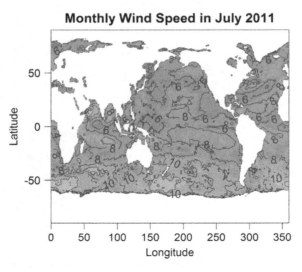

Figure 7.3 Monthly wind speed at 10 m ASL in July 2011.
Data reconstructed by NOAA, USA.

7.1.5   Sea ice and snow

The extent of cryosphere on the surface is an important indicator for a climate system. Figure 7.4 shows sea ice extent and snow cover on July 1, 2010 from NOAA, USA. It is seen that the north polar sea surface was covered with ice and, in high latitudes of the southern hemisphere, sea ice extended to 60°S. These ice sheets contain 80% of global fresh water, and any change in size will affect sea level. Meanwhile, the accumulated snow cover ranged from 1 to 323 kg m^{-2}, with an average value of

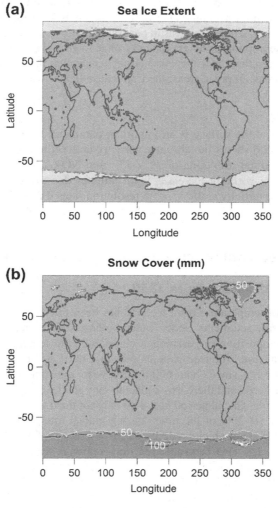

Figure 7.4 Sea ice extent (a) and snow cover amount (b) on July 1, 2010. Data reconstructed by NOAA, USA.

83.4 kg m^{-2}. It is seen that snow cover mainly existed over Greenland and high latitudes in the southern hemisphere.

7.1.6 Heights of PBL, maximum wind speed, and tropopause

The atmospheric vertical structure of temperature is closely related to the heights of PBL, maximum wind speed, and tropopause. Figure 7.5 shows these three heights on July 1, 2010 from NOAA, USA. The PBL height is sensitive to terrain and time of day, and ranged from 17 m in areas at night to 5405 m over a desert area. It is seen that the marine boundary layer height was ~500 m and the global average value of the PBL height was 634 m. The geopotential height with maximum wind speed ranged from 4.8 to 16.5 km, with an average value of 11.7 km. It is seen that the geopotential height with maximum wind speed was in general higher over tropical areas and lower in polar areas. The tropopause height ranged from 6 to 18.6 km, with an average value of 12.6 km. It is seen that the tropopause height was greater than 15 km over tropical and subtropical areas; in polar areas, the tropopause height was below 10 km.

7.1.7 Vorticity

Absolute vorticity at the surface is an important parameter for weather and climate modeling. Figure 7.6 shows the absolute vorticity at 1000 mb at 0 UTC on July 1, 2010, with a horizontal resolution of $1° \times 1°$, from NOAA, USA. While the zonal average of absolute vorticity increases with latitudes due to the apparent Coriolis force, the absolute vorticity at the $1° \times 1°$ resolution ranged from 0 over the Equator to over $5 \times 10^{-4} \text{ s}^{-1}$ at mid-latitudes, with an average value of $0.93 \times 10^{-4} \text{ s}^{-1}$.

7.2 Observed climate changes and greenhouse gases

7.2.1 Climate changes prior to 1850

Climate change of temperature prior to 1850 has been inferred geochemically from ice and sand cores, sediments under water bodies, fossils and living beings for isotopic or air composition or parameters correlated with temperature and precipitation in historical time. Inferred data suggest that the Earth has experienced both warmer and colder eras since about 600 million years ago, due to changes in orbital parameters of the Earth, and there was a glacial–interglacial cycle of ~100,000 years. Within each cycle, the glacial period lasted 80–90% of the time, while the interglacial period lasted only 10–20% of the time. During glacial time, atmospheric CO_2 was inferred to be ~190 ppm, which could be partly responsible for the cold climate.

Climate change has also been inferred on a regional scale, with a temperature increase of several degrees in a few decades. Such changes could have occurred during the last 10,000 years, which may be responsible for the formation of deserts. During historical warming periods, atmospheric circulations could have changed and polar areas could have become colder. Inferred data also suggest that a minor ice age occurred between 1350 and 1850, according to the IPCC.

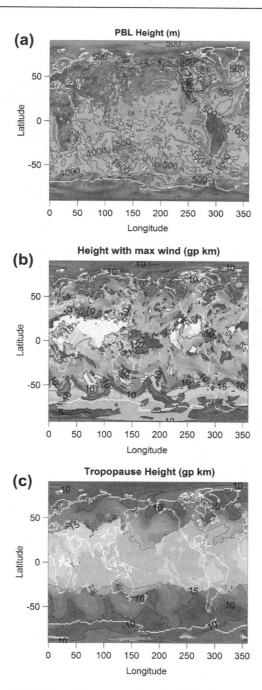

Figure 7.5 Heights of PBL (a), maximum wind speed (b), and tropopause (c). Data reconstructed by NOAA, USA.

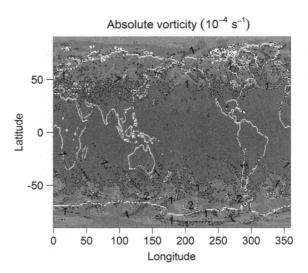

Figure 7.6 Absolute vorticity at 1000 mb at 0 UTC on July 1, 2010.
Data reconstructed by NOAA, USA.

7.2.2 Global warming since 1850

Surface temperature has increased rapidly in recent decades compared to 150 years ago, when temperature started to be recorded worldwide. According to the IPCC, surface temperature increased over large regions by 0.5 K between 1910 and 1940; since 1980, surface temperature has increased by 0.6 K, while changes in other periods since 1840 were smaller. The decade of 1995–2006 was the hottest since 1850, and the global warming rate from 1956 to 2005 was $0.13 \pm 0.03°C$ per decade. By comparison, the global warming rate from 1906 to 2005 was $0.74 \pm 0.18°C$ per century, or $0.07 \pm 0.02°C$ per decade. Thus, the trend of global mean surface temperature was nonlinear.

As most human activities occur over the land, global warming over the land was more rapid than over the ocean. For example, the urban island effect is about 2°C and temperatures in urban areas have increased rapidly in recent years compared to 150 years ago. The global warming rate over the land has been greater than 0.27°C per decade since 1979, which is about twice as much as over the ocean. Surface oceanic warming that affects atmospheric circulation showed hemispheric differences in the Pacific with the El Nino events and the Atlantic, while Indian Ocean warming has been steady.

Global warming trends show seasonality. The global warming rate was greatest during winter and spring in the northern hemisphere, which possibly reflects more rapid meridian transport of energy.

Climate change has occurred for extremes of temperature. The number of days with cold extremes and frost in mid-latitude regions has decreased, which is probably beneficial. Meanwhile, the number of days with warm extremes has increased in over

70% of the land regions observed. Cold nights have become rarer and warm nights have become more frequent in the last 50 years. For example, in the summer of 2003, there was an outstanding heat wave over Europe that was 1.4°C above the previous warmest in 1807 and probably the hottest since 1500. As a result, the need for air conditioning increases.

Climate change also occurs due to anthropogenic perturbations of temperature at higher levels. Global warming in the lower troposphere is expected to be a little larger than at the surface, partly due to the presence of black carbon. Due to global O_3 thinning and the O_3 hole, the stratosphere has been cooler in recent years than 150 years ago; as a result, the tropopause is expected to be higher than before.

Climate change in precipitation has not been well documented as yet. Limited data have suggested that, in recent years, more heavy precipitation events occurred over many land areas, including those with a decreased amount of total precipitation. Meanwhile, droughts have occurred more frequently over tropical and subtropical areas since 1970. Tropospheric water vapor increased over global oceans by 1.2 $\pm$ 0.3% per decade from 1988 to 2004.

Climate change in winds occurred in both hemispheres, as indicated by stronger mid-latitude westerly winds from the 1960s to the mid-1990s and more frequent tropical cyclones. The first recorded tropical cyclone in the South Atlantic occurred in March 2004.

7.2.3   Radiative force of long-lived greenhouse gases

Climate change may be best identified by looking at anthropogenic perturbations and their trends, as the relative climate changes of temperature, precipitation, and winds on a global scale are small. Anthropogenic perturbations to climate change are mainly through emissions of greenhouse gases and aerosols, as well as their precursors. According to radiative force at the top of the atmosphere, the main greenhouse gas of anthropogenic origin is CO_2, followed by CH_4, N_2O, halogen compounds, O_3, and aerosols. The average lifetime of aerosol pollutants in the sky is of the order of a week and their radiative forces are being actively researched (Henze et al., 2012; Lund et al., 2012); due to severe adverse effects on human health, aerosols will be the subject of Chapter 9. Long-lived greenhouse gases are the focus of this chapter.

Atmospheric mixing ratios of the long-lived greenhouse gases CO_2, CH_4, N_2O, and CF_4 are believed to have been 278 ppm, 715 ppb, 270 ppb, and 40 ppt respectively before 1750, when the industrial revolution began, according to geochemical studies. The atmospheric mixing ratio of CO_2 has increased rapidly in the last 50 years at a speed of ~20 ppm per decade, while the global annual anthropogenic emission of CO_2 has increased from 15 GT in 1970 to 45 GT in 2010. Meanwhile, the surface mixing ratio of N_2O has also increased rapidly at a speed of ~10 ppb per decade. Table 7.1 lists the mixing ratios of long-lived greenhouse gases in 2005 and their relative changes since 1998, according to the IPCC. It is seen that CO_2 and N_2O increased at ~2% year^{-1}, due to rapid increases in industrial and agricultural activities

Table 7.1 Mixing ratios (ppt) of long-lived greenhouse gases in 2005 and their percentage changes since 1998 (IPCC)

Species	M.R.	Change (%)	Species	M.R.	Change (%)
CO_2 (ppm)	379 ± 0.65	$+13$	CH_3CCl_3	19 ± 0.47	-72
CH_4 (ppb)	$1{,}774 \pm 1.8$	<1	CCl_4	93 ± 0.17	-7
N_2O (ppb)	319 ± 0.12	$+11$	HFC-125	3.7 ± 0.10	$+234$
CFC-11	251 ± 0.36	-5	HFC-134	35 ± 0.73	$+349$
CFC-12	538 ± 0.18	$+1$	HFC-152	3.9 ± 0.11	$+151$
CFC-113	79 ± 0.064	-5	HFC-23	18 ± 0.12	$+29$
HCFC-22	169 ± 1.0	$+29$	SF_6	5.6 ± 0.038	$+36$
HCFC-141	18 ± 0.068	$+93$	CF_4	74 ± 1.6	$-$
HCFC-142	15 ± 0.13	$+57$	C_2F_6	2.9 ± 0.025	$+22$

Table 7.2 Radiative force (W m^{-2}) of long-lived greenhouse gases

Species	R.F.	Change (%)
CO_2	1.66	$+13$
CH_4	0.48	<1
N_2O	0.16	$+11$
CFCs[*]	0.268	-1
HCFCs	0.039	$+33$
Montreal gases	0.320	-1
HFCs + PFCs + SF_6	0.017	$+69$
Halocarbons	0.337	$+1$
Total	2.63	$+9$

[*] CFCs include CFC-11, CFC-12, CFC-113, CFC-13, CFC-114, CFC-115, and the halons.

during the statistical period. CH_4 did not show a significant change, partly due to its relatively rapid oxidation in the atmosphere with a lifetime of ~10 years. CFC compounds and organic chlorides showed decreasing trends, due to international efforts to mitigate stratospheric O_3 hole. Meanwhile, it is seen that the use of HCFC, HFC, and PFC compounds has increased significantly as they are substitutes for banned CFC compounds for refrigeration, air conditioning, aerosol spray, foams, fire extinguishers, delicate electronic cleaners, etc.

The radiative force of a chemical is an important indicator as regards its perturbation to climate change. Table 7.2 shows the radiative forces of long-lived greenhouse gases in 2005 and the percentage change since 1998. It is seen that CO_2 had the largest radiative force (1.66 W m^{-2}) in 2005, and had increased by 13% since 1998. By comparison, CH_4 (0.48 W m^{-2}) and N_2O (0.16 W m^{-2}) had the second and third largest radiative forces in 2005. Halocarbons had a collective radiative force of 0.337 W m^{-2} in 2005. If all CFC compounds in the atmosphere had been removed in 2005, radiative force would have been reduced by 0.268 W m^{-2}. The radiative force of CFC substitutes, namely HCFCs, HFCs, PFCs, and SF_6, was an order of magnitude smaller, though they showed rapid increases from 1998 to 2005. Thus, reducing CFC emissions to mitigate the O_3 hole has a synergistic effect on mitigating climate change from the perspective of radiative force.

7.3 Simulating anthropogenic climate change

An important pathway of anthropogenic perturbation to climate change is by emitting greenhouse gases to exert radiative force to the energy budgets of the Earth. In this section, a scheme of energy budgets of the Earth will first be introduced. Then, radiative forces from anthropogenic greenhouse gases will be discussed based on atmospheric general circulation modeling. Finally, the results of state-of-science chemistry–climate models will be interpreted.

7.3.1 The Earth's energy budgets

The energy budgets of the Earth are illustrated in Figure 7.7 on an annual global average basis (Kiehl and Trenberth, 1997).

It is seen that the total incoming solar radiation (342 W m^{-2}), which changes little on a human timescale, is split into four parts: about half is absorbed by the surface (168 W m^{-2}), over 20% is reflected by atmospheric clouds, aerosols, and gases (77 W m^{-2}), nearly 20% is absorbed by the atmosphere, and nearly 10% is reflected by the surface (30 W m^{-2}). Important gases for solar radiation are the air molecules for Rayleigh scattering, stratospheric O_3 for absorbing ultraviolet radiation, and water vapor for absorbing at a wide spectrum. Important aerosol components for solar radiation include black carbon for absorption and sulfate for scattering. Important cloud properties for solar radiation include cloud water content, cloud

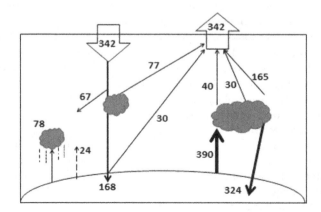

Figure 7.7 Global annual average energy budgets of the Earth (from Kiehl and Trenberth, 1997). Units are W m^{-2}.

droplet spectrum, and the phase of water in clouds. Important surface property for solar radiation is the surface albedo, which ranges from 0.05 over the liquid water surface to 0.3 over yellow grass to 0.7–0.9 over snow cover and ice sheets. The total long-wave radiation (390 W m^{-2}) from the surface of the Earth is mostly intercepted by the atmosphere, which reemits in the sky at cooler temperature: about 10% of the long-wave radiation in the atmospheric window escapes from the top of the atmosphere directly (40 W m^{-2}), clouds emit about 30 W m^{-2} long-wave radiation from the top of the atmosphere, and the atmospheric gases emit 165 W m^{-2} long-wave radiation from the top of the atmosphere. Greenhouse gases absorb long-wave radiation from the Earth's surface and reemit long-wave radiation above at lower temperatures in the troposphere. In the stratosphere, CO_2 emits at an in-situ temperature that is close to the temperature at the surface; thus, the addition of CO_2 to the upper stratosphere results in cooling. In addition, the surface releases about 78 W m^{-2} latent heat and 24 W m^{-2} sensible heat into the atmosphere. The atmospheric greenhouse gases and clouds emit 324 W m^{-2} long-wave radiation back to the surface. Thus, it is seen that solar radiation initiates the energy cycle by emitting incoming radiation energy of 342 W m^{-2} downward at the top of the atmosphere, and the atmospheric greenhouse gases and clouds magnify the radiation to the surface by emitting long-wave radiation to the surface to bring the total radiation energy to the surface up to 492 W m^{-2}. The upward radiation energy at the top of the atmosphere consists of 107 W m^{-2} reflected short-wave radiation and 235 W m^{-2} long-wave radiation emitted by atmospheric gases, clouds, and the surface. Note that the exact numbers shown in the diagram depend on the processes and feedback implemented in the model, but changes between models are small on a global annual average basis since climatological properties of the atmosphere and the surface change little over a decade.

The energy budgets on a zonal average basis differ significantly from the global average, due to uneven solar radiation at the top of the atmosphere. Over the tropics, incoming solar radiation is much stronger than over polar areas on an annual average basis. The differential heating of the surface and the atmosphere by solar radiation drives climatological circulations in the atmosphere and the surface water of oceans, namely the Hadley cell in tropical and subtropical atmosphere, the mid-latitude storm tracks, and the polar vortices; the clockwise circulations of surface waters in the northern hemisphere and the anticlockwise circulations of surface waters in the southern hemisphere. The meridian transport of heat due to general circulations of the atmosphere and oceans reduces the thermal gradient caused by uneven solar radiation.

At a latitudinal zone, annual average energy budgets vary with land cover and land use types that determine seasonal variations of surface albedos, atmospheric pressure patterns that define air mass, cloud and precipitation amounts that affect both long-wave and short-wave radiation, and anthropogenic activities that define other parameters, such as atmospheric black carbon, which are important for short-wave radiation budget calculations. Long-lived greenhouse gases are distributed almost uniformly in the atmosphere, with ~1% difference between the northern hemisphere and southern hemisphere, as most populations reside in the northern hemisphere.

7.3.2   Force and response from general circulation modeling

Assessment of anthropogenic perturbations to the climate system may be effectively conducted by the force and response method of general circulation modeling, due to the long-term nature of climate change and the associated complexity. The radiative force (F) of a greenhouse gas to the surface–atmosphere climate system is operationally defined as the change in the net downward flux of solar and Earth radiation at the tropopause, as defined in formula (F7.3.1), while holding all other parameters in a general circulation model constant:

$$F = \Delta E - \Delta S \qquad\qquad\qquad\qquad\qquad\qquad\qquad\qquad \text{(F7.3.1)}$$

In the above formula, S denotes net downward short-wave radiation from the Sun and E is the global mean emitted long-wave radiation energy; ΔE and ΔS represent responses in a general circulation model to the radiative force F. For example, in a historical climate, the absorbed solar energy and emitted infrared energy were both 240 W m^{-2} at the tropopause; an instantaneous doubling of atmospheric CO_2 reduced the emitted infrared energy to 236 W m^{-2}. Thus, doubling atmospheric CO_2 at that time produced a radiative force of 4 W m^{-2}.

The response of surface temperature is usually proportional to the radiative force from a greenhouse gas in clear-sky conditions:

$$\Delta T_s = k \cdot F \qquad\qquad\qquad\qquad\qquad\qquad\qquad\qquad\qquad \text{(F7.3.2)}$$

As $E = \varepsilon \cdot \sigma \cdot T_s^4$, it follows that $\Delta E/\Delta T_s = 4E/T_s = 3.3$ W m^{-2} K^{-1} in typical conditions. Thus, without any feedback in the atmosphere, the global sensitivity parameter k in formula (F7.3.2) is $\Delta T_s/\Delta E = 0.3$ K m^2 W^{-1}. Furthermore, the climate system would recover to energy balance by increasing surface temperature. Doubling atmospheric CO_2 would lead to an increase in surface temperature of $\Delta T_s = 0.3 \times 4 = 1.2°$C, to maintain the energy balance without any feedback.

7.3.3   Feedback

The atmosphere provides a considerable amount of feedback to a radiative force, such as doubling CO_2. For example, as the surface temperature increases due to doubling of the CO_2 mixing ratio in the atmosphere, more water evaporates from the surface into the air. As water vapor is a strong greenhouse gas, this positive feedback reduces $\Delta E/\Delta T_s$ from 3.3 W m^{-2} K^{-1} to ~2.2 W m^{-2} K^{-1}. In addition, water vapor absorbs solar radiation. According to general circulation modeling (Cess et al., 1989, 1990), this positive feedback results in $\Delta S/\Delta T_s \approx 0.2$ W m^{-2} K^{-1} on a global annual average basis. Together, water vapor feedback results in an increase of the global sensitivity parameter k from 0.3 to ~0.5 K m^2 W^{-1} in formula (F7.3.2). As a result, doubling atmospheric CO_2 would result in an increase of surface temperature of $\Delta T_s = 0.5 \times 4 = 2.0°$C, to maintain the radiative energy balance with the feedback by water vapor. Thus, water vapor feedback amplifies the initial global warming due to doubling of atmospheric CO_2 by a factor of 1.7.

Besides atmospheric water vapor, surface snow and ice may also provide significant feedback in response to a doubling of atmospheric CO_2. As snow and ice melt at the surface due to higher temperature, the surface absorbs more solar radiation. In addition, melt snow and ice interact with clouds and long-wave radiation. In a perpetual April simulation by an ensemble of atmospheric general circulation models (Cess et al., 1991), it was found that the feedback from snow cover over the land could increase the global sensitivity parameter k from 0.5 to 0.6 K m^2 W^{-1} in formula (F7.3.2) in clear-sky conditions. Thus, doubling atmospheric CO_2 would result in an increase of surface temperature of $\Delta T_s = 0.6 \times 4 = 2.4°C$, to maintain the radiative energy balance with the feedback from atmospheric water vapor and snow cover over the land. The sea ice feedback mechanism is similar to that for snow cover over the land, but is harder to simulate because the timescale for ocean circulation is orders of magnitude longer than for atmospheric circulation. Over permafrost lands, significant amounts of CO_2 and CH_4 are trapped in frozen land, these provide positive feedback to global warming at high latitudes. In polar spring, the feedback from snow and ice at the surface may reduce meridian transport from tropical areas to the polar vortex, which will trigger dynamic feedback in the real atmosphere.

Besides atmospheric water vapor and surface snow and ice, clouds provide important feedback to global warming due to their interaction with both short-wave and long-wave radiation, as shown in Figure 7.7. For example, on a global annual mean basis in the recent climate, clouds above the Earth increase the reflection of solar radiation by about -44 W m^{-2}. Meanwhile, clouds absorb the Earth's long-wave radiation and reemit long-wave radiation above back to the surface by 31 W m^{-2}. Together, the presence of clouds above the Earth results in a net cooling force of about 13 W m^{-2}. A number of satellites monitor climatological cloud distributions over the globe, and cloud optical depth is measured for precipitating and non-precipitating clouds. The optical properties of clouds are linked to cloud fraction, vertical extension, and droplet spectrum, which vary dramatically with time and space. Existing weather forecasts for these parameters rely heavily on observations. Thus, it is very hard for climate models to accurately simulate clouds and their feedback to a radiative force. Despite significant uncertainty, an ensemble of atmospheric general circulation models (Cess et al., 1991) estimated that the global sensitivity parameter k in formula (F7.3.2) could increase from 0.6 to 0.84 when the feedback from clouds is included. Therefore, doubling atmospheric CO_2 would result in an increase of surface temperature of $\Delta T_s = 0.84 \times 4 = 3.4°C$, to maintain the radiative energy balance with the feedback by atmospheric water vapor, snow cover over the land, and clouds.

7.3.4 Chemistry–climate modeling

Climate simulations involve many years of global changes governed by the atmosphere, the ocean, and the land surface as well as photochemical reactions of anthropogenic emissions. Chemistry–climate models have implemented important processes in the atmosphere, surface oceans, soil layers, and biosphere for interactive climate simulations of key feedback such as water vapor, surface snow and ice, clouds and aerosols, as well as chemical reactions of greenhouse gases. To simulate climate change, a number of climate facets have to be reproduced by chemistry–climate models before further analysis of model outputs.

Aerosols and clouds are important for temperature structure near the surface. A number of satellite observations are available for deriving aerosol and cloud parameters. As it is computationally challenging for global climate models to represent interactive aerosol processes, a simplified version of the sulfate aerosol formation from SO_2 and DMS together with chemical degradations of greenhouse gases has been implemented in the Community Climate System Model (CCSM3) by the National Center for Atmospheric Research, USA. An important indicator for correctly representing aerosols and clouds in the model is the amount of sunlight reflected from the top of the atmosphere, the annual mean of which is ~100 W m^{-2}, and the amount of long-wave radiation from the top of the atmosphere, which is 200–250 W m^{-2}. At most latitudes, climate model results agree with observations of outgoing short-wave radiation within 6% on an annual average basis. As temporal and spatial resolutions increase, the performance of individual climate models becomes increasingly degraded; multi-model mean climatology is much closer to observations than individual climate models, while no single climate model has consistently been closest to observations. The outgoing long-wave radiation at the top of the atmosphere depends on atmospheric temperature, humidity, clouds, and surface temperature. Climate models may simulate observed outgoing long-wave radiation within 5% on a zonal average basis with seasonality, and capture the relative minimum near the Equator with relatively high humidity and extensive cloud cover. A regional climate model could simulate the long-term effects of Saharan dust on African and European climate by implementing a desert dust module that includes emission, transport, gravitational settling, wet and dry removal, and calculations of dust optical properties (Zakey et al., 2006). A recent investigation with 40-year simulations using a conservative general circulation model, the GCM3 of the NASA Goddard Institute for Space Studies, USA, suggested that elimination of aerosols over the USA would increase air temperature in the eastern USA by 0.6°C on an annual average basis and by up to 2 K during summer heat waves (Mickley et al., 2012).

A state-of-science chemistry–climate model designed for regular PCs was able to reproduce many climate facets including observed weakening of Brewer–Dobson circulation and corresponding elevated temperature and O_3 mixing ratio in the lower tropical stratosphere when the polar night jet was strong (Egorova et al., 2005). In a century-long simulation, the chemistry–climate model included forces from sea surface temperature, sea ice, solar irradiance, stratospheric aerosols, quasi-biennial oscillation, changes in land properties, greenhouse gases, O_3-depleting substances, and emissions of CO and NO_x. By using an ensemble of nine members with perturbations of not more than 0.01% to atmospheric CO_2 in the first year of the simulations, the model was able to reproduce vertical O_3 distributions from the surface to the stratopause at selected stations. Despite significant differences between ensemble members over polar areas, probably due to poor Eliassen–Palm fluxes, the ensemble captured many observed climate facets in the twentieth century, such as the two-dimensional distribution of the ratio of zonally averaged column O_3 trend to equivalent effective stratospheric chlorine, zonal wind, and zonal temperature (Fischer et al., 2008).

7.3.5 Coupled atmosphere–ocean modeling

An ensemble of coupled atmosphere–ocean general circulation models that contributed to the last IPCC report were compared with detailed radiative transfer models to calculate radiative forces of well-mixed greenhouse gases CO_2, CH_4, N_2O, CFC-11, and CFC-12 under clear-sky conditions (Collins et al., 2006). Instantaneous perturbations were applied to well-mixed greenhouse gases for seven scenarios, and Table 7.3 lists the accurate results. It is seen that the increase of atmospheric CO_2 from the level in 1860 to the level in 2000 increased the radiative force at 200 hPa by 1.7 W m^{-2}, and doubling atmospheric CO_2 from the level in 1860 increased the radiative force at 200 hPa by 4.7 W m^{-2}; if water vapor increases by 20% from current conditions due to warmer weather at a CO_2 level of 574 ppm, an additional radiative force of 4.1 W m^{-2} will be produced. For comparison, the radiative force from CFC-11 and CFC-12 in 2000 was 0.4 W m^{-2}, which is only 13% of the radiative force caused by CH_4 and N_2O in 1860 and half of the radiative force caused by the increase of CH_4 from 1860 to 2000. Thus, anthropogenic global warming is mainly due to CO_2, CH_4, N_2O and feedback from water, with minor contributions from CFCs. Some of the compared coupled atmosphere–ocean general circulation models neglected the effects of CH_4, N_2O, and CFCs on short-wave radiation, and some others used radiative modeling methods that yielded different results from accurate calculations. The resulting difference from accurate calculations was within ~10%.

Shindell et al. (2012) analyzed simulated radiative forces from dozens of state-of-science climate models for anthropogenic emissions with a focus on aerosols. It was found that simulated radiative forces from anthropogenic aerosol components were on average about −0.4, 0.24, −0.04, 0.05, and −0.19 W m^{-2} for SO_4^{2-}, BC, OC, SOA, and NO_3^- respectively using the emission and concentration levels in 2000 compared to those in 1850. On a global average basis, the total radiative force of aerosols was −1.1 W m^{-2} in 2000 compared with 1850. As emission control

Table 7.3 Simulated radiative forces (W m^{-2}) at 200 hPa for seven scenarios

Case	F (200 mb)	Case	F (200 mb)
1. CO_2 (287 ppm $\rightarrow$ 369 ppm)	1.7	5. "2000": CFC-11/12	0.4
2. CO_2 (287 ppm $\rightarrow$ 574 ppm)	4.7	6. CFC: "2000" CH$_4$: "2000"–"1860"	0.8
3. "2000"–"1860"	2.6	7. CO_2 (574 ppm) H_2O ($\times$1.2)	4.1
4. "1860": CH_4 + N_2O	3.1		

Note: "1860" denotes conditions of CO_2 (287 ppm), CH_4 (806 ppb), and N_2O (275 ppb); "2000" denotes conditions of CO_2 (369 ppm), CH_4 (1760 ppb), N_2O (316 ppb), CFC-11 (267 ppt), and CFC-12 (535 ppt).

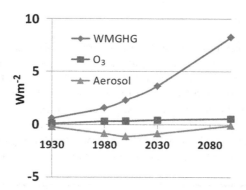

Figure 7.8 Global mean annual average radiative forces from anthropogenic emissions due to enhanced well-mixed greenhouse gases (WMGHG), O_3, and aerosol compared to 1850. Data are from Shindell et al. (2012).

measures are expected to reduce aerosol and O_3 pollution, anthropogenic radiative forces via enhanced aerosols and O_3 are predicted to decrease. Thus, anthropogenic perturbations to climate change in future will probably be via increasing well-mixed greenhouse gases. Figure 7.8 shows global mean annual average radiative forces of well-mixed greenhouse gases, O_3, and aerosols from anthropogenic emissions in 1930, 1980, 2000, 2030, and 2100, compared with those in 1850. Note that the values shown are the ensemble mean of a number of state-of-science climate models that showed a wide range of variations for each point in the figure.

7.4 Current issues on climate change

Relationships between air pollution and climate change are two-way. Air pollution mitigation may reduce climate forces by O_3, CH_4, and soot but may increase warming potential due to SO_2 reduction (Jacob and Gilliland, 2005; Mauzerall, 2011); Climate change may enhance air pollution in inland areas but may help ventilate fresh emissions in coastal areas (Mahmud et al., 2012).

Climate change simulations face two classic problems. Firstly, the time advancement from initial values in climate models with nonlinear and unstable governing equations is believed to have an upper limit. Secondly, the responses of the climate system to external forces have been found by ensemble simulations to depend on the initial state, which amplifies model uncertainties and errors. A number of model intercomparison exercises suggested that it is challenging to simulate cloud-radiation processes, the cryosphere, and atmosphere–ocean interactions. As climate models have started to implement full chemical and biogeochemical processes in a transition to simulate the full Earth system, model differences may be reduced; however, successful prediction of observed climate changes must be done to validate climate models.

Summary

In this chapter, the current states of important climate parameters are illustrated with a focus on the troposphere from an atmospheric chemical modeling perspective. Observed climate changes are described for periods prior to and after 1850. To analyze signals of climate change, the concept of radiative force is explained. Then, anthropogenic perturbations to the climate system are discussed using radiative forces of greenhouse gases, such as CO_2, CH_4, N_2O, CFCs and O_3, and aerosols. Climate modeling results are interpreted to understand important processes that regulate climate changes due to human activities. Further improvements on climate simulations are identified.

Exercises

1. A 20-year-old marine vessel cruises along a ship lane, where ocean water has abundant nutrition for biomass to grow. Assuming the productivity of the ocean water is 10 kg C m^{-2} day^{-1} along the ship lane, calculate how much anthropogenic CO_2 may be emitted daily into the atmosphere without causing long-term climate change, if the global annual average productivity of ocean water is 50% of that along the ship lane.

2. In a quasi-steady-state condition, the mixing ratio of SF_6 was 7 ppt in the northern hemisphere and 6 ppt in the southern hemisphere, while most emission sources are located in the northern hemisphere. If the lifetime of SF_6 in the northern hemisphere against interhemispheric transport to the southern hemisphere is 1 year, calculate the emission rate of SF_6 in the northern hemisphere. Survey the literature, and explain the difference between the emission rate in this problem and those from the literature.

3. In a historical year, global carbon reservoirs and annual fluxes are listed in Figure E7.1 for problems 3–5. In the figure, the unit of the sizes of the reservoirs and annual fluxes is 10^{15} gC. The percentage denotes $k(^{13}C)$, defined as $\{[^{13}C/^{12}C]/[^{13}C/^{12}C]_{standard} - 1\} \cdot 100\%$. Using the figure, calculate the atmospheric CO_2 mixing ratio and the value of $k(^{13}C)$ when all fossil fuel is used up by assuming constant fluxes.

4. Using Figure E7.1, estimate the corresponding pH value of the surface ocean in the historical year with the following assumptions: (1) the surface ocean was in equilibrium with the atmosphere where CO_2 was well mixed over the globe; (2) the only ion pair

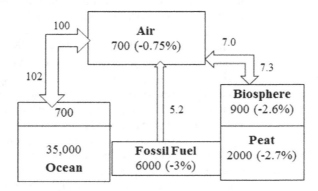

Figure E7.1

involving inorganic carbon in the ocean was $MgCO_3$; and (3) the activity coefficients of all ions in surface ocean obeyed the Davies equation. The global ocean is ~4000 m on average.

5. Estimate the equilibrium value of pH in the surface ocean after all fossil fuel is used up, assuming that the mixing between the surface ocean and deep ocean during the period is negligible. Then, if the diffusion coefficient between the surface ocean and the deep ocean is 1 cm^2 s^{-1} and the surface ocean and deep ocean were approximately in equilibrium in the historical year, recalculate the equilibrium pH value of the surface ocean when all fossil fuel is used up using a computer program.

8 Surface oxidants

Air oxidants are present at low levels in natural air. In modern air, human activities have elevated oxidants to harmful levels for humans and plants in some urban, rural, and park areas. Many nations have started to regulate ambient O_3 levels and to reduce its precursors to protect public health and welfare. Air quality modeling is often used to forecast regional ambient air quality status, to simulate observed O_3 fields, and to predict its response to changes in anthropogenic emissions of O_3 precursors. The details are presented below.

8.1   Observations of surface ozone and other oxidants

Ambient air contains oxidants (O_3, OH, HO_2, RO_2, Cl, NO_3, NO_2, H_2O_2) that may be harmful to humans and plants. O_3 is the most abundant oxidant, and thus is widely monitored for the protection of public health and welfare (Logan, 1985; Wang et al., 2001a, b; Wong et al., 2007; Zhang et al., 2009). For comparison, measurements of OH and H_2O_2 are still very limited, and their mixing ratios are of the order of 0.1 ppt and 1 ppb respectively in clear-sky conditions in daytime in summer; on rare occasions, the surface OH mixing ratio can reach 1 ppt (Hofzumahaus et al., 2009). No direct measurement of NO_3 is reported for ambient air conditions in the daytime, due to its rapid photolysis in sunlight; at night, it may become the major oxidant in

Chemical Modeling for Air Resources. http://dx.doi.org/10.1016/B978-0-12-408135-2.00008-2

ambient air with a mixing ratio of several ppt. NO_2 is both a primary pollutant and a precursor of O_3; it is widely measured despite significant uncertainty. Satellite data shows that its column concentration is of the order of 10^{15} molecules cm^{-2}. Providing that the column concentration of air at sea level is 2.13×10^{25} molecules cm^{-2}, the global average NO_2 mixing ratio is of the order of 100 ppt. In metropolitan and industrial regions, the NO mixing ratio often exceeds 100 ppb in early morning and evening, and the conversion of NO to NO_2 in the morning may increase the NO_2 mixing ratio to the order of 10 ppb despite its rapid photolysis in sunlight (Konovalov et al., 2005; Wang et al., 2009). Halogen radicals are also present in breathing zones, and the mixing ratio of chlorine radicals can approach that of OH in coastal areas; the increasing use of HCFCs and HFCs may make halogen radicals more important on a global scale in the future. Therefore, surface O_3 will be the focus below.

8.1.1   Surface O_3 episodes

Surface O_3 is routinely monitored in many countries nowadays, due to its potential adverse effects. In the USA, surface O_3 is measured at thousands of stations and the hourly averaged O_3 mixing ratio is reported to the US EPA for regulatory purposes. Based on potential adverse impacts on human health, environmental and ecological systems, the measured O_3 mixing ratio is grouped into six categories, namely "Good", "Moderate", "Unhealthy to sensitive groups", "Unhealthy", "Very unhealthy", and "Hazardous". Table 8.1 lists the corresponding surface O_3 mixing ratios in units of ppb, according to the O_3 air quality standard effective from 2009. It is seen that ambient air becomes unhealthy for sensitive groups, such as children and the elderly, when the 8-hour average O_3 reaches 76 ppb or the 1-hour average O_3 reaches 125 ppb. When the 8-hour average O_3 reaches 96 ppb or the 1-hour average O_3 reaches 165 ppb, ambient air becomes unhealthy. When the 8-hour average O_3 reaches 116 ppb or the 1-hour average O_3 reaches 205 ppb, ambient air becomes very unhealthy. In the 1960s, the 1-hour average O_3 in the Los Angeles area exceeded 600 ppb, which made the ambient air hazardous to residents and tourists.

Using the O_3 categories listed in Table 8.1, each measurement can be converted to a corresponding air quality index (AQI), using the linear formula:

$$I_p = (C_p - BP_{Lo}) \cdot (I_{Hi} - I_{Lo})/(BP_{Hi} - BP_{Lo}) + I_{Lo} \qquad \text{(F8.1.1)}$$

Table 8.1 Surface O_3 categories in the USA (US EPA, 2009)

USEPA-2012	AQI	O_3 (8-hour)	O_3 (1-hour)
Good	0–50	0–59	
Moderate	51–100	60–75	0–124
Sensitive	101–150	76–95	125–164
Unhealthy	151–200	96–115	165–204
Very unhealthy	201–300	116–374	205–404
Hazardous	301–500		405–604

where I_p = the air quality index for pollutant p, C_p = the rounded-up concentration of pollutant p, BP_{Hi} = the breakpoint that is greater than or equal to C_p, BP_{Lo} = the breakpoint that is less than or equal to C_p, I_{Hi} = the AQI value corresponding to BP_{Hi}, and I_{Lo} = the AQI value corresponding to BP_{Lo}. For example, if the 1-hour O_3 mixing ratio is 600 ppb, then ambient air is "hazardous" and the corresponding parameters in formula (F8.1.1) are listed in the last row of Table 8.1; the O_3 index is interpolated as $I_{O3} = (600 - 405) \cdot (500 - 301)/(604 - 405) + 301 = 496$. The air quality index is convenient for the dissemination of information. For example, Figure 8.1 shows converted surface O_3 at ground stations in the USA for August 13, 2011 and June 29, 2012. It is seen that, on August 13, 2011, surface O_3 reached an unhealthy level in inland areas of southern and central California. On June 29, 2012, surface O_3 was elevated in many eastern states to an unhealthy level and reached a very unhealthy level in eastern Tennessee, despite the incomplete data reported.

8.1.2 Long-term trends of surface O_3

Surface O_3 was measured at various ground stations in Europe a century ago. In Yugoslavia, surface O_3 was measured twice daily for 12 years from 1889 to 1900, and the original data were revisited a century later by Lisac and Grubišić (1991). During a 5-month warm period, the surface O_3 mixing ratio was believed to be 36 ppb during daytime and 30 ppb at night in the period 1893–1900. By 1975, the surface O_3 mixing ratio increased to 67 ppb during daytime and 56 ppb at night at the same station. Seasonality analysis showed that the day–night peak values occurred during the transition from spring to summer. In Greece, surface O_3 was measured for four decades at the beginning of the twentieth century, when the O_3 mixing ratio was 28 ppb during daytime and 27 ppb at night. By 1987–1990, the surface O_3 mixing ratio had increased to 57 ppb during the day and 48 ppb at night (Cartalis and Varotsos, 1994). In July and August, the surface O_3 mixing ratio increased by a factor of 1.8 during daytime but did not show significant change at night (Varotsos et al., 2001). Surface O_3 measurements in Switzerland during 1950s and from 1989 to 1991 revealed that the surface O_3 mixing ratio increased by a factor of 2 at a ground station over 40 years, and the relative increase was a factor of 3 in winter (Staehelin et al., 1994).

In the USA, anthropogenic emissions of the O_3 precursors NO_x and ROGs increased rapidly, from 7.4 and 17 million tons in 1940 to 21 and 31 million tons in 1970 respectively. By 1998, anthropogenic emission of NO_x slowly increased to 24 million tons, and anthropogenic emission of ROGs decreased rapidly to 18 million tons. The surface O_3 mixing ratio in the USA is usually elevated in May–September due to enhanced photochemical production. The observed O_3 mixing ratio at ground stations in contiguous USA over the period 1980–1998 was analyzed to identify the number of days when 1-hour O_3 exceeded 120 ppb and 8-hour O_3 exceeded 80 ppb in different regions, and to separate the effect of temperature (Lin et al., 2001). It was found that, at high temperature, a clear negative trend existed in the northeastern urban corridor, southern California, and the western bank of Lake Michigan, while a positive trend existed in Nashville, Tennessee. Except in southern California, O_3 air quality improvement was mild elsewhere from 1980 to 1998.

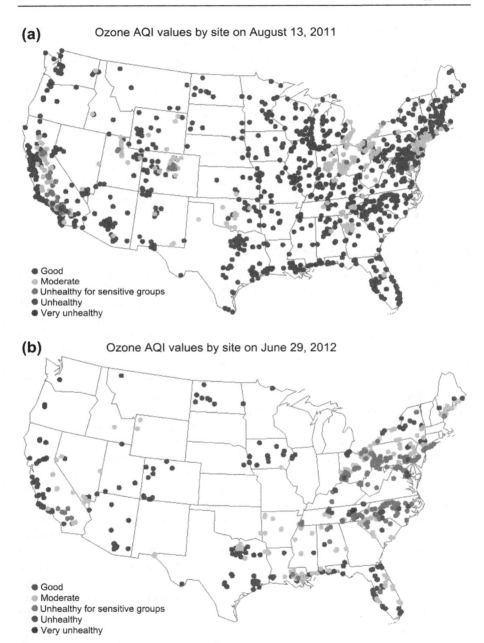

Figure 8.1 Surface O_3 air quality indices over the USA on August 13, 2011 (a) and June 29, 2012 (b).

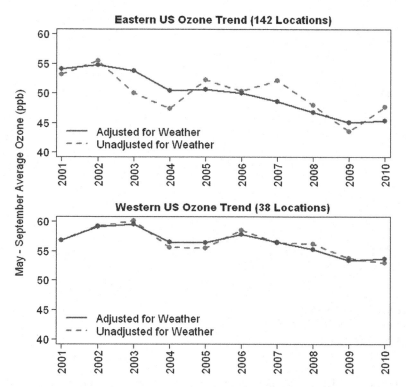

Figure 8.2 Trend of surface O_3 mixing ratio averaged from May to September in the eastern and western USA from 2001 to 2010.
Obtained from the US EPA (http://www.epa.gov, 2012).

Figure 8.2 shows trends of surface O_3 averaged at ground stations during the O_3 season from May to September in the eastern and western US respectively over the period 2001–2010. It is seen that, in the last decade, surface O_3 averaged over 142 stations in the eastern USA decreased by $\sim 10\%$, while the relative change in surface O_3 averaged over 38 stations in the western USA was small. As average O_3 within an air mass is closely related to emissions of O_3 precursors and weather conditions, the observed O_3 trend may be adjusted for weather conditions to show the effect of emission only. For example, the weather was persistently hot, dry, and stagnant in the eastern USA during the O_3 season in 2007; in 2009, the weather was cool, humid, and well ventilated. The weather effect was removed by statistical fitting conducted at the US EPA, and the adjusted O_3 trend is much smoother in the eastern USA than the observed O_3 trend that included all factors. In the western USA, the weather effect on average surface O_3 was relatively small over the period 2001–2010.

In California, USA, surface O_3 was found to be the major harmful component of smog in Los Angeles half a century ago (Haagen-Smit et al., 1951; McCabe, 1952;

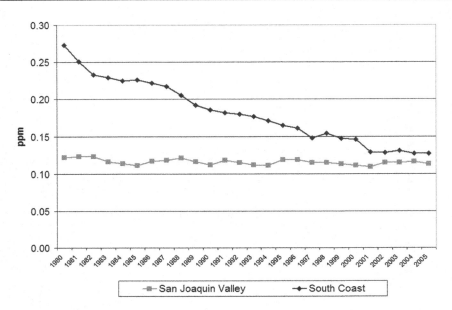

Figure 8.3 Trends of design values of 8-hour average O_3 in inland areas of southern and central California in the period 1980–2005.
From Larsen (2006).

Leighton, 1961). The peak 1-hour average O_3 mixing ratio was hazardous ($\geq$405 ppb) in the Los Angeles area in the 1960s and often very unhealthy ($\geq$205 ppb) in the San Francisco Bay area and San Diego in the 1970s. To combat severe air pollution, the State of California formed the Air Resources Board in early 1970 to coordinate statewide air pollution research and mitigation efforts. Anthropogenic emissions of NO_x and ROGs decreased from 4886 and 7058 tons per day in 1975 to 2981 and 2127 tons per day in 2010, mainly due to emission reductions in motor vehicles and industries. Currently implemented emission reduction measures will further reduce emissions of NO_x and ROGs to 2173 and 1950 tons per day respectively in 2020. Due to effective emission control, the San Francisco Bay area attained the national 1-hour average O_3 standard in 2005. Since 2005, the 8-hour average O_3 mixing ratio has become a major concern in California and the USA. Figure 8.3 shows the time series of "design values" for 8-hour average O_3 in two air districts of California, USA, from 1980 to 2005. In the USA, the design value for 8-hour average O_3 currently refers to the fourth highest daily maximum 8-hour O_3 concentration measured by an O_3 monitor; the design value in an air district refers to the peak design value from all monitors. It is seen that the design value of 8-hour average O_3 decreased significantly from 270 ppb in 1980 to 130 ppb in 2005 in southern California, while it was 110–120 ppb throughout the period in California San Joaquin Valley, where NO emission from agricultural fields may be significant (Fang and Mu, 2006, 2009; Pang et al., 2009). Table 8.2 lists the pollution status of 8-hour average O_3 in California, namely the

Table 8.2 Pollution status of 8-hour O_3 (ppb) in California

	Year	Ex. days	Peak 8-hour	Design 8-hour
Calif. South Coast	2009	113	129	118
	2010	102	123	112
	2011	106	137	107
San Joaquin Valley	2009	98	110	105
	2010	93	115	104
	2011	109	105	99

number of days that exceed the US national 8-hour O_3 standard together with peak and design values of 8-hour O_3 in California South Coast and San Joaquin Valley, from 2009 to 2011.

It is seen that the 8-hour O_3 level in California South Coast and San Joaquin Valley exceeded the USA national standard on most days of summer in 2009–2011, and peak 8-hour average O_3 was unhealthy for the general public in both air districts; however, the design value of 8-hour average O_3 has decreased in recent years.

Control of O_3 levels reduces personal exposure to surface O_3 and benefits health. Figure 8.4 shows the times series of personal exposure to O_3 in California South Coast and San Joaquin Valley from 1988 to 2007. It is seen that personal exposure to surface O_3 on the South Coast decreased from 33 ppm-hours per person in 1988 to 6.7 ppm-hours per person in 2007. In the San Joaquin Valley, personal exposure to surface O_3 changed little over the period 1989–2003, but decreased from 19 ppm-hours per person in 1988 to 5.2 ppm-hours per person in 2007.

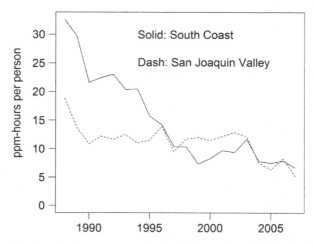

Figure 8.4 Trends of personal exposure to O_3 in California South Coast and San Joaquin Valley, USA in the period 1988–2007.

China started to investigate surface O_3 pollution in the early 1980s, when photo-chemical smog caused severe nuisance along an industrial corridor in Gansu province. In the late 1980s, the surface O_3 mixing ratio in suburban Guangzhou was measured to be ~ 60 ppb in summer. By 2004, the surface O_3 mixing ratio exceeded 90 ppb at 14 stations in the province (Zhang, Y.H. et al., 2008b). A 14-year continuous measurement at a coastal site in Hong Kong revealed that surface O_3 in the background air of southern China ranged from 10 ppb during unstable weather conditions with marine air flow in summer to 70 ppb during stable warm conditions with continental air stream in fall, on a monthly average basis. On an annual average basis, surface O_3 increased by 0.58 ppb per year over the period 1994–2007, probably due to rapid increases in column NO_2 in upwind regions (Wang et al., 2009). At a rural station downwind of metropolitan Hangzhou and Shanghai, China, the surface 8-hour average O_3 mixing ratio exceeded 80 ppb for over 3 weeks from June 1999 to July 2000 (Cheung and Wang, 2001). Biogenic ROG emissions are usually significant for O_3 formation in rural and remote areas, and may also be significant in urban areas (Xie et al., 2008). As NO_x emission continues to increase in China (Zhao et al., 2008), the O_3 level is expected to increase too.

8.2   Effects of oxidants on humans and plants

Humans and plants consist of reduced materials and are thus sensitive to oxidants. If the redox reaction rate exceeds a certain threshold beyond the repairing ability of biological systems, permanent damage will occur. In most ambient conditions, oxidant pollution is below the hazardous level nowadays, except in Mexico City occasionally, while O_3 pollution is serious in many large cities worldwide. In 1975, the maximum 1-hour mean NO_2 mixing ratio in California, USA reached a unhealthy level of 600 ppb; by 2007, it decreased to 130 ppb. Minor side-effects of O_3 pollution are summarized below.

8.2.1   Effects of O_3 pollution on humans

According to studies sponsored by the US EPA, an ambient O_3 mixing ratio less than 50 ppb on an 8-hour average basis is probably safe for humans. When it reaches 60–75 ppb, very sensitive individuals may experience respiratory symptoms, and unusually sensitive people should consider reducing prolonged or heavy outdoor exertion. When it reaches 76–95 ppb, active children and adults, such as athletes and physical laborers or people with lung disease, have a higher likelihood of suffering from respiratory symptoms and breathing discomfort, and should reduce prolonged or heavy outdoor exertion. When the level reaches 96–115 ppb, ordinary people may suffer respiratory symptoms and breathing difficulties, and should reduce prolonged or heavy outdoor exertion; sensitive people should avoid prolonged or heavy outdoor exertion. When the level reaches 116 ppb and above, ordinary people have an increased risk of suffering from respiratory symptoms and breathing difficulties, and

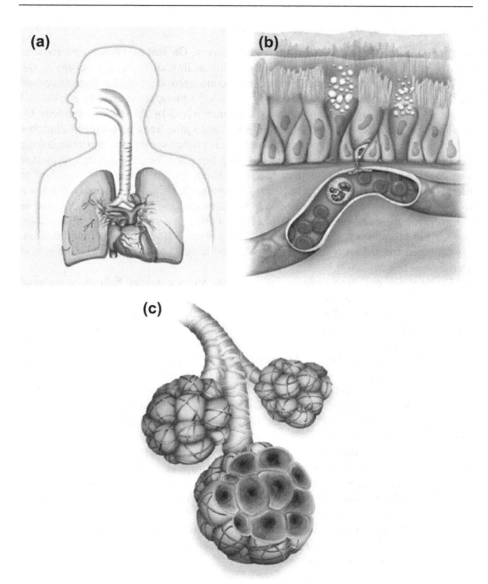

Figure 8.5 O_3 can damage human respiratory organs. (a) Chest. (b) Linings. (c) Alveoli. Pictures obtained from the US EPA (2012).

should reduce outdoor exertion; active children and adults as well as people with lung disease may suffer severe symptoms and impaired breathing, and should avoid all outdoor exertion. In particular, when the 1-hour average O_3 level reaches 405–600 ppb, ordinary people may suffer from severe effects on the chest, lung linings, and alveoli, as illustrated in Figure 8.5, and should avoid all outdoor exertion.

8.2.2   Effects of O_3 pollution on plants

Surface O_3 is deposited on plant trunks and leaves. On leaves, O_3 damage reduces photosynthesis and slows growth. As O_3 pollution in California was severe in the 1950s and 1960s, extensive studies have been conducted there on the damage caused by O_3 to plants. In the 1950s, ambient O_3 levels were found to be toxic to grapes, and to cause foliar symptoms on tobacco and pine trees. In the 1960s, ambient O_3 pollution was suspected to cause certain diseases to pine trees. Subsequent chamber experiments verified that O_3 pollution could cause foliar symptoms, and season-long O_3 exposure could reduce the growth of forest tree seedlings. O_3 damage to mature tree canopies was less severe than to young trees, but the damage is expected to increase with time (Karnosky et al., 2007). Ambient O_3 pollution in Mexico is known to have caused visible plant damage over 30 years ago, and foliar injury expressed as chlorotic mottling and premature defoliation on pines, a general decline of sacred fir, and visible symptoms on Mexican black cherry. Recent investigations found that, on O_3-damaged trees, root colonization by symbiotic fungi was limited (DeBauer and Hernández-Tejeda, 2007).

Ambient O_3 pollution reduces crop yields. Surface O_3 in East Asia reduced soybean yields by $\sim 25\%$ and the yields of wheat, rice, and corn by $\sim 5\%$ in 1990 (Wang and Mauzerall, 2004). It was estimated that, in 2006–2007, ambient O_3 pollution (with an O_3 mixing ratio over 40 ppb) of 7370–9150 ppb hour in rural India could reduce winter wheat and summer crop yields by 10% and 15% respectively (Debaje et al., 2010). In the Mediterranean and most of the Italian peninsula, experimental results have shown that, without water limitations, ambient O_3 pollution could reduce crop yields significantly; however, concurrent water and saline stress reduced additional damage caused by O_3 pollution (Fagnano et al., 2009). In Italy, ambient O_3 pollution was found to induce cell death of wheat, tomatoes, beans, and onions (Faoro and Iriti, 2009).

It should be noted that observations of the damage caused by ambient O_3 pollution were probably mixed with other environmental impacts, including climate change. Figure 8.6 illustrates O_3 damage on black cherry, a sensitive bio-indicator plant, according to the Department of Agriculture, USA. Also shown is the O_3 damage to leaf tissues. However, growth decline or increased susceptibility to other diseases, insects, harsh weather, invasive species, and other pollutants due to O_3 damage is not always visible. Thus, ambient O_3 pollution may reduce species diversity, habitat quality, and cycles of water and nutrients.

8.3   Efforts to control ozone

Surface O_3 pollution is caused by photochemical reactions of NO_x and reactive organic gases. Figure 8.7 illustrates the O_3 formation pathway and daytime budgets at 20 m AGL on the campus of Beijing University, China in August 2007 (Liu et al., 2012). To abate O_3 pollution in an area, sophisticated research and programs to reduce emissions of O_3 precursors need to be conducted.

Figure 8.6 O_3 damage to plant tissues.
Obtained from the US EPA (2012).

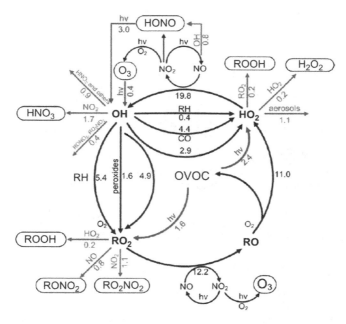

Figure 8.7 Formation pathway and budgets of surface O_3. Primary RO_x sources and sinks are shown in gray. The production and loss rates are in ppbv h^{-1} expressed as a daytime average pertinent to Beijing, China in August 2007.
Obtained from Liu et al. (2012).

8.3.1   Photochemical Assessment Monitoring Stations program

The system of Photochemical Assessment Monitoring Stations (PAMS) is an important program coordinated by the US EPA for monitoring O_3 pollution status in metropolitan areas with serious O_3 pollution. The PAMS program consists of two important parts: the siting and the speciation of ROG compounds. Figure 8.8 illustrates the siting standard. For a typical metropolitan area with serious O_3 pollution, at least five ground stations need to be chosen. The #1 station is situated in primary upwind locations of the metropolitan area, to monitor background concentrations of O_3 precursors. Two #2 stations are situated in primary and secondary downwind locations within urbanized fringe immediately outside the central business district with major emissions, to monitor elevated concentrations of O_3 precursors in urban plumes. The #3 station is

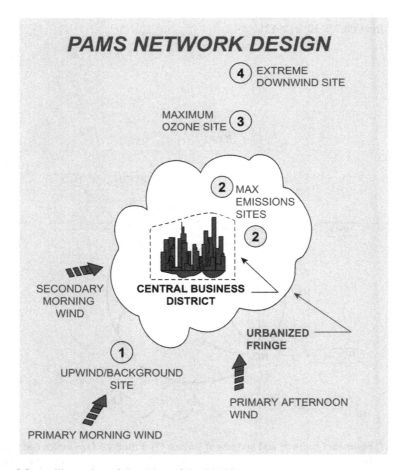

Figure 8.8 An illustration of the siting of the PAMS program.
Obtained from the US EPA (2012).

Table 8.3 Hydrocarbon compounds measured at PAMS stations

Ethylene	c-2-Pentene	2,2,4-Trimethylpentane	m-Ethyltoluene
Acetylene	2,2-Dimethylbutane	n-Heptane	p-Ethyltoluene
Ethane	Cyclopentane	Methylcyclohexane	1,3,5-Trimethylbenzene
Propylene	2,3-Dimethylbutane	2,3,4-Trimethylpentane	o-Ethyltoluene
Propane	2-Methylpentane	Toluene	1,2,4-Trimethylbenzene
Isobutane	3-Methylpentane	2-Methylheptane	n-Decane
1-Butene	2-Methyl-1-Pentene	3-Methylheptane	1,2,3-Trimethylbenzene
n-Butane	n-Hexane	n-Octane	m-Diethylbenzene
t-2-Butene	Methylcyclopentane	Ethylbenzene	p-Diethylbenzene
c-2-Butene	2,4-Dimethylpentane	m- and p-Xylenes	n-Undecane
Isopentane	Benzene	Styrene	Formaldehyde
1-Pentene	Cyclohexane	o-Xylene	Acetone
n-Pentane	2-Methylhexane	n-Nonane	Acetaldehyde
Isoprene	2,3-Dimethylpentane	Isopropylbenzene	
t-2-Pentene	3-Methylhexane	n-Propylbenzene	

situated outside the urbanized fringe at some distance (20–40 km) downwind of major emissions, to capture the maximum O_3 formed from urban plumes. The #4 station is situated outside the urbanized fringe at a further distance downwind of major emissions, to monitor the residual concentrations of O_3 precursors. The #4 station for one metropolitan area may be the #1 station for another metropolitan area. If two metropolitan areas are too close, they may be merged into one PAMS system.

At each PAMS station, O_3, NO_2, NO_x, HNO_3, total non-methane organic compounds, the sum of targeted hydrocarbons, and surface and upper air meteorological parameters are measured at hourly or 3-hourly intervals. Measured meteorological parameters include temperature, wind speed, wind direction, relative humidity, solar radiation, UV radiation, barometric pressure, and precipitation. Table 8.3 lists the targeted hydrocarbon compounds measured at PAMS stations in elution sequence. It is seen that measured alkanes include compounds with 2–11 carbons, and measured aliphatic alkenes include compounds with 2–6 carbons. Measured aromatic compounds contain a single benzene ring with 0–3 substitutes and 6–10 total carbons. In addition, ethyne, isoprene, HCHO, CH_3CHO, and CH_3COCH_3 are measured at PAMS stations. In general, alkanes and aromatics showed relatively higher mixing ratios in units of ppbC than alkenes and aldehydes, and the sum of targeted hydrocarbons accounted for $\sim 80\%$ of total non-methane hydrocarbons, which was often at the ppmC level at #2 PAMS stations. The NO_x mixing ratio usually exceeded 100 ppb at #2 stations in early morning hours.

Measured hydrocarbon compounds at PAMS stations may be analyzed to obtain statistically representative profiles of ROG species in photochemical mechanisms, such as the State Air Pollution Research Center (SAPRC) O_3 mechanism, which describes the formation of O_3 pollution in ambient air. Usually, measured ROG profiles at #2 stations during early morning hours are closer to fresh emissions than at noon, except for very reactive species. The average ROG profiles measured at ~ 50 metropolitan stations over the USA were used in the SAPRC O_3 mechanism implemented in the CMAQ model by the US EPA. Due to various emission mixtures in urban conditions, ambient measurements of the ROG profile may change significantly with time and space. Nevertheless, ROG measurements at PAMS stations provide a valuable validation for an emission inventory in metropolitan areas with serious O_3 pollution.

8.3.2 Emission programs

The precursors of surface O_3 pollution, NO_x and ROGs, may be emitted by natural sources and anthropogenic sources. Anthropogenic emissions may be classified as stationary sources, area-wide sources, and mobile sources. Stationary source emissions are usually estimated from specific facilities at a fixed location by facility operators or air pollution control districts. Area-wide sources include small individual sources, such as residential fireplaces, and widely distributed sources that cannot be tied to a single location, such as dust from unpaved roads. Mobile sources include on-road cars, trucks and buses, off-road recreational vehicles, trains, boats, and aircrafts. Natural sources include wildfires, natural wind-blown dust, geogenic and biogenic hydrocarbons. Besides stationary sources, all other sources (natural, area-wide, and mobile sources) are usually estimated by researchers employed or sponsored by state and local government agencies in the USA. For example, the California Air Resources Board has published an emission and air quality almanac each year since 1999, and the US EPA updates national emission inventory every 3 years.

NO_x and ROGs are precursors of O_3 pollution in modern air. In the global troposphere, NO_x are believed to be the limiting precursor for O_3 formation; in surface air, NO_x are usually abundant in urban airsheds due to emissions from high-temperature fuel combustion in mobile sources and stationary sources, and ROGs become the limiting precursor there after evaporative and fugitive sources are well controlled. On an urban scale, controlling the limiting precursor may be more cost-effective for reducing peak O_3 pollution if feasible measures exist (Wei et al., 2011). In practice, regional planning is pivotal to controlling O_3 levels.

California in the USA has a successful story of reducing emissions of O_3 precursors in the past two decades. On-road mobile sources emitted 2510 and 2654 tons NO_x per day in California in 1975 and 1990; the increase from 1975 to 1990 was driven by diesel vehicles. By 2005, NO_x emission by on-road mobile sources decreased to 1823 tons NO_x per day in California; the decrease of NO_x emission by gasoline vehicles from 2197 tons per day in 1975 to 754 tons per day in 2005 exceeded the increase of NO_x emissions by diesel vehicles from 312 tons per day in 1975 to 1069 tons per day in 2005. Implemented emission control measures are expected to reduce NO_x emissions by diesel and gasoline vehicles to 435 and 263 tons per day in 2020. Meanwhile, NO_x emissions from stationary sources increased from 1161 tons per day in 1975 to 1170 tons per day in 1980, and then decreased to 372 tons per day in 2005. NO_x emissions from stationary sources are expected to increase to 387 tons per day in 2020 due to increasing demands on energy and other goods, and resulting power generation and industrial activities.

Significant efforts to reduce emissions of O_3 precursors in other areas of the USA, such as the eastern states and Texas, as well as California, contributed to national emission reduction trends in recent years, as shown in Figure 8.9. It is seen that anthropogenic emissions of NO_x increased from 7 million tons in 1940 to over 20 million tons in 1970. Between 1980 and 1998, anthropogenic emissions of NO_x stabilized at ~ 24 million tons; by 2008, this decreased to 19 million tons, below the level in 1970. It is noteworthy that reducing NO_x emissions is often costly, as reductants, catalytic techniques, fuel modifications, and changes in the design of combustion conditions are necessary to reduce NO_x emissions into ambient air without hindering industrial production. In contrast, reducing anthropogenic emissions of ROGs may result in fuel recovery and increased energy efficiency if the ROGs in exhaust gas are not merely engineered to be burnt. Thus, it is seen that anthropogenic emissions of ROGs in USA decreased from 31 million tons in 1970 to 18 million tons in recent years, which is close to the level in 1940. By comparison, biogenic emissions of ROGs from the USA were estimated to be 32 million tons in 2008, 77% higher than anthropogenic ROGs.

8.4   Regulatory ozone modeling

Computer simulation of observed O_3 episodes and their responses to changes in emission distribution and quantity has been required in the USA since 1990 for attainment of acceptable O_3 levels in regions with serious surface O_3 pollution. Regulatory O_3 modeling involves sophisticated research efforts, to be discussed below.

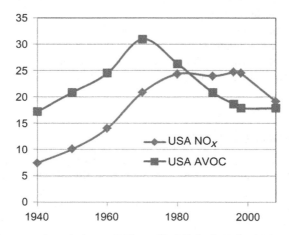

Figure 8.9 Anthropogenic emissions of NO_x and VOCs in the USA in the period 1940–2008. Units are million tons. Biogenic VOC emission was estimated to be 31.7 million tons in 2008. Data from the US EPA (2012).

8.4.1 Chamber experiments on O_3 formation

Ambient O_3 formation involves too many parameters to measure and resolve. To investigate mechanism of formation of O_3, smog chamber experiments under controlled conditions are indispensable to study the possible O_3 photochemical mechanisms. Typically, an experiment with a smog chamber starts from an initial condition with a high mixing ratio of a precursor. After the chamber air is exposed to natural or simulated summer sunlight from 6 a.m. to 6 p.m., a time series of O_3, NO, NO_2, ROGs, and other important compounds is measured. Thousands of these experiments were used to construct a credible photochemical mechanism, such as CBM, SAPRC, and RADM, for O_3 simulation.

An example is given below to illustrate how to evaluate a new photochemical mechanism, such as MCM (Liang and Jacobson, 2000b). Table 8.4 lists initial mixing ratios of important compounds for O_3 simulation at nine sites in the greater Los Angeles area, California, USA to provide a comparison of several photochemical mechanisms.

Using the initial conditions listed in Table 8.4, box-model simulations were conducted from 6 a.m. to 6 p.m. using SMVGEAR II (Jacobson, 1998a) with MCM and two versions of CBM photochemical mechanisms (CBIV and ACBM). To apply CBM mechanisms, explicit species concentrations need to be partitioned to carbon-bond lumping groups using partitioning fractions provided by the California Air Resources Board. Figure 8.10 shows time-series comparisons of NO_x, O_3, HNO_3, HCHO, PAN, and H_2O_2 at one site. It is seen that the MCM consumes NO_x more slowly but produces more O_3 compared with CBM mechanisms. The agreement between CBIV and ACBM is close in terms of the daily maximum O_3 concentration, but CBIV and ACBM differ by 50% for the daily maximum concentration of H_2O_2. For the daily maximum concentrations of HCHO and PAN, CBIV and ACBM agree within 30%. The daily maximum concentrations of O_3, HCHO, PAN, and H_2O_2 predicted by MCM differ from those predicted by CBIV and ACBM by up to 50%, 40%, 40%, and 80% respectively at all nine sites. To fix the problem with a new mechanism, additional simulations were conducted by replacing all aromatic reactions in the version of MCM (292 reactions) with those in ACBM (26 reactions). The agreement on O_3 time series became close after modifying the aromatic reactions in the MCM mechanism.

8.4.2 Multicolumn O_3 simulation

Multicolumn simulation refers to massive one-dimensional simulations over a region without horizontal transport processes but with realistic vertical transport processes. The effectiveness of regional O_3 mitigation efforts is sensitive to whether the major sources of O_3 precursors are well understood. To understand the nature of emission sources, multicolumn simulation is useful for diagnosing O_3 production potentials from local emissions over a region.

An example is given below to illustrate how to apply the multicolumn simulation method to evaluate the effects of emission control efforts on O_3 in California, USA. The

Table 8.4 Initial mixing ratios at nine sites in the greater Los Angeles area

Species	ANAH	AZUS	BURK	CELA	CLAR	HAWT	LBCC	RIVR	SNI
CO	1575.3	1896.5	3561.5	2839	2871.7	809.3	1516	2595	154.5
CH_4	2306	2391	2654	2341	2268	1949	2171.3	3323.5	1710
C_2H_6	20.3	12.9	88.4	26	35.4	19.4	45.6	37.2	0.9
C_2H_4	16.2	15.1	52.8	19.1	39.9	14.9	30.6	33.4	0.7
C_2H_2	10.9	12.5	46.1	11.7	32.1	12.8	17.3	41.1	1.9
C_3H_8	45.4	47.6	169.4	60.8	56.2	39	92.6	60.2	1.8
C_3H_6	5.4	10.1	27.4	15.5	13.4	9	16.3	23.3	0.5
i-C_4H_{10}	23.2	17.9	34.8	20.7	21.4	15.9	30.1	20.6	0.6
n-C_4H_{10}	52.9	45	84.4	57.3	52.5	26.8	73.8	52.3	1.5
trans-2-Butene	0.8	1	2.9	1.1	1.3	0.4	2	2.2	0
cis-2-Butene	1.1	1.1	2.5	1.6	1.2	0.5	1.9	1.5	0
3-Methyl-1-butene	0.6	0.6	1.3	1.2	0.8	0.4	0.9	0.9	0.2
i-C_5H_{12}	56.7	62.2	110.6	80.1	77.2	32.1	79.8	80.6	4.5
1-Pentene	2	3.3	6.6	2.7	2.1	2.5	3.9	3.4	0
2-Methyl-1-butene	1.8	2.3	5.6	3.6	2.2	6.8	3.5	2.8	0.2
n-C_5H_{12}	31.6	28.5	45.5	35.2	33.8	11.9	40.8	37.5	1.6
trans-2-Pentene	1.9	2.5	5.8	3.7	2.5	0.7	3.5	3.2	0.2
cis-2-Pentene	1	1.2	3.2	2.1	1.5	0.3	2.1	1.8	0.2
2,2-Dimethylbutane	0	0	0	0.9	1.1	0	0	0.9	0
2,3-Dimethylbutane	4.6	5.7	9.9	7.2	6.8	2.5	5.8	7.5	0.3
2-Methylpentane	18	21.6	36	26.9	27	9.3	20.8	28.4	1.3
3-Methylpentane	12.8	14.1	25	17.5	17.5	6.3	14.2	18	1.3
n-C_6H_{14}	12.6	15	23.2	17.2	19.5	5.5	14.9	17.5	0.7
Benzene	18	22.4	40.7	30.9	30.6	9.1	19.4	30.6	1.2
Cyclohexane	3.4	3.8	6.2	5.2	4.4	1.6	4.5	4.2	0.2
2-Methylhexane	6.2	8.6	14.5	10.9	11.9	3.2	6.8	10.6	0.6

(Continued)

Table 8.4 Initial mixing ratios at nine sites in the greater Los Angeles area—*cont'd*

Species	ANAH	AZUS	BURK	CELA	CLAR	HAWT	LBCC	RIVR	SNI
3-Methylhexane	7.6	10.2	16.5	12.7	16	3.9	8.2	11.9	1
n-C_7H_{16}	7.3	9.4	14.1	11.6	10.8	3	6.5	8.4	0.2
Toluene	48.7	76.5	126.4	90.2	91.6	25.5	46.5	80.9	3.8
n-C_8H_{18}	2.4	5.6	5.2	4.7	4.3	0.8	3.3	3.6	0.1
Ethylbenzene	8.4	16.5	18.5	14.6	14.6	3.8	7.7	14	0.6
m-Xylene	31.6	51.9	74.2	56.2	54.8	14.6	28.9	53.8	3.8
o-Xylene	11.9	18	26.8	19.8	19.6	6.5	10.2	19.9	2.3
n-C_9H_{20}	1.4	5.4	3.4	2.9	2.2	0.9	1.5	2.5	0
n-Propylbenzene	1.8	3.3	4.2	3.6	3.8	0.8	1.7	2.8	0
m-Ethyltoluene	7.1	10.9	16.9	13.2	13.1	3.3	6.5	11.9	0
p-Ethyltoluene	3.5	5.8	8.7	6.6	6.7	1.7	3.3	5.9	0.4
o-Ethyltoluene	3.3	5.7	7.7	5.8	6.5	1.5	3.1	5.1	0
1,2,4-Trimethylbenzene	11.6	16.2	25.7	21.8	20	4.9	10.6	17.7	1
HCHO	11	4.3	6.9	5.2	8.9	3.5	7.2	12.1	4.1
CH_3CHO	12.3	8.5	9.5	7.1	12.3	5.7	7.5	21.4	6.5
CH_3COCH_3	17.8	23.1	32	16	23.2	7.9	9.4	34.1	2.4
C_2H_5CHO	3.4	0.6	3	2.2	2.9	2.7	2.3	5.4	2.7
Methyl ethyl ketone	4.9	15.8	12	10.4	8.8	1.3	6.4	23.2	2
C_3H_7CHO	2.8	0.4	3.8	0.2	0.7	1.1	0	2.8	1.4
C_4H_9CHO	9	4	5	3.8	8.5	4	9.5	17	8
NO	110	30	160	90	43	20	82	120	8
NO_2	60	50	50	50	57	30	46	20	3.1
O_3	10	10	10	0	3	10	5	0	30

Note: Initial mixing ratios (ppb) are the averages from measurements at 6 a.m. on August 27–29, 1987. The temperature and pressure were assumed to be 298 K and 1013 mb respectively. Values of NO, NO_2, and O_3 at SNI were assumed, due to the lack of measurements.

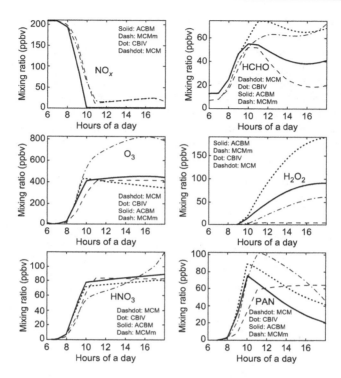

Figure 8.10 Time series of NO_x, O_3, HNO_3, HCHO, PAN, and H_2O_2 using the CBIV, ACBM, and MCM mechanisms, with initial conditions measured at site BURK. MCMm is the same as MCM except that the aromatic scheme of MCM is replaced by that of ACBM.
From Liang and Jacobson (2000b).

US EPA CMAQ model is a three-dimensional air quality model. By turning off horizontal transport operators in the code, a multicolumn O_3 model is created. A regulatory emission inventory in 2005 was produced from the database at the California Air Resources Board for mobile sources, stationary sources, area sources, and biogenic sources gridded to a horizontal resolution of 4 km $\times$ 4 km with plume rises calculated from meteorological model outputs. The Mesoscale Meteorological model version 5 (MM5, Grell et al., 1995) was employed to simulate California weather conditions in July 2000 and to provide inputs to the multicolumn version of CMAQ. CMAQ was configured to use SMVGEAR (Jacobson and Turco, 1994) to solve the SAPRC photochemical mechanism (Carter, 2000). The additional OH source (Li et al., 2008; Wennberg and Dabdub, 2008; Liang and Jacobson, 2011) was not included in the chemical mechanism. The horizontal model domain was configured at a resolution of 12 km $\times$ 12 km to cover the state of California and its surrounding area to limit influences of boundary conditions. Vertical layers were configured to be expanding from the surface to 10,000 Pa in the terrain-following pressure coordinate with a bottom layer thickness of $\sim$30 m. Initial and boundary conditions of O_3 and $PM_{2.5}$ (precursor) concentrations were taken from a global simulation (Horowitz, 2006). The CMAQ modeling system was then applied to simulate the O_3 mixing ratio over California using the regulatory emission inventory in 2005 and a hypothetical emission inventory (2005-BAU) assuming no emission control measure had been enacted in the greater Los Angeles area in the period 1975–2005.

Figure 8.11 shows the modeled O_3 mixing ratio from the multicolumn version of CMAQ with the regulatory emission inventory in 2005 and the 2005-BAU emission scenario. The ratios of ROGs and NO_x emissions in the 2005-BAU scenario to that in the regulatory emission inventory are also shown. It is seen that, using the regulatory emission

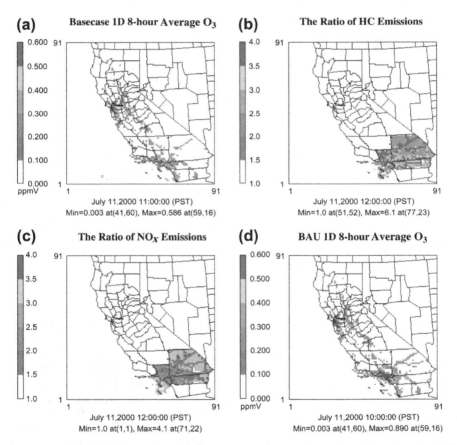

Figure 8.11 Multicolumn simulation of O_3 production in California. (a) O_3 from base-case emission inventory in 2005 and meteorological parameters on July 11, 2000. (b) Ratios of reactive HC emissions in the 2005-BAU scenario to that in 2005. (c) Ratios of NO_x emissions in the 2005-BAU scenario to that in 2005. (d) O_3 from the 2005-BAU emission scenario. The 2005-BAU scenario refers to a hypothetical emission inventory assuming that no control measure was implemented in the greater Los Angeles area in the period 1975–2005. From Livingstone et al. (2010).

inventory in 2005, peak O_3 formation potential was located over Los Angeles area, while elevated O_3 formation potential was located over San Diego area, San Francisco Bay area, and Sacramento area. It is also seen that emission control measures implemented in the greater Los Angeles area in 1975–2005 decreased ROG and NO_x emissions by several times. Without the significant emission reduction due to the control measures implemented from 1975 to 2005, peak O_3 formation potential could be over 50% higher over Los Angeles, and areas with serious O_3 pollution would also be much larger.

8.4.3 Three-dimensional O_3 simulation

Simulation of the surface O_3 mixing ratio requires three-dimensional modeling that represents the chemical transformation of O_3 precursors emitted and transported from various sources. Both Eulerian and Lagrangian methods may be applied to obtain three-dimensional O_3 fields. The Lagrangian method is also termed trajectory modeling using a box or a column of boxes moving with winds while exchanging with background air (e.g. Cheng et al., 2010). The first attempt to simulate the three-dimensional distribution of the surface O_3 mixing ratio due to anthropogenic emissions with fixed grids (Eulerian method) was made by Reynolds et al. (1973) for the urban airshed of Los Angeles, California. Subsequent Eulerian air quality models were developed to implement more realistic representations of the chemical and physical processes important for surface O_3 (Fiore et al., 2009). Depending on target parameters, a three-dimensional simulation of surface O_3 can be very challenging as the atmospheric boundary layer contains all phases of versatile chemicals including aqueous-phase radical reactions in cloud and fog droplets (Jacob, 1986; Lelieveld and Crutzen, 1990; Liang and Jacob, 1997). In practice, effects of condensed phases on surface O_3 may be split into two parts: radiative effect and chemical effect. The radiative effect of precipitation, cloud and fog may be approximated by Mie theory (Liou, 1992), and the chemical effect may be evaluated by including detailed heterogeneous reactions. Table 8.5 lists modeled budgets of O_3 and NO_x in the continental boundary layer of contiguous USA in a GCM January and July.

Table 8.5 shows model results for the US boundary layer obtained for (a) a "gas–aerosol–cloud" simulation, (b) a "gas–aerosol" simulation ignoring aqueous-phase chemistry, (c) a "gas-only" simulation ignoring also reactions of NO_3 and N_2O_5 in aerosols, and (d) a "gas–cloud" simulation including aqueous-phase chemistry in clouds but ignoring reactions of NO_3 and N_2O_5 in aerosols. The "gas–aerosol–cloud" simulation was conducted for the baseline non-precipitating cloud abundances and pH value, and also for sensitivity cases with tripled cloud volume and increased pH value. It is seen that O_3 in the US boundary layer has a chemical lifetime of about a week in July and over 2 weeks in February; photochemical production of O_3 in the US boundary layer during July is about five times faster than in February. It is seen that the monthly average mixing ratio of NO_x in the boundary layer of USA in February is about four times that in July. It is also observed that changes in the O_3 mixing ratio between different simulations are small, except for conditions corresponding to a marine boundary with a significant amount of cloud. Three-dimensional chemical transport modeling has been conducted to simulate effects of anthropogenic emissions

Table 8.5 Budgets of O_3 and NO_x in the US boundary layer

Simulation	Production of O_x (ppb day^{-1})	Loss of O_x (ppb day^{-1})	O_3 (ppb)	NO_x (ppb)
July				
Gas-only	12.8	7.34	49.1	0.24
Gas–cloud	12.7	7.34	48.8	0.24
Gas–aerosol	12.2	7.18	48.1	0.22
Gas–aerosol–cloud				
Standard (pH 4.5)	12.1	7.18	47.8	0.22
fx3, pH 4.5	11.9	7.19	46.9	0.22
fx3, pH 7	12.0	8.17	43.3	0.23
February				
Gas-only	2.61	1.99	38.3	0.87
Gas–cloud	2.47	2.00	37.9	0.89
Gas–aerosol	2.48	2.02	37.9	0.79
Gas–aerosol–cloud				
Standard (pH 4.5)	2.33	2.03	37.5	0.81

All quantities are monthly averages for the US boundary layer calculated with a continental-scale three-dimensional model (Liang and Jacob, 1997). The "gas-only" simulation includes no aerosol or cloud reactions. The "gas–cloud" simulation includes aqueous-phase cloud chemistry but no aerosol reactions. The "gas–aerosol" simulation includes reactions of N_2O_5 and NO_3 in aerosols but no cloud chemistry. The "gas–aerosol–cloud" simulation includes both cloud chemistry and reactions of N_2O_5 and NO_3 in aerosols; results of sensitivity calculations with increased cloud water pH and increased cloud volume fraction (f) are also shown.

over North America and Europe on surface O_3 in Asia, and it was found that regional emissions were responsible for over 90% of peak O_3 in Asia (Holloway et al., 2008).

California experiences surface O_3 pollution frequently in summer, partly due to terrain-induced poor ventilation of surface air. The terrain of California has coastal areas in the west, for example the Los Angeles and San Diego areas, and a mountain range in the east, with the highest mountain in the USA, which provides a natural border between California and adjacent states. In central California, there is a large valley extending nearly 1000 km from northwest to southeast. On a typical summer day, California land areas receive abundant solar radiation with little cloud in the sky; significant clouds are only present over the adjacent Pacific Ocean. Due to land–sea circulation, surface temperatures along the coast and around bays are mild; in valleys and desert areas, surface temperatures could exceed 313 K. Wind speed of sea breeze is typically about 5 m s^{-1}; in valley areas, wind speed is slow. The mixing height of the atmospheric boundary layer is within 1 km, and often below 500 m, over coastal and valley areas. Figure 8.12 illustrates modeled weather conditions at noon on July 11, 2000 (Livingstone et al., 2010).

California surface O_3 distribution and its response to emission changes are typically simulated using simulated weather parameters, such as those described above, and regulatory emission scenarios. For example, Figure 8.13 shows simulated distributions of the 8-hour average O_3 mixing ratio at ~ 15 m above ground level,

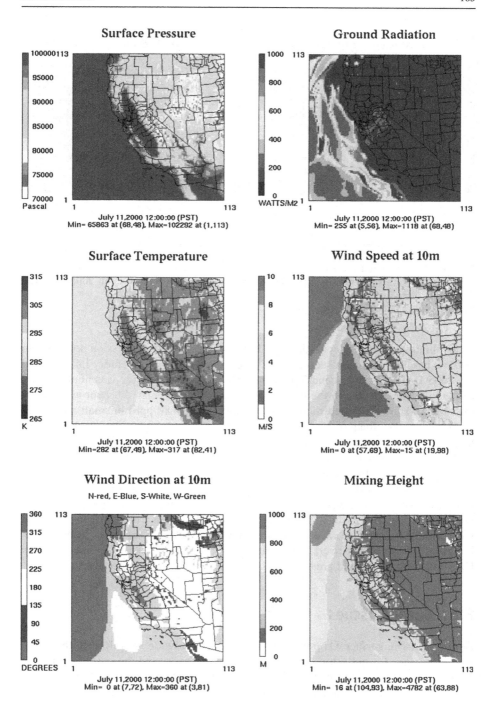

Figure 8.12 California weather conditions at noon on July 11, 2000.
From Livingstone et al. (2010).

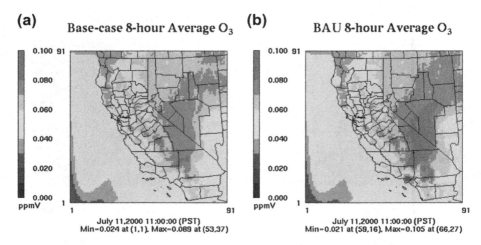

(a) Base-case 8-hour Average O_3 **(b)** BAU 8-hour Average O_3

July 11,2000 11:00:00 (PST)
Min=0.024 at (1,1). Max=0.089 at (53,37)

July 11,2000 11:00:00 (PST)
Min=0.021 at (59,16). Max=0.105 at (66,27)

Figure 8.13 Simulated surface 8-hour average O_3 with weather conditions on July 11, 2000 and two emission scenarios in 2005. (a) Regulatory emission inventory in 2005. (b) The 2005-BAU emission scenario.
From Livingstone et al. (2010).

using simulated weather conditions on July 11, 2000 and two emission scenarios in 2005: a baseline emission scenario from regulatory emission inventory in 2005, and the 2005-BAU emission scenario that assumes no control measure was implemented in the Los Angeles area in the period 1975–2005. The simulated 8-hour average O_3 distribution shows elevated mixing ratios (pink color, >80 ppb) in small areas over the southern San Joaquin Valley and San Bernardino county, with a peak value of 89 ppb over central California under weather conditions in July 11, 2010. In the 2005-BAU emission scenario, areas with elevated mixing ratios (pink color, >80 ppb) expanded significantly to cover a much larger portion of the San Bernardino county of California, the majority of the Inyo County of Nevada, and about half of the Nye county of Nevada, with a peak value of 105 ppb over southern California. Thus, it is seen that emission control in the greater Los Angeles area from 1975 to 2005 had significantly reduced the excess magnitudes and areas of surface O_3 air pollution, as indicated by the 8-hour average O_3 mixing ratio at ~15 m above ground level, in California and downwind areas, on a summer day with fair weather conditions.

8.5 Current issues

As an air toxin, surface O_3 pollution has drawn worldwide attention. In the USA, serious surface O_3 pollution occurs in California, Texas, and the eastern states. In Mexico, serious O_3 pollution occurs in the metropolitan areas of Mexico City. Surface O_3 pollution in China has also drawn public attention. For example, China's national ambient standards for O_3 and NO_2, as listed in Table 8.6, designate non-

Table 8.6 Ambient standards for O_3 and NO_2 in China (China MEP, 2012)

		Class I ($\mu g\ m^{-3}$)	Class II ($\mu g\ m^{-3}$)	Class I (ppbv)	Class II (ppbv)
NO_2	Annual	40	40	19	19
	24-hour	80	80	39	39
	1-hour	200	200	97	97
O_3	8-hour	100	160	47	75
	1-hour	160	200	75	93

attainment areas with an 8-hour average O_3 mixing ratio of 47 ppb for scenery and other specially protected areas and 75 ppb for urban, industrial, and rural areas; the corresponding designations for the 1-hour average O_3 mixing ratio are 75 and 93 ppb respectively.

In addition, maximal 1-hour and 24-hour average NO_2 mixing ratios are 97 and 39 ppb in all attainment areas. It follows that, in the worst case, the total mixing ratio of surface oxidants ($O_3 + NO_2$) is ∼ 100 ppb in attainment areas of China, which is lower than in the USA. However, according to the latest available official emission inventory, China emitted 22.7 million tons of NO_x in 2010 while USA emitted only 15.7 million tons of NO_x in 2008. Therefore, spatial distributions of NO_x emission could play an important role in mitigating surface oxidant pollution (Liang and Jacobson, 2000b).

Summary

In this chapter, the distribution and formation of surface oxidants in ambient air are discussed. As oxidant precursors are emitted worldwide, oxidant pollution is a ubiquitous phenomenon from a global perspective; as surface oxidants are produced on a timescale of within a week, regional control is effective in reducing oxidant pollution. Observations of surface O_3 episodes and long-term trends in various locations of the world are discussed. The effects of surface O_3 pollution on human health, vegetation, and crops are summarized. Ozone control efforts in the USA, especially in California, are presented using examples of the PAMS system, emission programs, and regulatory O_3 modeling. The latter is elaborated by using examples evaluating a new chemical mechanism with chamber experiments on O_3 formation, multicolumn O_3 simulations, and three-dimensional O_3 simulations. Current issues on surface oxidants are considered.

Exercises

1. Hydroxyl radical (OH) is believed to be the dominant oxidant in the troposphere that degrades reduced compounds such as CO, CH_4, and hydrocarbons with more carbons. Using Figure 8.7, write a reaction pathway to show that photochemical O_3 production could be limited by the availability of NO_x.

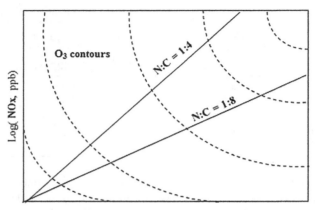

Log(**Reactive Organic Gases, ppmC**)

Figure E8.1

2. Repeat exercise 1, but show that photochemical O_3 production could be limited by the availability of reactive organic gases.

3. Figure E8.1 illustrates the dependence of peak surface O_3 on the mixing ratios of O_3 precursors, and is commonly referred to as an EKMA diagram. Regions between the horizontal axis and the line of N:C = 1:8 are NO_x limited, and regions between the vertical axis and the line of N:C = 1:4 are ROG limited; the exact boundaries may vary with space and time. Survey the emission and observational data in your region, to evaluate whether surface O_3 production during stagnant summer daytime is NO_x limited or ROG limited.

4. Ozone production efficiency is defined as the number of O_3 molecules produced per NO_x molecule lost to form HNO_3, and is commonly used to evaluate the photochemical reactivity of NO_x in NO_x-limited regions, such as rural and suburban areas. Apply a box-model computer program that solves photochemical reactions represented by a CBM mechanism, to calculate the ozone production efficiency of an air parcel with initial chemical conditions defined by the column "ANAH" of Table 8.4. Assume the air parcel was a closed system during daytime.

5. Repeat exercise 4, but use the column "RIVR" of Table 8.4. Write a brief note to compare the ozone production efficiencies in the two air parcels, in the context of the EKMA diagram and the CBM mechanism. (a) Make an EKMA diagram using a box model, the CBM mechanism, and the ROG profile of the column "RIVR" of Table 8.4, with the mixing ratio of ROG in the range 100 ppb C to 10 ppm C and the mixing ratio of NO_x in the range 10 ppb to 2 ppm. (b) Traditional vehicle exhaust contains NO_x at ppm level; in the exhaust, NO_2 is probably the most important oxidant. Add the oxidation reactions of unsaturated hydrocarbons by NO_2 into the CBM mechanism and construct a second EKMA diagram.

9 Particulate matter

Particulate matter (PM), also known as aerosol, refers to condensed phases suspended in the air. Besides molecules, radicals, atoms, and precipitations, particles with diameters larger than ~ 10 nm and smaller than ~ 50 μm are collectively called total suspended particles, aerosol or PM. PM is mostly observed near the surface, and its mechanism of formation differs from gases. When inhaled, some PM can cause severe damage to the lungs and other organs, but other PM sprays are marketed for medical use. To protect special professionals and the public from PM pollution, substantial efforts have been made to reduce PM levels in ambient air and built environments. Simulating PM in the atmosphere is one of the most challenging tasks for environmental professionals. The details are elaborated below.

9.1   Physical and chemical characteristics of atmospheric PM

Atmospheric PM is varies widely near the surface. Primary PM, such as dusts, pollens, and sea salt sprays, is mechanically emitted into the air from the ground,

Chemical Modeling for Air Resources. http://dx.doi.org/10.1016/B978-0-12-408135-2.00009-4

plants, and ocean, while secondary PM is formed or transformed in the air. Cloud and fog droplets belong to secondary PM, but are typically treated separately by meteorologists due to their importance to weather (e.g. Shi et al., 2008). PM measurements have been widely conducted in developed countries and developing countries for many decades. Number concentrations in various size ranges may be monitored continuously using light and electrical methods, such as scanning mobility particle spectrometers, while mass concentrations and associated speciation at certain size ranges may be measured via filter methods, such as the micro-orifice uniform deposit impactor and other size-segregated impactors. For filter methods, the minimum interval of sampling time depends on the pollution level and detection limit of instruments. For chemical identifications, ion chromatography has been widely used for ionic composition, and a number of PM organic compounds have been identified by gas chromatography–mass spectrometry and high-pressure liquid chromatography–mass spectrometry. In addition, proton-induced X-ray emission and neutron activation analysis methods as well as inductively coupled plasma–mass spectrometry have been widely used to analyze elemental composition, while thermal methods have been used for carbon analysis to distinguish between elemental and organic carbons.

9.1.1 Size representations

In modern air, polluting PM particles are typically classified as ultrafine ($D_p < 0.1$ μm), fine ($D_p = 0.1$–2.5 μm), and coarse ($D_p > 2.5$ μm) PM. In urban air, PM number concentration often exceeds $100,000$ cm^{-3}, and most particles are ultrafine PM; in the free troposphere, it is orders of magnitude smaller, and most particles are fine PM. While PM size distribution measurements depend on the instruments used, two methods have been widely used to represent the PM size distribution in models. The first method is called the sectional method, which uses a discrete representation within a number of size intervals; for example, the logarithm of the center of discrete size intervals may be evenly distributed. The second method is called the modal method, which typically uses three modes to approximate a continuous size distribution, though different numbers of modes have been used in the literature. For each mode, a log-normal distribution may be assumed:

$$\mathrm{d}N/\mathrm{d}(LD) = \left[N \Big/ \left(\sqrt{2\pi} \cdot L\sigma_g \right) \right] \cdot \exp\left\{ - (LD - LD_g)^2 \Big/ \left(2L\sigma_g{}^2 \right) \right\} \quad \text{(F9.1.1)}$$

The log-normal distribution is defined by three parameters, namely the total number concentration of PM over all sizes in the mode (N), the mean of the logarithm of the diameter in the mode (LD_g), and the standard deviation of the logarithm of the diameter in the mode ($L\sigma_g$). Note that both LD_g and $L\sigma_g$ denote single parameters, and calculations of these two parameters from measurements are best achieved by first converting all diameters (D) into logarithms (LD). The typical value of $L\sigma_g$ is 0.3 and the typical value of D_g is ~ 0.1 for the mode representing combustion exhaust and ~ 0.6 for the mode representing dust in PM$_{2.5}$. For the mode representing particles with diameters in the range of 2.5–10 μm, D_g is $\sim 5 \ \mu m$. Of course, these parameters may vary with time and location of measurements, and other distributions with similar shape may also be fitted from measurements.

9.1.2   Size and chemical profiles of PM sources

Human activities emit a variety of PM particles directly into the air. For example, various combustions using gas, liquid, and solid fuels emit ultrafine, fine, and coarse PM particles. With control devices, the fraction of coarse PM may be reduced. Evaporation of paint and coating materials emit PM with unique compositions. Smelting, chemical manufacturing, agricultural production and food processing, construction-related activities, and other industrial activities emit various types of PM particles. Waste management, road dusts of natural and anthropogenic origin, tire and brake wear emit PM particles with a significant fraction in coarse mode. Table 9.1 lists measured size profiles of various PM sources in the USA, which were used for PM air quality modeling in support of control measures in California, USA.

Knowledge of the chemical compositions of PM sources is important for effective mitigation of PM pollution. Particles emitted by combustion sources contain a significant fraction of non-volatile carbon due to incomplete combustion (e.g. Cao et al., 2006; Qin and Xie, 2011), though modern combustion efficiency often exceeds 99%. Depending on fuel sulfur content, particles emitted by combustion sources may contain a significant fraction of sulfate. In addition, ammonium nitrate has also been measured as a minor fraction from emissions of various combustion sources. However, secondary formation of sulfate, ammonium, and nitrate from the gas phase in ambient air is usually much larger than their primary emissions. Trace elements are emitted by a number of PM sources, and have been used for source identification and apportionment. Table 9.2 lists chemical profiles of $PM_{2.5}$ from 20 selected sources in the USA.

It is seen that $PM_{2.5}$ particles emitted by various sources have characteristic elemental compositions. For example, the metal industry profile contained significant fractions of toxic heavy metals Cd, Cr, Pb, As, Ni, and Zn, and the coal combustion profile contained significant fractions of Al and P as well as Si, Fe, Ca, K, and Ti. A number of profiles contained significant fractions of Fe, Cl, and Si, and the profiles of biomass burning, fossil fuel combustion, and industrial emissions contained a significant fraction of K. The brake wear profile contains significant fractions of Fe, Mg, Si, and Ba.

9.1.3   Mixing states of ambient PM

Ambient PM has been represented as both externally and internally mixed particles by PM modelers. As freshly emitted PM particles from different sources have distinct physical and chemical characteristics, they may be represented by unique model species; PM represented by this scheme is usually termed externally mixed PM, and both Eulerian and Lagrangian models may implement this scheme depending on the purpose and computational resources.

Table 9.1 Size profiles of PM sources (California Air Resources Board, 2012)

Profile	$<PM_{10}$	$<PM_{2.5}$	$<PM_1$	$>PM_{10}$
Agric. burning – Field crops	0.9835	0.9379	0.8816	0.0165
Agricultural burning	0.88	0.85	0.78	0.12
Agricultural tillage dust	0.45	0.1	0.03	0.55
Agricultural tilling dust	0.4543	0.0681	0.035	0.5457
Aircraft – jet fuel	0.976	0.967	0.96	0.024
Almond prunings burning	0.9829	0.9303	0.87	0.0171
Aluminum foundry	0.95	0.903	0.86	0.05
Asbestos	0.5	0.5	0.5	0.5
Asphalt roofing manufacture	0.98	0.945	0.91	0.02
Asphaltic concrete batch plant	0.4	0.333	0.3	0.6
Barley straw burning	0.9899	0.956	0.9031	0.0101
Basic oxygen furnace – Steel	1	1	0.95	0
Brake wear	0.98	0.42	0.14	0.02
Brake wear (replaced by 473)	0.98	0.42	0.14	0.02
Cadmium	1	1	1	0
Calcination of gypsum	0.88	0.495	0.22	0.12
Cement prod./concrete batching	0.92	0.62	0.34	0.08
Chemical fertilizer – Urea	0.96	0.95	0.94	0.04
Chemical manufacturing	0.9	0.89	0.88	0.1
Chrome: Hexavalent chromium	1	1	1	0
Clay and related products mfg.	0.56	0.513	0.48	0.44
Coal/coke combustion	0.4	0.15	0.03	0.6
Coating material evaporation	0.96	0.925	0.9	0.04
Coffee roasting	0.62	0.61	0.6	0.38
Construction dust	0.4893	0.0489	0.0385	0.5107
Corn residue burning	0.985	0.9438	0.8926	0.015
Cotton ginning	0.62	0.08	0.04	0.38
Diesel vehicle exhaust	1	0.92	0.86	0
Douglas fir slash burning	0.9646	0.8417	0.7562	0.0354
Electric arc furnace	0.83	0.6	0.45	0.17
EPA avg: Chemical manufacturing	0.505	0.279	0.279	0.495
EPA avg: Food and agriculture	0.49	0.14	0.14	0.51
EPA avg: Gray iron foundries	0.925	0.835	0.835	0.075
EPA avg: Industrial mfg.	0.574	0.407	0.407	0.426
EPA avg: Metal mining – General	0.51	0.15	0.15	0.49
EPA avg: Mineral products	0.545	0.33	0.33	0.455
EPA avg: Petroleum industry	0.691	0.396	0.396	0.309
EPA avg: Primary metal production	0.644	0.464	0.464	0.356
EPA avg: Pulp and paper industry	0.608	0.486	0.486	0.392
EPA avg: Secondary metal production	0.633	0.474	0.474	0.367
EPA avg: Solid waste	0.19	0.13	0.13	0.81

Table 9.1 Size profiles of PM sources (California Air Resources Board, 2012)—*cont'd*

Profile	$<PM_{10}$	$<PM_{2.5}$	$<PM_1$	$>PM_{10}$
EPA avg: Steel foundry – General	0.86	0.765	0.765	0.14
EPA avg: Steel production	0.6	0.52	0.52	0.4
Evaporation	0.96	0.925	0.9	0.04
Feed and grain operations	0.29	0.01	0	0.71
Fiberglass forming line	0.994	0.992	0.99	0.006
Fireplaces	0.92	0.87	0.8	0.08
Fireplaces and woodstoves	0.935	0.9001	0.8392	0.0651
Forest management burning	0.961	0.8544	0.7579	0.039
Fuel combustion – Distillate	0.976	0.967	0.96	0.024
Fuel combustion – Residual	0.87	0.76	0.66	0.13
Gaseous material combustion	1	1	1	0
Gasoline vehicles – Catalyst	0.97	0.9	0.88	0.03
Gasoline vehicles – No catalyst	0.9	0.68	0.53	0.1
Glass melting furnace	0.98	0.963	0.95	0.02
Grain drying	0.54	0.4	0.3	0.46
Grass/woodland fires	0.9825	0.9316	0.8745	0.0175
Hexavalent, trivalent chromium	1	1	1	0
Incineration – Gaseous fuel	1	1	1	0
Incineration – Liquid fuel	0.976	0.967	0.96	0.024
Incineration – Solid fuel	0.3	0.2	0.13	0.7
Landfill dust	0.55	0.378	0.28	0.45
Landfill dust	0.4893	0.0734	0.0385	0.5107
Lime manufacturing	0.3	0.117	0.07	0.7
Liquid material combustion	0.976	0.967	0.96	0.024
Livestock dust	0.48	0.06	0.05	0.52
Livestock operations dust	0.4818	0.055	0.048	0.5182
Marine vessels – Liquid fuel	0.96	0.937	0.92	0.04
Mineral process loss	0.5	0.075	0.02	0.5
Open burning	0.9825	0.9316	0.8745	0.0175
Orchard heaters	0.976	0.967	0.96	0.024
Orchard prunings burning	0.9814	0.9252	0.8675	0.0186
Other waste combustion	0.997	0.927	0.737	0.003
Paint application – Oil based	0.96	0.925	0.9	0.04
Paint application – Water based	0.68	0.62	0.58	0.32
Paved road dust	0.46	0.08	0.03	0.54
Paved road dust (before 1997)	0.4572	0.0772	0.0302	0.5428
Paved road dust (1997 onwards)	0.4572	0.0686		0.5428
Petroleum heaters – Gas	0.95	0.93	0.91	0.05
Petroleum refining	0.61	0.555	0.51	0.39
Pine slash burning	0.9573	0.8672	0.7596	0.0427
Planned/unplanned forest fires	0.88	0.85	0.78	0.12
Pulp and paper mills	1	0.76	0.4	0
Range improvement burning	0.9825	0.9316	0.8745	0.0175

(Continued)

Table 9.1 Size profiles of PM sources (California Air Resources Board, 2012)—*cont'd*

Profile	<PM$_{10}$	<PM$_{2.5}$	<PM$_1$	>PM$_{10}$
Residential – Natural gas	1	1	1	0
Rice straw burning	0.9758	0.9186	0.8581	0.0242
Road and building construction dust	0.64	0.37	0.24	0.36
Rock crushers	0.1	0.075	0.01	0.9
Rock screening and handling	0.5	0.075	0.02	0.5
Solid material combustion	0.997	0.927	0.737	0.003
Stat. IC engine – Diesel	0.96	0.937	0.92	0.04
Stat. IC engine – Dist./diesel	0.976	0.967	0.96	0.024
Stat. IC engine – Gasoline	0.994	0.992	0.988	0.006
Stat. IC engine – Solid fuel	0.997	0.927	0.737	0.003
Stat. IC engine – Gas	0.994	0.992	0.988	0.006
Steel abrasive blasting	0.86	0.79	0.74	0.14
Steel heat treatng – Salt quench	0.96	0.86	0.74	0.04
Steel open hearth furnace	0.98	0.93	0.87	0.02
Steel sinter plant	0.98	0.97	0.96	0.02
Timber and brush fires	0.961	0.8544	0.7579	0.039
Tire wear	0.4	0.32	0.2	0.6
Tire wear	1	0.25	0.1	0
Tire wear (replaced by 472)	1	0.25	0.1	0
Unpaved road dust	0.61	0.13	0.05	0.39
Unpaved road dust (before 1997)	0.5943	0.126	0.0516	0.4057
Unpaved road dust (1997 onwards)	0.5943	0.0594		0.4057
Unplanned structural fires	0.98	0.914	0.8	0.02
Unspecified	0.7	0.42	0.25	0.3
Utility boilers – Residual	0.97	0.953	0.94	0.03
Vehicular sources – Diesel	0.96	0.937	0.92	0.04
Vehicular sources – Gasoline	0.994	0.992	0.988	0.006
Walnut prunings burning	0.9799	0.9202	0.8649	0.0201
Waste burning	0.9825	0.9316	0.8745	0.0175
Weed abatement burning	0.9835	0.9379	0.8816	0.0165
Wheat straw burning	0.9834	0.9334	0.8726	0.0166
Windblown dust – Agric. lands	0.4543	0.0786	0.035	0.5457
Windblown dust – Desert lands	0.5937	0.1131	0.0476	0.4063
Windblown dust – Agricultural	0.5	0.1	0.03	0.5
Windblown dust – Unpaved areas	0.5	0.12	0.06	0.5
Windblown dust – Unpaved rd/area	0.5943	0.0786	0.0516	0.4057
Wood operation – Resawing	0.4	0.283	0.2	0.6
Wood operation – Sanding	0.92	0.885	0.86	0.08
Wood waste combustion	0.997	0.927	0.737	0.003

Table 9.2 Chemical profiles (%) of PM$_{2.5}$ from selected sources in the USA (California Air Resources Board)

ID	OC	EC	Cd	Cr	Pb	As	Ni	Zn	Fe
(a) Species 1–9									
1		15	0.05	0.55	0.55	0.55	0.05	0.55	
2		6		0.55		0.05	0.55		4
3		20		0.05			0.05	0.05	0.05
4		14							9
5		30	0.05	0.05	0.05	0.05	0.55	0.55	0.55
6		55							0.55
7		33	0.05				0.55	0.05	0.55
8		8							
9									2
10		30		0.05	0.05		0.05	0.05	0.55
11		8	0.05	0.55	0.05		2	0.05	9
12		0.1	0.05	0.55	0.05		0.55		15
13		20	0.05	2	0.55	0.05	0.55		10
14							0.05		0.55
15	47	22			0.02			0.53	0.5
16	11	2.6		0.1			0.1	0.03	28.7
17	54	17							
18	69	26						0.03	
19	33	16						0.02	0.01
20	3.2	0.3	2.8	0.4	11.7	9.9	0.17	6.2	4.6

ID	Cl	Si	Al	K	Ca	Mg	P	Ti	V	Ba
(b) Species 10–19										
1										
2										
3	7									
4		23	20	3	5	2	8	1		
5	2	10		1	1					
6					0.6			3		
7	11			0.6	0.6					
8		10		2	20					
9		10			30					
10	15			0.6	0.1					
11	30			0.6	0.6					
12					0.6					
13				5	0.6				0.6	
14		40								
15	0.8				0.1					
16	0.2	6.8			0.1	8.3			0.1	5.4
17	0.8			1.7						
18		0.1								
19	18	0.2		16						
20	1.8	5.6	4.3	3.1	1.3	0.2			0.1	

(*Continued*)

Table 9.2 Chemical profiles (%) of $PM_{2.5}$ from selected sources in the USA
(California Air Resources Board)—*cont'd*

ID	Source description
(c) Description of selected sources	
1	Liquid material combustion
2	Fuel combustion – Residual
3	Stat. IC engine – Gasoline
4	Coal/coke combustion
5	Wood waste combustion
6	Coating material evaporation
7	Chemical fertilizer – Urea
8	Cement prod./concrete batching
9	Lime manufacturing
10	Fiberglass forming line
11	Steel heat treating – Salt quench
12	Steel abrasive blasting
13	Steel open hearth furnace
14	Rock crushers
15	Tire wear
16	Brake wear
17	Fireplaces and woodstoves
18	Diesel vehicle exhaust
19	Agric. burning – Field crops
20	US EPA avg: Primary metal production

Statistical thermodynamics suggests that molecules and other particles collide frequently in ambient air (e.g. Seinfeld and Pandis, 1998). The mean free paths of an air molecule (L_a) and a PM particle (L_p) in ambient air depend on their "diameters" (d and D), air dynamic viscosity (μ), temperature (T) and pressure (P):

$$L_a = k \cdot T / (4.443 dP)$$

$$L_p = \left[C_s \Big/ \left(6\mu \cdot \sqrt{3} \right) \right] \cdot \sqrt{\rho K T D} \tag{F9.1.2a}$$

with

$$C_s = 1 + \{2.514 + 0.8 \cdot \exp[-0.55 D/L_a]\} \cdot L_a/D \tag{F9.1.2b}$$

where k is Boltzmann's constant (1.38×10^{-23} J K^{-1}) and $d \approx 4 \times 10^{-10}$ m for an "air molecule". C_s denotes a factor that corrects for non-continuum effects when a PM particle is close to or smaller than the mean free path of air molecules. For example, at the mean sea level, the mean free path of air $L_a = 0.05$ μm at 298.15 K from formula (F9.1.2a). For a particle with diameter 0.3 μm, the non-continuum correction factor

$C_s = 1.4$. Assuming air dynamic viscosity $\mu \approx 2 \times 10^{-5}$ Pa $\cdot$ s and particle density $\rho = 1.5 \times 10^3$ kg m^{-3} (e.g. Hu et al., 2012), the mean free path of the particle $L_p = 9$ nm.

Two distinct particles may also collide and combine to form a single particle in the air, which is termed coagulation. The coagulation processes in combination with the condensation of water and soluble gases result in a pseudo-homogeneous composition of PM particles in ambient air, especially in the global atmosphere. To represent a chemical in particles within a size interval without distinguishing its origins by one model species is termed the internally mixed method; both Eulerian and Lagrangian models may implement this scheme. In practice, the internally mixed PM scheme is used for general research purposes and the externally mixed PM scheme is useful for investigating source–receptor relationships on urban and regional scales.

Figure 9.1 illustrates a cluster of fine PM particles formed from incomplete combustion of diesel fuels, and a wet particle droplet that contains a core of black carbon was illustrated in Figure 2.1. The first type of particle is most abundant over highways and in air parcels containing fresh plumes of fossil fuel combustion and biomass burning. The second type of particle is widespread and responsible for PM episodes observed in urban, suburban, and even rural areas; it is conceivable that this type of particle also plays an important role in urban hazes observed over many metropolitan areas, such as Los Angeles (California, USA) and Shanghai (China).

Figure 9.1 A cluster of fine particles from diesel exhaust. Obtained from California Air Resources Board (2012).

9.1.4 Ambient PM composition

Ambient PM particles contain versatile chemicals. For example, in the urban air of London, over 10,000 organic species were present in 10-μg PM samples, though complete molecular identifications were impossible due to instrumental limitations (Hamilton et al., 2004). Ambient PM composition may change with time and location, due to changes in source characteristics. Particles in rural northern China were found to be relatively clean before China's economic boom (Winchester et al., 1981). Fine PM samples collected in southern China in March 1987 were analyzed with PIXE for a major size bin during a study to investigate PM transport pathways in the region. Despite the myth of absolute units, Table 9.3 lists average elemental mass concentration profiles at suburban, rural hilltop, and mountain stations. It is seen that fine PM at the suburban station was several times higher than at the rural hilltop and mountain stations, and sulfur was the dominant element in all stations. At the suburban station, potassium, zinc, chlorine, sulfur, lead, and arsenic were much higher than at the rural hilltop and mountain stations.

While it is intriguing to sample the average ionic composition of PM in a column of air, soluble PM precipitates in rainwater; the latter may be conveniently collected at the surface. Table 9.4 lists the average ionic composition of 71 rain samples collected during a campaign to survey acid rain pollution in southern China from February 24 to March 1, 1986. Most samples were collected at various urban and suburban stations in

Table 9.3 Historical profiles of fine PM in a major size bin in southern China

PM_1	Suburban	Rural hilltop	Mountain (Lechang)
K	13.80	3.92	8.05
Ca	19.15	12.99	28.64
V	0.09	0.06	0.10
Mn	1.28	0.49	0.97
Fe	14.91	9.26	11.44
Co	0.09	0.12	0.14
Ni	0.89	0.41	0.50
Cu	0.65	0.25	0.51
Zn	15.35	1.36	1.09
Se	0.10	0.04	0.02
Br	0.10	0.09	0.12
Si	18.01	10.24	14.97
Cl	58.68	9.97	30.63
Ti	1.76	0.95	1.29
Cr	1.97	1.01	1.23
S	294.38	88.64	105.16
Sr	0.23	0.04	0.07
Pb	38.38	1.01	1.12
As	5.02	0.26	0.31
Total	**484.85**	**141.12**	**206.33**

Table 9.4 Average ionic composition of rainwater samples

Ions	Mean (μeq L^{-1})	Mean (ppmw)
F$^-$	7.77	0.15
Cl$^-$	48.30	1.71
NO$_3^-$	32.30	2.00
SO$_4^{2-}$	252.00	12.10
Na$^+$	30.40	0.70
NH$_4^+$	118.00	2.12
K$^+$	29.70	1.16
Ca^{2+}	137.00	2.75
Mg^{2+}	16.40	0.20
H$^+$	79.70	0.08
Total	**751.57**	**22.97**
C/A ratio	**1.21**	**0.44**

Guangdong province, and only seven samples were collected at various ground stations in the provincial capital of Guangxi province. As rainfall amount varies with stations in the two provinces, the standard deviation was of about the same in magnitude as the mean. Nevertheless, it is seen that soluble PM in the two provinces in February 1986 consisted mainly of sulfate (52% by mass); nitrate and sodium chloride concentrations were similar in the rainwater samples at that time.

A number of regional and global field campaigns have been conducted in recent years to investigate PM characteristics in urban, rural, and recreational areas (Cheng et al., 2000; He et al., 2001; Hu et al., 2002; Yao et al., 2002; Li, 2006; Vingarzan and Li, 2006; Hagler et al., 2007; Park et al., 2007; Streets et al., 2007; Eichler et al., 2008; Zhang et al., 2008a, b; Cohen et al., 2010; Deng et al., 2012; Guo et al., 2012). In a polluted environment, diurnal amplitude is significant for anthropogenic chemicals. For example, hourly averaged concentration reached 300, 5000, and 100 ppbv for NO, CO, and SO$_2$ respectively, and exceeded 350 μg m^{-3} for PM$_{2.5}$ in metropolitan Guangzhou in October, 2004 (Zhang et al., 2008b). Table 9.5 lists sample ambient concentrations of PM and its components over Europe, America, and Asia. Maximum 24-hour average values are given whenever possible. It is seen that both size distribution and composition of PM differed significantly between samples from the three continents. Nevertheless, sulfate, nitrate, and ammonium are still major ionic components of fine PM, while contributions from EC and OC (not shown) are also significant. High values of PM concentrations in America originated from a widespread wildfire in California, USA in summer 2008, and that in Asia was measured in the Yangtze River Delta of China during a pollution episode in winter–spring 2007 (Fu et al., 2008).

In modern air, heavy metals account for only a minor fraction of fine PM mass, due to effective emission reduction efforts in past decades. Table 9.6 lists sample PM heavy metal concentrations in Chicago in 1964 (Winchester and Nifong, 1971) and elsewhere in recent years (Wang et al., 2010; Moreno et al., 2011; Yang et al., 2011). It is seen that zinc, titanium, lead, copper, and manganese were relatively abundant, while arsenic and cadmium concentrations differed by an order of magnitude between

Table 9.5 Sample ambient PM concentrations in recent years ($\mu g\,m^{-3}$)

	Europe	America	Asia
NH_4^+	4.6	2	24
NO_3^-	10	4	54
SO_4^{2-}	30	5	62
EC	11	10	28
PM_1	50	33	100
$PM_{2.5}$	58	400	400
PM_{10}	110	500	623

Table 9.6 Sample PM heavy metal concentrations ($ng\,m^{-3}$)

	Tokyo	Taiwan	China	Mexico	Spain	Chicago
As	2	50	66	7	1.4	35
Cr	6	20	40	192	9.8	24
Zn	233	380	1409	482	208	1260
Sr		10	110	38	4.1	
Pb	64	90	616	253	36	900
Ni	5	50	75	9	11	272
Co	1		6	1.6	0.4	<8.5
Cd	2	10	70	3	1.3	<45
Mn	30	70	312	83	20	175
V	6	10	50	50	30	35
Cu	27	250	170	140	21	80
Ti	40		742	440	33	35

sample locations. Heavy metal pollution in China in recent episodes was more severe than the measurements in Chicago in 1964, except for Pb and Ni, despite significant efforts to mitigate air pollution in metropolitan Beijing, Shanghai, and Guangzhou (Chan and Yao, 2008).

9.2 Mechanism of formation of PM

9.2.1 Secondary organic aerosols

Photochemical oxidation of organic compounds in the atmosphere produces a variety of intermediate compounds before the formation of CO_2, H_2O, and HNO_3 (e.g. Tkacik et al., 2012). Nitrate and many intermediate compounds may stay in condensed phases, which were observed in smog chamber experiments designed to evaluate the O_3 mechanism. The term "PM yield" was used to denote the ratio of total secondary organic aerosol (SOA) mass produced to the mass of targeted VOCs reacted during a smog chamber experiment (e.g. Pandis et al., 1992).

It is conceivable that "PM yield" depends on the chemical and physical environments of specific smog chamber experiments. For an SOA molecule, SOA_i, its formation from a gas-phase precursor GOP_i in the presence of an existing PM mass concentration may be described by reaction (R9.2.1), and the evaporation of SOA_i back to its gas-phase precursor GOP_i may be described by reaction (R9.2.2):

$$GOP_i + PM \rightarrow SOA_i + (1 + f_i)PM \tag{R9.2.1}$$

$$SOA_i \rightarrow GOP_i - f_i \cdot PM \tag{R9.2.2}$$

where f_i is simply a constant that converts the units of GOP_i to the units of PM. Reactions (R9.2.1) and (R9.2.2) are similar to cross-phase reactions in interstitial clouds, and the rate coefficients may be denoted as $k(c_i)$ for reaction (R9.2.1) and $k(e_i)$ for reaction (R9.2.2). Let $k(p_i)$ denote the equilibrium constant for reactions (R9.2.1) and (R9.2.2), then:

$$k(p_i) = k(c_i)/k(e_i) = [SOA_i]/\{[PM] \cdot [GOP_i]\} \tag{F9.2.1}$$

In the literature on PM formation, $k(p_i)$ is also called the PM partitioning coefficient of species i, which is similar to Henry's constant for a soluble gas. The rate of reaction (R9.2.1) is determined by kinetic and diffusion processes, and its coefficient $k(c_i)$ is given by formula (F9.2.2) if all PM particles are spherical with the same radius r:

$$k(c_i) = \frac{3\dfrac{[PM]}{\rho(PM)}}{r\left\{\dfrac{r}{D(g,\ i)} + \dfrac{4}{[v(i) \cdot \gamma(i)]}\right\}} \tag{F9.2.2}$$

where $\rho(PM)$ denotes the bulk density of PM and $D(g, i)$ is the diffusion coefficient of molecule i in the gas phase; $v(i)$ denotes the mean speed of molecule i in the gas phase and $\gamma(i)$ is the sticking coefficient of molecule i to PM particles. Thus, accurate simulation of an SOA molecule requires its molecular weight, partitioning and sticking coefficients, as well as mass concentration, density, and effective radius of existing PM. It is noteworthy that a small variation in molecular weight and sticking coefficient may result in a small change in the partitioning of an SOA molecule between the gas and PM phases, based on formula (F9.2.2). Nevertheless, the partitioning coefficient is determined by molecular identity.

Degradation of an aromatic compound in ambient air may produce a number of SOA molecules. Figure 9.2 illustrates various possible pathways used in the SAPRC photochemical mechanism for the formation of SOA molecules during the degradation of an aromatic molecule, such as m-xylene (Carter et al., 2012).

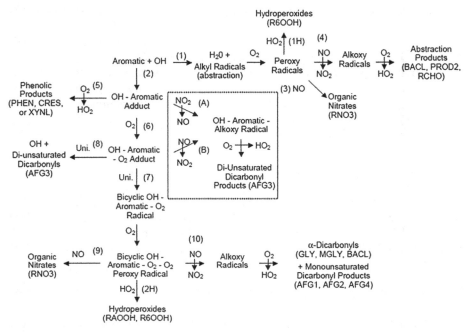

Figure 9.2 Possible pathways to form SOA molecules during photochemical degradation of an aromatic compound in ambient air. Boxed pathways were recently added to the SAPRC mechanism.
From Carter et al., (2012).

It is seen in Figure 9.2 that photochemical reactions of an aromatic compound may form at least 15 SOA model species, namely R6OOH, RNO3, BACL, PROD2, RCHO, PHEN, CRES, XYNL, RAOOH, GLY, MGLY, AFG1, AFG2, AFG3, and AFG4; all notation follows the SAPRC mechanism. Thus, for each aromatic compound, such as *m*-xylene, accurate simulation of its associated SOA formation requires molecular weights, partitioning and sticking coefficients of at least 15 corresponding SOA species. In fact, a nearly explicit gas-phase photochemical mechanism, MCM, uses 4675 gas species to represent the degradation of 135 volatile organic compounds with up to 12 carbons. The application of MCM in global-through-urban nested three-dimensional simulation of climate-air pollution indicated that O_3 simulation could be simplified by using the CBM mechanism, but MCM provided unique feedback to meteorological variables and also to gases themselves (Jacobson and Ginnebaugh, 2010). Of course, many more SOA molecules are explicitly represented in MCM than in the CBM mechanism.

A reliable SOA mechanism of formation needs to be evaluated with specially designed smog chamber data, similar to the widely applied O_3 mechanisms that are based on thousands of smog chamber experiments conducted in the USA, Europe, Australia, and Asia. In an effort to construct an SOA mechanism suitable for regulatory

Table 9.7 Aromatics studied for SOA formation

	NO_x	H_2O_2	CO or ROG
Benzene	x	x	x
Toluene	x	x	x
Ethylbenzene	x	x	
o-Xylene	x	x	
m-Xylene	x	x	x
p-Xylene	x	x	
n-Propylbenzene	x	x	
Isopropylbenzene	x	x	
o-Ethyltoluene	x	x	
m-Ethyltoluene	x	x	
p-Ethyltoluene	x	x	
1,2,3-Trimethylbenzene	x	x	
1,2,4-Trimethylbenzene	x	x	
1,3,5-Trimethylbenzene	x	x	
Phenol	x	x	
o-Cresol	x	x	
m-Cresol		x	
p-Cresol		x	
2,4-Dimethylphenol	x	x	
2,6-Dimethylphenol		x	
3,5-Dimethylphenol		x	
Catechol		x	

air quality modeling, hundreds of experiments were recently conducted in California, USA using the US EPA smog chamber at the University of California at Riverside (Carter et al., 2012). Table 9.7 lists aromatic compounds investigated for SOA formation. Reactions of aromatic compounds with NO_x or H_2O_2 at various concentration levels, under dry conditions, at 300 K, without seed aerosol, were observed in a temperature-controlled, 85 m^3, smog chamber irradiated by black lights. Effects of concurrent reactants in ambient air were observed in a few experiments with added CO in some aromatic–H_2O_2 experiments or with an additional reactive organic compound in some aromatic–NO_x experiments, mostly for toluene or m-xylene.

9.2.2 Aerosol mechanisms

The SAPRC photochemical mechanism has traditionally been built by using the least number of reactions and parameters to simulate observed data in smog chamber experiments. In earlier versions of the SAPRC mechanism, a unique "PM yield" was used to represent the formation of SOA surrogates from each of the model precursors, such as alkanes, aromatics, isoprene, and terpenes. In a recent update, a dozen more surrogate molecules were added to improve the fitting of aromatic degradation data obtained from several hundred recent smog chamber experiments. The typical error of fitting was ~50% for PM mass yields in experimental conditions. As laboratory data for building a reliable, nearly explicit SOA mechanism for regulatory air quality modeling is still very limited, dramatic improvements of the SAPRC–SOA mechanism from a scientific perspective are likely in the long term.

Nearly explicit photochemical mechanisms, such as MCM and those representing permutation reactions of organic peroxy radicals, contain tens of thousands to millions of reactions (Madronich and Calvert, 1990; Aumont et al., 2005; Szopa et al., 2005). These mechanisms make use of ~5000 to 360,000 species, and are ideal for SOA simulation. For example, Johnson et al. (2006) analyzed ~2000 model species in MCM for potential SOA formation, and found that 70–100 of them could account for most SOA mass. While scientifically sound, the use of these mechanisms requires estimates of many unknown parameters (Jacobson and Ginnebaugh, 2010). Thus, there is no guarantee that the resulting uncertainty is significantly less than by using a condensed photochemical mechanism, such as CBM or SAPRC, which has been thoroughly evaluated with data from smog chamber experiments.

9.3   Health hazards of PM

The PM in breathing zones is a unique malicious pollutant to humans. Excessive exposure to hazardous particles may darken lungs, as shown in Figure 9.3, and clog the air–blood exchange surfaces of respiratory organs.

9.3.1   Career miners' health

Miners are traditionally exposed to much higher levels of particles than most people. Besides high risks of accidental death, occupational risks of death for miners were almost one per 1000 workers each year in the USA between 1955 and 1978, according

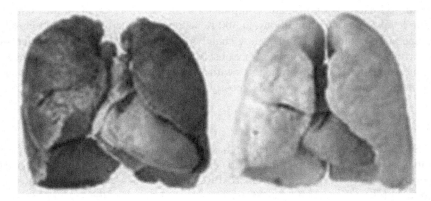

Figure 9.3 A lung damaged by PM pollution. The left lung is polluted by smoking, the right lung is unpolluted.
Pictures downloaded from http://www.baidu.com.

to the National Safety Council of the USA. This risk level is equivalent to the risk of a smoker who smokes 16 cigarettes a day (Crouch and Wilson, 1982), and may correspond to the current risk levels of miners in developing countries.

9.3.2 Ambient air quality standard

The effects of ambient PM on humans have been investigated via large-scale human epidemiological surveys, when ambient PM pollution was severe in the USA and other developed countries during the 1960s and 1970s. Data analyses showed that the mortality rate in US cities and the incidence of bronchitis in Japanese cities were proportional to PM sulfate concentration, and health effects were significant when annual average PM sulfate concentration was as low as 10 $\mu g\ m^{-3}$, while the annual average PM sulfate concentration in eastern US cities was $\sim$20 $\mu g\ m^{-3}$. The average death risk of US residents due to air pollution was estimated to be 2.4 $\times$ 10^{-4} at that time. Compared with national average conditions, the risk due to air pollution is much higher in pollution hot spots, such as in ports and along goods movement corridors where diesel fuels are often used. In early 2000, the total health risk due to air pollution over major highways in California was calculated to be equivalent to that from smoking continuously. In the years 2000–2006, coal-combustion-related components (e.g. selenium) in New York City were still statistically associated with cardiovascular disease (CVD) mortality in summer and CVD hospitalizations in winter, whereas elemental carbon and NO_2 showed associations with these outcomes in both seasons (Ito et al., 2011).

Current ambient PM standards in the USA and China are health based. Table 9.8 lists various levels of 24-hour average PM concentrations in ambient air and corresponding health effects and guidance.

It is seen that respiratory symptoms can be triggered on very sensitive people by fine PM with a 24-hour average concentration of 15–40 $\mu g\ m^{-3}$, or by PM_{10} with a 24-hour average concentration of 50–150 $\mu g\ m^{-3}$. As the daily fine PM concentration increases to an unhealthy level of 65–150 $\mu g\ m^{-3}$, increased respiratory effects may be observed in the general population; in addition, older adults and people with cardiopulmonary disease may suffer increased risks of heart or lung diseases and premature mortality. When the daily fine PM concentration reaches levels of 150–250 $\mu g\ m^{-3}$, such as during a PM episode observed in central California from December 25, 2000 to January 7, 2001, significant increases in respiratory diseases in the general population may be observed, as may heart or lung disease and premature mortality in people with cardiopulmonary disease and older adults. While it is rare in modern air, the daily $PM_{2.5}$ concentration occasionally reaches a hazardous level of 250–500 $\mu g\ m^{-3}$. For example, in the summer of 2008, a number of wildfires occurred concurrently over mountain ranges surrounding the California Central Valley in the USA. The PM emitted by wildfires elevated the ambient daily $PM_{2.5}$ concentration in the valley to 400–500 $\mu g\ m^{-3}$ for several days. At that time, the sky in the California Central Valley was covered with dense smoke, and the general population there were at serious risk of respiratory effects; older adults and people with cardiopulmonary disease were at serious risk of heart and lung diseases and premature mortality.

Table 9.8 The 24-hour PM standards and health effects in the USA

	$PM_{2.5}$	PM_{10}	Effects	Guidance
Good	0–15	0–50	None.	None.
Moderate	15–40	50–150	Respiratory symptoms possible in unusually sensitive individuals, possible aggravation of heart or lung disease in people with cardiopulmonary disease and older adults.	Unusually sensitive people should consider reducing prolonged or heavy exertion.
Unhealthy for sensitive groups	40–65	150–250	Increasing likelihood of respiratory symptoms in sensitive individuals, aggravation of heart or lung disease and premature mortality in people with cardiopulmonary disease and older adults.	People with heart or lung disease, older adults, and children should reduce prolonged or heavy exertion.
Unhealthy	65–150	250–350	Increased aggravation of heart or lung disease and premature mortality in people with cardiopulmonary disease and older adults; increased respiratory effects in general population.	People with heart or lung disease, older adults, and children should avoid prolonged or heavy exertion; everyone else should reduce prolonged or heavy exertion.
Very unhealthy	150–250	350–420	Significant aggravation of heart or lung disease and premature mortality in people with cardiopulmonary disease and older adults; significant increase in respiratory effects in general population.	People with heart or lung disease, older adults, and children should avoid all physical activity outdoors. Everyone else should avoid prolonged or heavy exertion.
Hazardous	250–500	420–600	Serious aggravation of heart or lung disease and premature mortality in people with cardiopulmonary disease and older adults; serious risk of respiratory effects in general population.	Everyone should avoid all physical activity outdoors; people with heart or lung disease, older adults, and children should remain indoors and keep activity levels low.

9.4 PM pollution control

9.4.1 SIP procedures

To reduce ambient PM pollution on a regional scale, the use of State Implementation Plans (SIPs) in the USA may serve as an example. The procedures of a successful state implementation plan in California are illustrated in Figure 9.4.

It is seen that, for a region with ambient PM pollution at an unhealthy level, in order to reduce the ambient PM pollution to a moderate level, four key elements need to be conducted coherently. Firstly, current emissions responsible for ambient PM pollution need to be compiled or modeled and evaluated. Secondly, in addition to a regular ambient monitoring network, additional air quality monitoring needs to be conducted to evaluate an emission inventory and to capture changes in ambient PM concentration due to the improved distribution of emissions or reduced emissions. Thirdly, quantifiable measures need to be implemented to change the distribution or intensity of emissions responsible for ambient PM pollution. Fourthly, air quality models need to be applied to demonstrate the achievement of the goal. If the goal is not expected to be achieved by existing measures, additional measures need to be implemented; thus the third and fourth elements may need continuous monitoring before the goal is achieved. This scheme has been used in California and other states of the USA to attain acceptable levels of O_3 and PM in ambient air in the last 20 years.

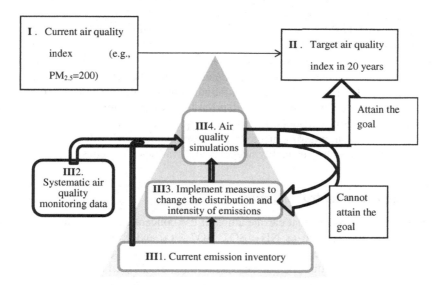

Figure 9.4 Procedures of a State Implementation Plan.

9.4.2   Monitoring network

The primary purpose of a systematic air quality monitoring network is to distinguish between areas where pollutant levels violate an ambient air quality standard and areas where they do not. As health-based ambient air quality standards are set at levels of pollutant concentrations that result in adverse impacts on human health, evidence of levels exceeding an ambient air quality standard in an area requires a public air quality agency to mitigate the corresponding pollutant. In other words, strategies, technologies, and regulations need to be developed to achieve the necessary reduction in pollution. The secondary purpose of a systematic monitoring network is to document the success of this sophisticated endeavor, either to record the rate of progress towards attaining the ambient air quality standard or to show that the standard has been achieved.

A systematic monitoring network in support of a successful implementation plan to reduce regional air pollution exposure must represent most of the population, and cover a diverse range of topography, meteorology, emissions, and air quality in the region. If possible, multiple pollutants and precursors need to be monitored at the same locations at the same time. For example, California requires annual evaluation of its air quality monitoring network, which consists of over 700 monitors at 250 sites, in order to document the current status and improvement of ambient air quality (CARB, 2012). The monitoring network represents various geographical areas, such as coastal areas, desert areas, interior valleys, mountain areas, and border areas, with more monitors in populated and polluted areas than in remote and clean areas. For individual monitors, spatial scales, whether they are microscale, middle scale, neighborhood scale, urban scale, or regional scale, are important considerations. In addition, diverse functions may be served by a monitor for the benefit of researchers, industries and media, as well as the general public. Table 9.9 lists known functions of air quality monitoring stations in California, USA.

It is seen that, besides the general purpose of documenting population exposure to air pollution, both national and state bodies have established monitoring stations in California for special purposes. A subset of monitoring stations may be used by health researchers, environmentalists, business interests groups, the general public, regulatory agency staff, and others. In the USA, a minimum number of monitors are required for areas with certain populations considering the level of air pollution severity and for areas near major traffic corridors. To ensure data quality, change in a monitoring station has to be approved, monitoring methods are recommended, and collocated monitoring has been coordinated by air quality agencies with various hierarches.

9.4.3   Emission control measures

Anthropogenic emissions that elevate ambient PM concentration include primary PM and precursor gases, such as NH_3, SO_x, NO_x, CO, and ROGs. Classifications of various emission sources are somewhat arbitrary, depending on the purpose. For chemical modeling, anthropogenic emissions may be classified as stationary sources that have fixed geographic locations and important parameters for estimating plume rises, area-wide sources that include small stationary sources as well as disperse

Table 9.9 Functions of air quality monitoring stations in California, USA

Code	Purposes
General	Population exposure, population representation, spatial representation, geographical representation, public reporting
USA1	Peak concentration; may have multiple sites due to seasonal meteorology
USA2	Representative concentrations
USA3	Impact of a stationary or mobile source
USA4	Pollutant-specific background levels
USA5	Pollutant transport between (populated, agricultural, wildland) areas
USA6	Measure air pollution impacts on visibility, vegetation damage, or other welfare-based impacts
CA1	Monitoring at expected high concentration sites relative to California Ambient Air Quality Standards
CA2	Support agricultural prescribed burning decisions
CA3	Support residential burning program
CA4	Spatial and temporal trend analyses
CA5	Support California area designation for a specific pollutant
CA6	State Implementation Plan maintenance requirement
CA7	Geyser Air Monitoring Program for geothermal H_2S
CA8	Special purpose monitoring

sources, and mobile sources that include onroad and offroad vehicles, boats, trains, and aircraft. These emission sources vary dramatically with time and location in modern air. For example, stationary sources may dominate SO_x emission in areas near major coal-fired power plants, and area-wide sources may dominate NH_3 emission in dairy farms, while mobile sources may dominate NO_x emission in ports and airports. Table 9.10 provides a summary of anthropogenic emission inventories of PM and precursors in California, USA in 2008.

It is seen that, in California, USA in 2008, primary emissions of PM_{10} and $PM_{2.5}$ were dominated by area-wide sources. In particular, mechanical forces were responsible for PM_{10} emissions in California; for $PM_{2.5}$ emissions, both mechanical forces and non-industrial burning, such as biomass burning and cooking, were responsible. Anthropogenic SO_x emission was mainly by boats and stationary sources, and livestock sources dominated NH_3 emission. ROGs and NO_x are precursors of both PM and O_3; while mobile sources dominated anthropogenic NO_x emissions, area-wide and stationary sources contributed significantly to anthropogenic ROG emissions, as well as the major contribution by mobile sources. CO is itself a toxin, and is marginally a precursor for O_3 and thus PM. Anthropogenic CO emissions were mainly from mobile sources with relatively small engines and from biomass burning in residential areas and in agricultural fields (e.g. Yan et al., 2006), while stationary sources contributed less than 3%, which reflects the high combustion efficiency of fuels used in boats, power plants, and industrial activities.

Table 9.10 Anthropogenic emissions of PM and precursors in California, USA in 2008

Major category	ROGs	CO	NO_x	SO_x	PM_{10}	$PM_{2.5}$	NH_3
Stationary sources	**427**	**317**	**368**	**109**	**161**	**95**	**91**
Fuel combustion	32	245	262	39	34	31	14
Waste disposal	53	4	3	1	1	1	58
Cleaning and surface coatings	142	0	0	0	1	1	0
Petroleum production and marketing	135	12	8	40	4	3	2
Industrial processes	65	56	94	30	121	58	17
Area-wide sources	**652**	**1968**	**95**	**6**	**1791**	**448**	**610**
Solvent evaporation	408	0	0	0	0	0	34
Miscellaneous processes	243	1968	95	6	1791	448	575
Mobile sources	**1135**	**9042**	**2747**	**166**	**160**	**133**	**55**
Light-duty passenger vehicles	231	2207	189	2	16	9	25
Light- and medium-duty trucks	231	2568	300	2	19	13	27
Heavy-duty trucks	119	796	1020	1	40	34	2
Other on-road	51	529	75	0	2	2	1
Aircraft and trains	47	313	198	5	13	13	
Ocean-going vessels and commercial harbor craft	14	40	315	154	23	22	
Pleasure crafts	137	740	40	0	9	7	
Recreational vehicles	69	192	2	1	1	1	
Off-road equipment	190	1546	503	0	30	27	
Other off-road	45	113	104	0	6	6	0
Total sources	**2213**	**11,327**	**3209**	**281**	**2112**	**677**	**756**

Note: The units are tons per day, and natural sources are not included.

To achieve the goal of reducing serious PM pollution in a region, a number of emission reduction measures may be applied after a thorough emission inventory has been compiled. For example, to alleviate the PM_{10} pollution in central California, USA from serious non-attainment status, California San Joaquin Valley Unified Air Pollution Control District (SJVAPCD) implemented a successful plan in 2003. Major elements of the plan included: (1) demonstration of attainment at the earliest practicable date; (2) implementation of the best available control measures/technology, approved by the US EPA, for all significant sources of PM_{10} and (limiting) precursors; (3) provision of annual reductions of at least 5% of PM_{10} or (limiting) precursor emissions based on the most recent inventory until attainment; (4) provision of quantitative milestones for reasonable further progress; and (5) adoption of contingency measures to assure that emission reductions are in place that can be implemented if a milestone is not achieved on schedule.

In order to plan successfully, a \$30 million research project, the California Regional Particulate Air Quality Study, was carried out by a number of research institutions, industrial interests groups, and public air quality agencies involving national and international specialists. Investigations showed that constituents of

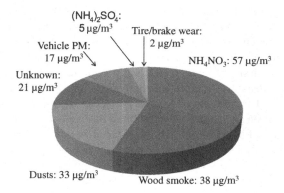

Figure 9.5 Average PM_{10} composition in Fresno city, California, USA in the period January 1–4, 2001.

elevated PM_{10} concentrations in ambient air quality monitoring stations were mainly ammonium nitrate, wood smoke, road and agricultural dusts, and vehicle exhaust particles; Figure 9.5 shows the average PM_{10} composition in metropolitan Fresno in the period of January 1–4, 2001. During the planning process, air quality modeling identified the limiting precursors and important emission sources for unhealthy PM_{10} concentrations in the region. A number of existing control measures for reducing NO_x and ROG emissions were analyzed; new feasible and cost-effective control measures for reducing PM_{10} and SO_x emissions and unique control measures for reducing NH_3 emission from dairy farms were also implemented. The resulting emissions reductions in this huge district are listed in Table 9.11, and air quality modeling identified the earliest practicable attainment date as 2010; in reality, this region achieved the goal a few years earlier than the model projection, partly due to the control of wood burning in urban areas during winter.

9.4.4   Exposure reduction

To reduce the risk of human exposure to ambient PM pollution, reducing emissions of PM and its precursors is the most effective method, though personal exposure to PM pollution may be significantly reduced by using quality PM filters in built environments, such as at home, in offices, and in vehicles. However, due to limitations in

Table 9.11 Anthropogenic emissions (tons per day) in SJVAPCD

	1990	2001	2010	Cut-off level
PM_{10}	450	465	284	0.9
NO_x	796	547	363	1.3
ROGs	625	472	344	2.8
SO_x	86	46	27	2.5

Note: Sources with emission rates below the cut-off level were not controlled.

available technology and other resources in addition to the pressure of needs and demands of consumers and regional economies, optimizing the spatial distribution of airborne PM and precursors according to atmospheric carrying capacity and population density is an effective alternative to attain health-based ambient PM air quality standards. Major emission sources are usually built in well-ventilated areas with a sparse population density, if it is feasible and cost-effective. Microscale adjustments may also be useful. The addition of highway walls in metropolitan areas is a literal example: total emission and peak concentration of PM are not reduced, but population-weighted exposure to PM pollution may be reduced besides the benefits of reducing noise and accidental risks. The use of low-emission hybrid school/transit buses is another example: this reduces vehicle emissions in populated areas while increasing emissions due to electric and battery productions elsewhere, so population-weighted exposure may be reduced. Other examples include the use of electric vehicles in ports and the use of clean fuels for auxiliary engines in marine vessels: both measures reduce local emissions while increasing emissions in electric and refinery plants, but population-weighted exposure may be reduced. Similarly, the use of highway bypasses around populated urban centers, the use of carpool lanes and electronic toll passes in special sections of highways, the use of electric furnaces instead of natural gas furnaces or biomass stoves, and the use of air conditioning instead of biomass fireplaces during cold winter may significantly reduce population-weighted exposure to PM pollution, while total emission to and peak concentration in ambient air may not improve at all. Due to systematic measures to reduce human exposure to PM pollution, the population-weighted annual average $PM_{2.5}$ concentration in California was only 8.7 μg m^{-3} in the period 2000–2006 (Mahmud et al., 2012). Chemical modeling has been applied to optimize spatial distributions of various sources, in order to reduce human exposure to ambient PM pollution.

9.5   Challenges of simulating PM in the atmosphere

Atmospheric PM concentration and composition may be simulated by Eulerian or Lagrangian three-dimensional models. Inorganic components of PM are relatively well understood despite significant disparities between models and observations (Jacob, 1986; Chang et al., 1987; Liang and Jacobson, 1999; Livingstone et al., 2009; Wang et al., 2010; Deandreis et al., 2012), but organic components of PM are more intriguing due to the complexity of condensable organic compounds (Pandis et al., 1992; Wexler et al., 1994; Jacobson and Turco, 1995; Liang et al., 1997; Lurmann et al., 1997; Saxena and Hildemann, 1997; US EPA, 1999; Livingstone et al., 2009; Carter et al., 2012). Details are elaborated below.

9.5.1   Simulating a PM episode

The California Central Valley in the USA is the second most polluted and studied area in California in recent decades. During a 14-month field campaign, an extended PM

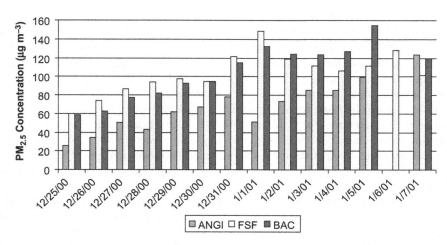

Figure 9.6 The CRPAQS winter PM$_{2.5}$ episode.

episode was captured in the southern San Joaquin Valley from December 25, 2000 to January 7, 2001. The daily PM$_{2.5}$ concentrations at a rural station ("ANGI") and two metropolitan stations ("FSF" and "BAC") during the episode are shown in Figure 9.6. It is seen that the daily PM$_{2.5}$ concentration exceeded 50 μg m^{-3} every day during the 2-week period at urban stations, and also occurred on eight of the 13 days when measured at the rural station. The peak daily PM$_{2.5}$ concentration exceeded 150 μg m^{-3} at the southern end of the valley.

To simulate the observed episode, US EPA air quality model CMAQ was configured for the spatial domain shown in Figure 9.7. The horizontal domain consists

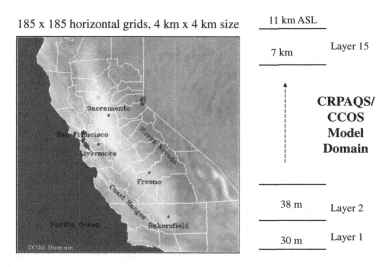

Figure 9.7 The model domain for the CRPAQS/CCOS programs in California, USA.

of 185 × 185 grids and the size of each grid is 4 km × 4 km; the vertical domain consists of 15 terrain-following expanding layers, with the bottom layer thickness of ~30 m above the surface. Configured photochemical mechanisms include the SAPRC mechanism for gas-phase reactions and SOA yields, the RADM mechanism for aqueous-phase reactions, ISORROPIA for partitioning inorganic components of PM, and the ideal solution for SOAs. Meteorological fields were simulated by a mesoscale non-hydrostatic meteorological model MM5, and a regulatory emission inventory was compiled at California Air Resources Board. Initial and boundary conditions were based on measurements and global model results. About a hundred simulations were conducted; in each simulation, the model ran from December 17, 2000 to January 7, 2001, with the first 8 days for model spin-up to reduce effects of initial and boundary conditions.

As the default setup in US EPA CMAQ was mainly based on typical conditions in the eastern USA, California-specific information was implemented in the model. The simulated surface distribution of sulfate and total inorganic components of $PM_{2.5}$ is shown in Figure 9.8 for both cases. It is seen that the total inorganic components of $PM_{2.5}$ in Fresno city in the initial base case were close to the observed values for $PM_{2.5}$, which reflects a significant overestimate, while values in the updated case are closer to observations. The simulated peak concentration of $PM_{2.5}$ sulfate in the updated base case is close to observations, while that in the initial base case is an order of magnitude too high.

As PM concentration is very sensitive to ambient relative humidity, modeled PM components contain propagated uncertainty from meteorological inputs, compared with observations that also contain significant variations. For the CRPAQS study, the model performance regarding insoluble $PM_{2.5}$ components, mainly total carbon, was significantly better than for soluble components, based on hourly time series. As the average time increases from an hour to a day, the disparity between model and observations decreases significantly. Figure 9.9 shows the observed and simulated daily $PM_{2.5}$ concentration at three stations during the episode. When averaged over the 2-week episode, ammonium nitrate in the model and from observations was even closer than daily average $PM_{2.5}$; for example, at the metropolitan Bakersfield station, the difference was ~4%. This phenomenon suggests that the model simulated long-term averages better than short-term variations for surface $PM_{2.5}$.

9.5.2 Simulating seasonal PM

Ambient PM concentration often shows substantial seasonal variations for several reasons. Firstly, the concentration of ammonium nitrate is much higher when relative humidity is ~80% than when relative humidity is below the deliquescent point, which depends on PM composition. Secondly, the mixing height in summer is usually significantly higher than in winter, which would result in a lower concentration of primary PM if emissions were constant throughout the year. Thirdly, anthropogenic and biogenic emissions of PM and precursors change with the seasons, as well as locations and control measures (e.g. Penner et al., 1993; Chang et al., 2012; Song et al., 2012). To simulate seasonal variations in a region, a nested-grid modeling

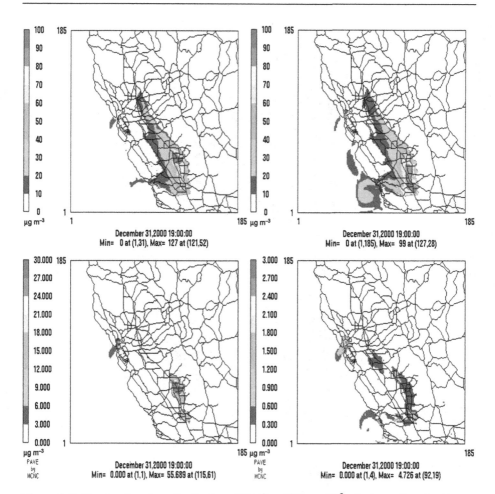

Figure 9.8 Simulated surface distribution of $NH_4^+ + NO_3^- + SO_4^{2-}$ (top panel) and sulfate (bottom panel) in $PM_{2.5}$. Left: initial base case. Right: updated base case.

method is necessary to balance the needs for representing long-term, large-scale variations in meteorological and chemical fields and for representing regional details.

Seasonal variations of PM in central California were investigated in a modeling exercise (Fahey et al., 2006). General circulation was simulated by the Community Climate System Model (National Center for Atmospheric Research, Colorado, USA), which provided one-way inputs to a mesoscale meteorological model (MM5). The MM5 model has been developed to simulate California weather features for several decades. Two one-way nested, downsizing grids were simulated using MM5, which provided meteorological fields at a horizontal resolution of 12 km × 12 km for the exercise. The PM air quality model, US EPA CMAQ, was applied to simulate seasonal patterns of the 14-month observations from December 1999 to January 2001.

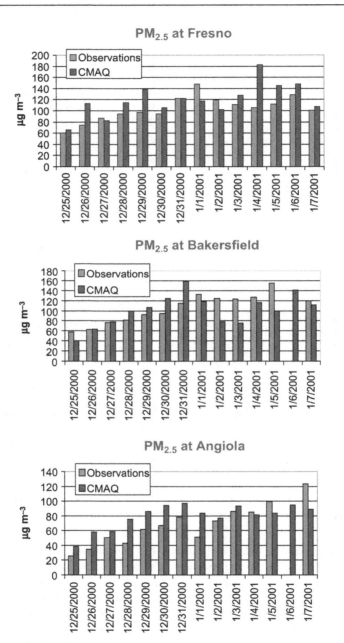

Figure 9.9 Observed and simulated daily PM$_{2.5}$ at three stations in California, USA.

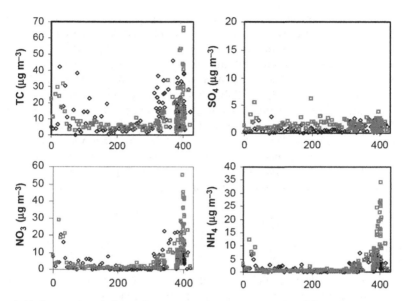

Figure 9.10 Seasonal variations of $PM_{2.5}$ constituents in central California, USA. Diamonds: model; squares: observations.
Obtained from Fahey et al. (2006).

Simulated and observed time series of $PM_{2.5}$ components are shown in Figure 9.10. It is seen that total carbon and ammonium nitrate peaked in winter months with the summed concentration over 150 μg m^{-3} in December 2000. Sulfate concentration was below 5 μg m^{-3} and showed much less amplitude in seasonal variations, due to two factors: firstly, sulfuric acid was mainly partitioned in the PM phase in all seasons; secondly, the aqueous-phase formation of sulfate during wet winter months compensated the slower formation rate in the gas phase than in dry summer months.

9.5.3 Simulating climate effects on PM

As PM pollution is sensitive to weather conditions, as many other forms of air pollution, the effect of anthropogenic climate change on PM pollution is an interesting research topic. For example, California has a long coastline in the west and a long valley in central California; the land–sea circulation provides an important natural air conditioning for the California Central Valley in summer. If sea surface temperature increases, the land–sea circulation is expected to weaken, which may intensify the stagnancy of air over California Central Valley in summer and in turn worsen PM pollution there. The effect of 50-year anthropogenic global climate change on PM pollution in California was assessed recently, using climate modeling results from the IPCC with regional meteorological and air quality models (Mahmud et al., 2012). The IPCC applied a parallel climate model to simulate global climate change from 1995 to 2099 due to constant increase in atmospheric CO_2 mixing ratio by 1% per year. Using the IPCC climate modeling results as boundary and initial conditions, the Weather Research and Forecast model (USA) was applied to simulate regional meteorological parameters in the periods 2000–2006 and 2047–2053, with a horizontal resolution of 4 km × 4 km. A source-oriented regional PM air quality model was applied to simulate PM constituents in California in selected periods of the 14 meteorological years, using the California regulatory emission inventory in 2000. The investigation suggested that 50-year global climate change is not expected to worsen bulk PM pollution in the greater Los Angeles area, but may intensify PM episodes in the California Central Valley. In particular, population-weighted exposure to NH_4NO_3, diesel engines particles, and wood combustion particles in San Joaquin Valley could increase by 21%, 32%, and 14% respectively due to 50-year global climate change driven by increasing atmospheric CO_2 mixing ratios at the rate of 1% per year.

Summary

In this chapter, the physical and chemical characteristics of atmospheric PM have been discussed stepwise by presenting size representations in computer models, size and chemical profiles of various PM sources, mixing states of PM in ambient air, and observed ambient PM compositions in sequence. Photochemical mechanisms of formation for secondary organic aerosol compounds are discussed from a modeling perspective. Health hazards of PM pollution have been introduced for professionals and the general public. A systematic method to control PM pollution has been elaborated, which includes procedures to implement a successful plan, the functions of a monitoring network, measures to control emissions, and proven methods to reduce human exposure. The challenges of simulating PM in the air are illustrated by the simulation of a PM episode, seasonality of PM constituents, and climate effects on PM pollution.

Exercises

Story: A graduate student went to an international conference on PM, and heard two research presentations by two distinguished professors. The first professor described findings by his group of researchers on the aerosol effects on climate change; the graduate student was puzzled by his statement that all aerosols in the research were "internally mixed". The second professor presented the findings of her group on sources responsible for a PM episode observed in a valley; the graduate student was puzzled by her statement that PM was treated as "externally mixed". Please help the graduate student by answering the following questions.

1. Assume the average composition of aerosols with significant radiative force is 50% NH_4NO_3 and 50% $(NH_4)_2SO_4$ for inorganic components, while organic components consist of two-thirds oxygenated carbon and one-third "black carbon". The mass ratio of inorganic to organic components is 4:1. Illustrate the size and chemical profiles of aerosols

Table E9.1 Mean fine PM composition at Rubidoux, Riverside, CA, USA in 1982

	$\mu g\ m^{-3}$		$ng\ m^{-3}$
$PM_{2.5}$	42.1	C_{23}–C_{34} n-alkanes	50
Sulfate	6.3	C_9–C_{30} n-alkanoic acids	262
BCs	2.6	C_3–C_9 aliphatic diacids	312
Others	10.5	Woodsmoke markers	15
Organics	6.2	Aromatic polyacids	106
Nitrate	10.5	PAHs	3.7
NH_4^+	5.9	Nitrogen-containing organics	1.9

used by the first professor, using molar units for chemicals and a modal representation for sizes with typical assumptions described in this chapter or other literature.

2. At a suburban ground station downwind of Los Angeles, the measured annual mean fine PM composition in 1982 is shown in Table E9.1. Though not shown, oxygenated organic compounds tend to be concentrated in particles with larger diameters than primary hydrocarbons. It is seen that 12% of the $PM_{2.5}$ organic compounds were identified. Among identified organic compounds, let us assume primary PM organics (PPOs) consisted of N-containing organics, PAH compounds, woodsmoke markers, and C_{23}–C_{34} n-alkanes; secondary organic aerosols contained C_9–C_{30} n-alkanoic acids, C_3–C_9 aliphatic diacids, and aromatic polyacids. (a) Calculate the mass fraction of identified SOAs from the right column. (b) Calculate the molar fraction of identified SOAs by using a surrogate compound to represent all identified SOA compounds and another surrogate compound to represent all identified PPO compounds.

3. Assume there were two groups of PM particles at Rubidoux station in 1982: the first group consisted of "BC" core coated with PPO; the second group consisted of "others" core coated with SOAs and inorganic ions, as well as liquid water. In the second group, the aerosol liquid water concentration is determined by molar concentrations of solutes at a fixed relative humidity. Survey the literature and estimate the average aerosol liquid water concentration at Rubidoux station in 1982, when the average relative humidity was 80% and 40% respectively. Assume that water activity in the second group of PM particles was exactly the relative humidity.

4. Change the molecular weights of the surrogate compounds by a factor of 2 and repeat the above calculations.

5. The number concentration of PM particles at Rubidoux in 1982 could exceed 100,000 cm^{-3}. Assume that the two groups had the same number concentrations of 100,000 cm^{-3} and the same log-normal distribution as described by formula (F9.1.1), but with different mean diameter. Calculate the LD_g for each group, using a computer program or analytical formula.

6. Discuss the implications of the above calculations for PM simulations. What would you like to do differently from the two distinguished professors in their projects?

10 Other toxins in the air

Chapter Outline

Besides oxidants and PM, a number of other toxic chemicals have been identified in ambient air. These additional air toxics may be classified as toxic metal compounds, aromatic compounds, and non-aromatic toxic compounds. Most toxic chemicals are emitted from fossil fuel combustion, biomass burning, industrial production, solvent evaporation, and waste incineration (e.g. Sorooshian et al., 2012; Tian et al., 2012a). A significant fraction of these chemicals are traditional pesticides and herbicides, which are typically persistent organic pollutants. Exposure in built environments is becoming more and more important, due to social advancements. It is necessary to understand the sources, transformation, transport, and fates of these additional air toxins in order to mitigate the risks of human exposure. The details are discussed below.

Chemical Modeling for Air Resources. http://dx.doi.org/10.1016/B978-0-12-408135-2.00010-0

10.1 Toxic metals

Metals are usually toxic if inhaled. Studies of the toxicities of heavy metals have been conducted on rats, mice, rabbits, monkeys, etc., while toxicities to humans are based on statistical analyses of tragedies involving occupational workers and neighboring residents of some industrial plants in the early to middle twentieth century, as well as criminal acts in historical wars. Figure 10.1 illustrates the extreme levels of air pollution in industrial and urban areas of the USA in the twentieth century; the black fumes from coal burning contained toxic metals, such as arsenic, and the smog from primitive automobiles contained toxic organics, such as polycyclic aromatic hydrocarbons (PAHs).

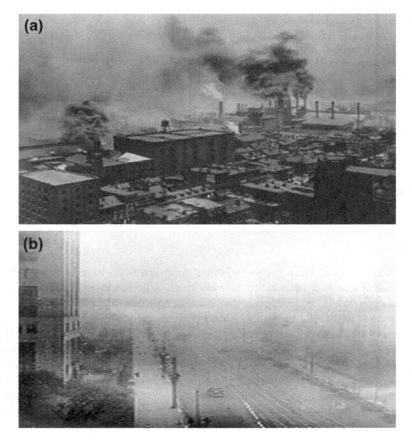

Figure 10.1 Illustrations of air pollution. (a) Coal burning fumes in Pittsburgh, PA, USA at 11 a.m. in June, 1906. Obtained from the US EPA. (b) Smog in Los Angeles, CA, USA on July 26, 1943. Obtained from South Coast Air Quality Management District, California, USA.

10.1.1 Levels for exposure assessment

There are several concentration levels important for assessing an air toxin, besides ambient air quality standards if they exist. The inhalation reference concentration (RfC) of a toxin is an estimate, with uncertainties of an order of magnitude, of a safe level; daily exposure to a toxin below this level is expected to have no observable adverse effect on humans during their lifetime. It is assumed that there is a threshold for toxic effects to occur. The lowest observed adverse effect level (LOAEL) of a toxin is the lowest concentration or amount of a toxin, found by experiment or observation, which causes an adverse alteration of morphology, functional capacity, growth, development, or life span of a target organism distinguishable from normal (control) organisms of the same species and strain under defined conditions of exposure, according to the International Union of Pure and Applied Chemistry. The immediately dangerous to life and health (IDLH) concentration for a toxin corresponds to the exposure that is likely to cause death or immediate or delayed permanent adverse health effects or prevent escape from such an environment, as defined by the National Institute of Occupational Safety and Health. The value of IDLH for a toxin is an occupational exposure limit for protecting the life of workers at their workplace. The values of LOAEL and RfC for a toxin constitute its ambient air quality standard.

10.1.2 Beryllium

Occupational workers and people in the neighborhood of beryllium alloy manufacturing plants have suffered from lung cancer due to elevated exposure to Be and its compounds. Studies showed that the inhalation toxicity of Be depended on the solubility of compounds and the size of the host particles. According to the Integrated Risk Information System of the US EPA, inhalation of As, Be, and Ni at an ambient concentration of 1 $\mu g\ m^{-3}$ could cause 4.3, 2.4, and 0.24 cases of lung cancer per 1000 people during their lifetime respectively. Inhalation of Cd and Cr may cause cancers in various organs, and inhalation of Co, Hg, Mn, Pb, and Sb may also cause severe damage to the human body.

10.1.3 Cadmium

Cadmium is a toxic heavy metal emitted from industrial production, fossil fuel combustion, and biomass burning (e.g. Tian et al., 2012b). Animal experiments have shown that inhalation of cadmium chloride resulted in excess lung cancer in rats, and toxic effects of cadmium on humans were confirmed by a historical survey of a group of 602 workers. These workers had worked in cadmium smelters for at least 6 months during the years 1940–1969, and were followed to the end of 1978; a two-fold excess risk of lung cancer was observed in the cohort. Table 10.1 lists health levels and regulatory numbers for cadmium concentration in breathing zones; also listed are concentration levels of cobalt for risk assessment.

It is seen that the 50% lethal dose for rats is 500 mg m^{-3} and the lowest observable adverse effect level is 0.02 mg m^{-3}. The California EPA of the USA recommends a reference level of 0.0001 mg m^{-3} for the public, and the safe level with 1 millionth

Table 10.1 Cadmium (cobalt) health levels and regulatory numbers

Health	mg m^{-3}
LC$_{50}$ (rats)	500 (165)
LOAEL	0.02 (0.2)
CalEPA	0.0001 (5×10^{-6})
Safe level*	6×10^{-7} (5×10^{-8})
Occupational	
NIOSH IDLH	9 (20)
OSHA PEL (dusts)	0.2 (0.1)
OSHA PEL (fumes)	0.1 (0.1)
ACGIH TLV (dusts)	0.01 (0.02)
ACGIH TLV (respiratory dusts)	0.002 (0.02)

* The safe level of cobalt is assumed to be 1% of the reference exposure level by the California EPA, USA. For comparison, the ambient concentration of cobalt in the USA is ~0.4 ng m^{-3}.

lifetime cancer risk is 0.6 ng m^{-3}. Occupational risk levels and standards are also shown. The IDLH concentration for Cd published by the National Institute for Occupational Safety and Health (NIOSH) of the USA is 9 mg m^{-3}, and the permissible exposure limits set by the Occupational Safety and Health Administration (OSHA) of the USA are 0.2 mg m^{-3} for dusts and 0.1 mg m^{-3} for fumes. The American Conference of Governmental Industrial Hygienists (ACGIH) set a threshold limit value of 0.01 mg m^{-3} for dusts and 0.002 mg m^{-3} for respiratory dusts. The lower limits for finer particles reflect the effective penetration by finer particles.

10.1.4 Lead

Gasoline used to be added with tetraethyl-lead and aromatics as anti-knocking agents to increase octane number in many countries and, as a result, vehicle emission of lead dominated other sources, such as coal burning and industrial emissions, until recent decades. For example, China emitted 0.2 Tg Pb in the period 1990–2009, and 60% was from motor vehicle gasoline combustion (Li et al., 2012); while the total emission of Pb increased from 2.7 Gg in 1980 to 13 Gg in 2008, industrial emission accounted for over 80% of total emission in 2008 (Tian et al., 2012b). After airborne lead was identified as an air pollutant, gasoline standards in many nations have reduced or eliminated organic lead while increasing aromatic content to maintain octane number. For example, vehicle emission of lead in mainland China exceeded 12,000 tons in 2000; in 2001, the amount decreased to within 200 tons due to the phase-out of organic lead additive in gasoline (Li et al., 2012). A dramatic reduction of lead in gasoline in the USA occurred in 1986. Gasoline in Califorina, USA contained added methyl t-butyl ether for several years in the late 1990s, and is currently supplemented with ethanol to reduce toxicity. However, as ethanol production may compete with food sources, innovative use of biomass, otherwise burnt, for ethanol production needs to be pursued cautiously to avoid the depletion of plant nutrients. Currently, coal combustion and industrial production are major sources of lead pollution, and ambient air quality standards regulate seasonal average lead concentration to be within 1 μg m^{-3} in China and 0.15 μg m^{-3} in the USA.

European countries conducted significant efforts to reduce anthropogenic emissions of Cd and Pb. In the period 1990–2009, emissions of Cd and Pb were reduced by 50% and 80%, respectively in Europe. As a result, atmospheric concentrations of Cd and Pb decreased by 80% and 90% at a network of stations (Tørseth et al., 2012); the additional yield is presumably due to changes in the spatial distribution of emission sources.

10.1.5 Mercury

Airborne mercury may be emitted by non-ferrous metal smelting, coal combustion, battery and fluorescent lamp production, cement production, and solid waste incineration (e.g. Streets et al., 2005; Hu et al., 2012). Despite significant efforts to reduce anthropogenic emissions of mercury in developed countries, gas-phase mercury is rather persistent in ambient air after emission. A recent cross-year measurement found that airborne mercury was present at 1.6 ng m^{-3} at a remote station in Canada, similar to levels at rural (up to 1.8 ng m^{-3}) and industrial stations (up to 5 ng m^{-3}) in other parts of North America (Cheng et al., 2012). By comparison, measured Hg concentration ranged from 1.8–3.8 ng m^{-3} at background sites to 120 ng m^{-3} at polluted stations in Asia in 2007–2008 (Nguyen et al., 2011). The concentration of PM Hg was measured as 0.07–1.45 ng m^{-3} in Shanghai, China in 2004–2006 (Xiu et al., 2009).

10.1.6 Summary of toxic metals

Table 10.2 lists sample sources, toxicities, safe levels, and unit risks of 11 metals. Metal Se is listed here due to wide public concerns: in some local areas of China, a deficiency of Se intake was responsible for a local disease while overdoses of Se have also been shown to be harmful.

10.2 Toxic aromatics

Aromatic compounds constitute a significant fraction of fossil fuels, and vehicle exhausts are a major source of aromatic compounds in metropolitan areas and nearby traffic corridors. Aromatic compounds have also been used as flame retardants and solvents in consumer products and industrial production, and thus may be present in significant amounts in built environments (e.g. Venier et al., 2012). Toxic aromatic compounds with a single benzene ring include benzene, toluene, ethylbenzene, styrene, xylenes, and mono- and poly-substituted benzenes containing other functional groups, such as –OH, –NO$_2$, –Cl, –Br, and –NH$_2$.

10.2.1 Polychlorinated biphenyls

An important class of biphenyl toxin is polychlorinated biphenyls (PCBs), which is one of the 12 organic pollutant groups regulated by the Stockholm Protocol

Table 10.2 Eleven selected metals in ambient air

	Sample sources	Toxicity	Safe level (ng m^{-3})	Unit risk [/(μg m^{-3})]
As	Metal production	Lung cancer	0.2	4.30E-03
Be	Dusts, coal combustion, Be–Al alloy	Lung cancer	20	2.40E-03
Cd	Metal production	Carcinogen	0.6	1.80E-03
Co	Metal/paint industry, oil, asphalt	Hard metal disease, pneumonia, asthma	0.05	
Cr	Ferrochrome	Carcinogen	0.08	1.20E-02
Hg	Coal, solid waste, metal production	Hand/memory	300	
Mn	Fuel, metal production, dusts	Neural	50	
Ni	Fuel, metal production	Lung cancer	4	2.40E-04
Pb	Solid waste, fuel, metal production	Neural	150	
Sb	Fuel, metal production, solid waste	Pulmonary inflammation	200	
Se	Fuel, glass, metal production	Liver	?	

coordinated by the United Nations Environmental Program. Industrial PCBs comprise 209 synthetic compounds used as insulating materials in electrical transformers and capacitors, but that have been banned from industrial production in developed countries since the 1970s due to their toxicity to humans. However, due to their stability in the environment, their residues persist and are detectable in tissues of most residents in developed countries. As PCBs are lipophilic, consumption of fatty fish from contaminated waters may be a major source of human exposure in coastal and lake areas, which can cause harm to infants via mother's milk (Jacobson and Jacobson, 1996). Animal studies have reported an increase in liver tumors in rats and mice exposed orally to all tested PCB formulations, according to the US EPA.

Airborne PCBs may also be significant as regards human exposure. According to research results coordinated by the US EPA, chronic exposure to some PCB formulations by inhalation in humans results in respiratory tract symptoms, gastrointestinal effects, mild liver effects, and effects on the skin and eyes such as chloracne, skin rashes, and eye irritation; the association between inhalation exposure and liver cancer was suggested from human exposure studies. The US EPA estimated a unit risk of 1.0×10^{-4} (μg m^{-3})$^{-1}$ for inhalation of evaporated PCB congeners, and a safe level of 10 ng m^{-3} when the lifetime cancer risk is within one millionth. Occupational limits of PCBs are 0.5–1 mg m^{-3} for the OSHA-permissible exposure limit and ACGIH threshold limit value for most workers, and 1 μg m^{-3} for the NIOSH-recommended exposure limit; these limits are for an 8- or 10-hour time-weighted average exposure a day and 40 hours a week.

In a study in China, PCBs were found to be highly enriched in computer dust, with concentrations as high as 72 µg g^{-1}. By comparison, the average PCB concentration measured in various parts of the world in recent decades ranges from 9 to 1400 ng g^{-1} for indoor dust, with the highest value reported in the northeastern USA; in outdoor dust, the average PCB concentration ranged from 4 to 1023 ng g^{-1}, with the highest value reported in Germany (Li et al., 2010). It is conceivable that atmospheric PCB compounds mainly reside in fine and ultrafine particles. Measurements found that the the total concentration of 28 polychlorinated dibenzo-p-dioxins and dibenzofurans (PCDD/F) ranged from 2.6 to 120 ng m^{-3} at 14 stations in southern China in the period 2007–2009 (Chen et al., 2011), and that the total concentration of seven organochlorine pesticides and three PCB compounds was 44 pg m^{-3} in Tibet, China in 2006–2007 (Gong et al., 2010).

10.2.2 Polycyclic aromatic hydrocarbons

Polycyclic aromatic hydrocarbons (PAHs) are a special group of air toxins regulated by the United Nations Environmental Program. Over 100 PAH compounds are formed or released from cigarette smoke, incomplete combustion of fossil fuels, woods and other forms of biomass, coke ovens, metal processing plants, coal tar and asphalt processing, incinerators, consumer and electronic products, and foods that are smoked, grilled, or charcoal broiled. Unlike ROG or VOC molecules, most PAH compounds are found in the particulate phase due to low vapor pressure and resulting condensation, unless in a very clean environment. There are three reasons why vehicle exhausts may contain PAHs: (a) synthesis from aromatic compounds in the fuel; (b) storage in engine deposits and subsequent emission of PAH already in the fuel; (c) pyrolysis of lubricant; more PAH mass is formed at higher engine load, greater high-boiling aromatic content of fuel, and lower air to fuel ratios (Baek et al., 1991). Before emission control, emissions of a number of PAH compounds in the USA were almost equally contributed by residential heating, open burning, non-catalyzed mobile sources, and industrial processes, as shown in Figure 10.2. These have reduced after several decades of the emission control program in the USA, and in 2004 consumer products were the major source category for the 16 PAH compounds prioritized by the US EPA, while global emission of the 16 PAH compounds in 2004 was estimated to be 0.52 Tg, with major contributions by biofuel (57%), wildfire (17%), and consumer products (7%) (Zhang and Tao, 2009). China (0.114 Tg), India (0.09 Tg), and the USA (0.032 Tg) were estimated to be the top three emitters in 2004. Figure 10.3 illustrates estimated total emission rates of the 16 PAH compounds in selected nations and regions in 2004, based on modeled data from Zhang and Tao (2009). It is seen that Asia, mainly China and India, emitted several times more PAHs than Africa, Europe, the USA, and South America in 2004. Vehicle emissions of PAH compounds differed significantly between nations. For example, vehicles in Mexico emitted about 0.01 g particle-phase PAH compounds for every kg fuel combusted, while the emission factor in Switzerland was 0.002 and 0.007 g kg^{-1} for light-duty and heavy-duty vehicles respectively, and the corresponding emission factor values in California, USA were 9 × 10^{-5} and 0.0023 g kg^{-1} (Jiang et al., 2005). For comparison, burning 1 kg rice and bean straws was

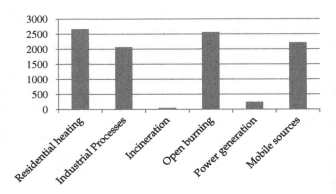

Figure 10.2 Emissions of PAHs from various sources in the USA in 1980. The units are tons, and the numbers are from Baek et al. (1991).

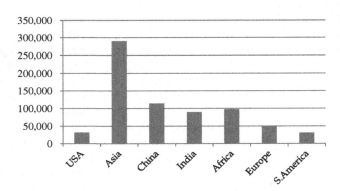

Figure 10.3 Estimated global emissions of 16 PAH compounds in 2004. The annual total emission rate was 520,000 tons globally, and the 16 PAH compounds are: naphthalene, acenaphthylene, acenaphthene, fluorene, phenanthrene, anthracene, fluoranthene, pyrene, benz[a]anthracene, chrysene, benzo[b]fluoranthene, benzo[k]fluoranthene, benzo[a]pyrene, dibenz[a,h]anthracene, indeno[1,2,3-cd]pyrene, and benzo[g,h,i]perylene. The numbers are from Zhang and Tao (2009).

measured to emit a total of 0.003–0.05 g PAH compounds (Lu et al., 2009). Due to the dramatic reduction in emissions of PAH compounds in developed countries and the gradual increases in China and India in recent decades, Asia accounted for over half of the global emissions of PAH compounds in recent years, while the total emission of PAHs and background PM concentrations in China have started to decrease (Zhang et al., 2011; Yan et al., 2012).

PAH compounds are present in ambient air at trace levels except near sources, where lifetime excess risk of lung cancer via inhalation can exceed one per 1000 people (Wang et al., 2012). For example, the naphthalene mixing ratio was 3 ppb in a diluted aircraft plume, and a few hundred ppt over Hong Kong airport. In Houston and New York City,

USA, the naphthalene mixing ratio was generally below 100 ppt, but occasionally exceeded 250 ppt in the summers of 2000 and 2001 (Martinez et al., 2004). The transport of PAH compounds from Asia may elevate the mixing ratio of PAH compounds up to 100 pg m^{-3} over the northern Pacific Ocean and several pg m^{-3} over western North America, on an annual average basis. A study along a traffic artery in China revealed that PAH compounds with four and five rings were removed by dry deposition, and PAH compounds with three rings were mainly removed by wet deposition, while PAH compounds with two rings were not efficiently removed by deposition. Particles on the traffic artery were enriched with PAH compounds, and the concentration of 12 identified PAH compounds in dusts was 648 ng g^{-1} on average. Due to traffic dust, the PAH concentration in runoff water was 3272 ng L^{-1}, which was significantly higher than in rainwater with a PAH concentration of 2157 ng L^{-1} (Wang and Zhu, 2005).

Among a number of identified PAH compounds, benzo[a]pyrene (BaP), a five-ring PAH compound with molecular formula $C_{20}H_{12}$ and structure shown in Figure 10.4, is the most potent carcinogen with unit risk of 0.087 at 1 μg m^{-3}. BaP crystals are light yellow and needlelike, with a faint odor. BaP digestion by humans and animals produces other forms of toxic compounds; together, they combine with human DNA to form BaP–DNA adducts that may interfere with or alter DNA replication and thus increase the risk of several forms of cancer. According to the US EPA, children and pregnant women may inhale BaP emitted from mobile sources, industrial sources (e.g. coke ovens, metal processing plants), first-hand and second-hand cigarette smoking; a diet of grilled or charcoal-broiled foods and children's hand-to-mouth activities may lead to oral exposure. A number of European countries set limits for ambient concentration of BaP on an annual average basis, as shown in Table 10.3. Due to effective control, BaP concentration in central London decreased from 46 ng m^{-3} in 1950, to 4 and 0.8 ng m^{-3} in 1972 and 1985, on an annual average basis. Annual mean BaP concentration in Germany was 2 ng m^{-3} in 1978–1979 in urban areas dominated by traffic emissions, and ranged from 2.4 to 13 ng m^{-3} in areas affected by cokery and in urban and traffic areas in 1985; by 1999, annual mean BaP concentration decreased to below 1 ng m^{-3} in urban and traffic areas (European Commission, 2001). BaP concentration in metropolitan Los Angeles and urban New Jersey was 0.1–0.3 ng m^{-3} in summer and 0.6–1.3 ng m^{-3} in winter in 1975–1982; in rural areas, BaP concentration was much lower. China also regulates BaP with an ambient air quality standard of 1 and 2.5 ng m^{-3} respectively on annual and 24-hour average bases.

Table 10.4 lists 65 selected toxic aromatic species detected in ambient air. Their safe levels and unit risks are not listed here, and interested readers are referred to the Integrated Risk Information System of the US EPA.

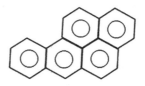

Figure 10.4 Molecular structure of benzo[a]pyrene ($C_{20}H_{12}$).

Table 10.3 Ambient BaP standards in a number of nations

	ng m^{-3}
Belgium	1
China	1
Croatia	2
France	0.7
Germany	1.3
Italy	1
Netherlands	1
Sweden	0.1
UK	0.25

10.3   Other air toxins

Table 10.5 lists an additional 103 inorganic and non-aromatic organic toxic air pollutants. These compounds are mostly hydrocarbons containing the functional groups –OH, –C=O, –C(O)OH, –NH$_2$, –CN, –X, and –S. Inorganic toxins include CS$_2$, COS, CCl$_4$, Cl$_2$, HF, HCl, H$_2$S, P, TiCl$_4$, and inorganic cyanides. Radon, other radionuclides, mixtures of fine mineral fibers and coke oven emissions are also toxic.

10.4   Pollutants in built environments

10.4.1   Transport from outdoors

Indoor air pollution has been investigated concurrently with outdoor air pollution. When outdoor air is polluted, ventilation transports pollutants from outdoors to the indoor environment. For pollutants transported into the indoor environment, the major differences between indoor and outdoor environments are: (1) the deposition surface area in the indoor environment is relatively large; (2) photochemical reactions usually cease but chemical reactions continue to proceed in the indoor environment. Thus, the effects of outdoor air pollution on the indoor environment depend on ventilation rate and the mixing ratios and lifetimes of outdoor air pollutants. For example, when outdoor O$_3$ is elevated, indoor cleaning using limonene may elevate SOAs, mainly organic nitrates, to 20 µg m^{-3} (Carslaw et al., 2012).

10.4.2   Natural indoor sources

There are unique sources of air pollution in the indoor environment, even when the outdoor environment is clean. Radon gas has been identified to exceed safe levels in poorly ventilated houses with floors not well sealed, as it is naturally emitted from

Table 10.4 Sixty-five selected toxic aromatic species in ambient air

CAS	HAP aromatics	CAS	HAP aromatics
	Polycylic organic matter	100027	4-Nitrophenol
	Xylenes	101688	Methylene diphenyl diisocyanate
95954	2,4,5-Trichlorophenol	118741	Hexachlorobenzene
88062	2,4,6-Trichlorophenol	100414	Ethylbenzene
108883	Toluene	122667	1,2-Diphenylhydrazine
95807	2,4-Toluene diamine	534521	4,6-Dinitro-o-cresol and salts
584849	2,4-Toluene diisocyanate	51285	2,4-Dinitrophenol
95534	o-Toluidine	121142	2,4-Dinitrotoluene
8001352	Chlorinated camphene	60117	Dimethyl aminoazobenzene
120821	1,2,4-Trichlorobenzene	106467	1,4-Dichlorobenzene
100425	Styrene	1319773	Cresols/Cresylic acid
96093	Styrene oxide	108907	Chlorobenzene
1746016	2,3,7,8-Tetrachlorodibenzo-p-dioxin	120809	Catechol
1336363	Polychlorinated biphenyls	71432	Benzene
82688	Pentachloronitrobenzene	92875	Benzidine
87865	Pentachlorophenol	98077	Benzotrichloride
91203	Naphthalene	100447	Benzyl chloride
98953	Nitrobenzene	92524	Biphenyl
92933	4-Nitrobiphenyl	92671	4-Aminobiphenyl
53963	2-Acetylaminofluorene	98862	Acetophenone
117817	Bis(2-ethylhexyl)phthalate	510156	Chlorobenzilate
119904	3,3-Dimethoxybenzidine	85449	Phthalic anhydride
84742	Dibutylphthalate	62737	Dichlorvos
91941	3,3-Dichlorobenzidene	119937	3,3'-Dimethylbenzidine
131113	Dimethylphthalate	91225	Quinoline
1582098	Trifluralin	114261	Propoxur
56382	Parathion	98828	Cumene
101779	4,4'-Methylenedianiline	101144	4,4-Methylene bis(2-chloroaniline)
532274	2-Chloroacetophenone	133904	Chloramben
63252	Carbaryl	72435	Methoxychlor
123319	Hydroquinone	90040	o-Anisidine
62533	Aniline	302012	Hydrazine
121697	N,N-Dimethylaniline		

soil. In nature, radon gas is produced from the radioactive decay of uranium in soil, and decays to form Pb in the air, as shown in Figure 10.5.

Thus, natural air contains trace amounts of Pb. The number of deaths from lung cancer due to indoor radon pollution in the USA was estimated to be 20,000 in 2010, using an average Rn decay rate of 1.3 pCi L^{-1} or 0.048 decays per s · L or 15,842 atoms L^{-1}. The ambient Rn concentration in the USA is about 30% of indoor Rn concentration.

Table 10.5 One hundred and three selected inorganic and non-aromatic
organic toxic air pollutants

CAS	Name	CAS	Name
75070	Acetaldehyde	822060	Hexamethylene-1,6-diisocyanate
60355	Acetamide	680319	Hexamethylphosphoramide
75058	Acetonitrile	110543	Hexane
107028	Acrolein	7647010	Hydrochloric acid
79061	Acrylamide	7664393	Hydrogen fluoride
79107	Acrylic acid	7783064	Hydrogen sulfide
107131	Acrylonitrile	78591	Isophorone
107051	Allyl chloride	58899	Lindane (all isomers)
542881	Bis(chloromethyl)ether	108316	Maleic anhydride
75252	Bromoform	67561	Methanol
106990	1,3-Butadiene	74839	Methyl bromide
156627	Calcium cyanamide	74873	Methyl chloride
105602	Caprolactam	71556	1,1,1-Trichloroethane
133062	Captan	78933	2-Butanone
75150	Carbon disulfide	60344	Methyl hydrazine
56235	Carbon tetrachloride	74884	Methyl iodide
463581	Carbonyl sulfide	108101	Methyl isobutyl ketone
57749	Chlordane	624839	Methyl isocyanate
7782505	Chlorine	80626	Methyl methacrylate
79118	Chloroacetic acid	1634044	Methyl tert-butyl ether
67663	Chloroform	75092	Dichloromethane
107302	Chloromethyl methyl ether	79469	2-Nitropropane
126998	Chloroprene	684935	*N*-Nitroso-*N*-methylurea
334883	Diazomethane	62759	*N*-Nitrosodimethylamine
132649	Dibenzofurans	59892	*N*-Nitrosomorpholine
96128	1,2-Dibromo-3-chloropropane	75445	Phosgene
111444	Bis(2-chloroethyl)ether	7803512	Phosphine
542756	1,3-Dichloropropene	7723140	Phosphorus
111422	Diethanolamine	1120714	1,3-Propane sultone
64675	Diethyl sulfate	57578	β-Propiolactone
79447	Dimethyl carbamoyl chloride	123386	Propionaldehyde
68122	Dimethyl formamide	78875	1,2-Dichloropropane
57147	1,1-Dimethyl hydrazine	75569	Propylene oxide
77781	Dimethyl sulfate	75558	1,2-Propylenimine
123911	1,4-Diethyleneoxide	106514	Quinone
106898	l-Chloro-2,3-epoxypropane	79345	1,1,2,2-Tetrachloroethane
106887	1,2-Epoxybutane	127184	Tetrachloroethylene
140885	Ethyl acrylate	7550450	Titanium tetrachloride
51796	Ethyl carbamate	79005	1,1,2-Trichloroethane
75003	Chloroethane	79016	Trichloroethylene
106934	Dibromoethane	121448	Triethylamine
107062	1,2-Dichloroethane	540841	2,2,4-Trimethylpentane

Table 10.5 One hundred and three selected inorganic and non-aromatic organic toxic air pollutants—*cont'd*

CAS	Name	CAS	Name
107211	Ethylene glycol	108054	Vinyl acetate
151564	Ethylene imine (aziridine)	593602	Vinyl bromide
75218	Ethylene oxide	75014	Vinyl chloride
96457	Ethylene thiourea	75354	1,1-Dichloroethylene
75343	1,1-Dichloroethane		Coke oven emissions
50000	Formaldehyde		Cyanide compounds
76448	Heptachlor		Glycol ethers
87683	Hexachlorobutadiene		Fine mineral fibers
77474	Hexachlorocyclopentadiene		Radionuclides (including radon)
67721	Hexachloroethane		

^{238}U(4.5e9 y) → $\quad$ ^{234}Th(24.5 d) → $\quad$ ^{234}Pa(1.14 m) → $\quad$ ^{234}U(2.3e5 y)

$\qquad\qquad\qquad\qquad\qquad\qquad\qquad\qquad\qquad\qquad\qquad\qquad$ ↓

^{218}Po(3.05 m) $\qquad$ ← ^{222}Rn(3.82 d) $\quad$ ← ^{226}Ra(1.6e3 y) $\quad$ ← ^{230}Th(8.3e4 y)

↓

^{214}Pb(26.8 m) → $\quad$ ^{214}Bi(19.7 m) → $\quad$ ^{214}Po(164 µs) → $\quad$ ^{210}Pb(22.3 y)

$\qquad\qquad\qquad\qquad\qquad\qquad\qquad\qquad\qquad\qquad\qquad\qquad$ ↓

$\qquad\qquad\qquad$ ^{206}Pb(stable) $\qquad$ ← ^{210}Po(138 d) $\qquad$ ← ^{210}Bi(5.01 d)

Figure 10.5 Natural productions of Rn and Pb from radioactive uranium.

10.4.3 Anthropogenic indoor sources

Besides Rn and Pb, other major sources of pollution identified in the indoor environment include cigarette smoking, mixtures of fine mineral fibers, emissions from kerosene and ethanol lamps, wood-burning fireplaces, coal, coke and biomass ovens as well as stir-fry and deep-fry cooking. Building materials and applications to carpets and vinyl may also emit significant amounts of pollutants in the indoor environment (Shin et al., 2012). For example, asbestos had been used as roofing material in the USA before it was identified as a carcinogen for human lungs a few decades ago. Fire-resistant organic fibers contain halogens, which are used in sofas and walls of many woodhouses and may evaporate into the indoor environment during hot summers if not well sealed. Paints and coatings of furniture, electronics, and toys may emit significant amounts of HCHO and other organic compounds into commercial, office, and residential indoor air. A recent comparative investigation suggested that indoor

VOC pollution could cause excess lifetime cancer risk of 23–210 per million people in two cities in China and Japan in 2006–2007, depending on smoking incidence and the use of paints or wallpapers in houses (Ohura et al., 2009). Children's habits, such as chewing and licking toys and hands, increase their exposure to indoor risks.

10.4.4 Indoor humidity

Humidity in the indoor environment can cause severe health problems. Low humidity with high sulfur concentration may result in respiratory irritation. Excess moisture may cultivate germs that are carcinogenic for liver and skin, besides its direct effect on skin temperature and thus on the health of joints and bones. Leprosy is still present in the coastal USA and parts of China, and its association with bacterial infection was recorded in the literature thousands of years ago. Most incidents of leprosy in modern times are associated with human-to-human transmission of the bacteria *Mycobacterium leprae* via nasal droplets in the air. Application of organic paints with low toxicity to coat inside walls of woodhouses is common practice in many places, to maintain the indoor environment. However, according to the US EPA, molds may grow on virtually any organic substance, such as wood, paper, carpet, foods, and insulation, with the presence of moisture and oxygen (Figure 10.6). While it is challenging to eliminate all molds and mold spores in the indoor environment, mold growth may be effectively reduced by controlling indoor moisture.

10.4.5 Inside vehicles

Besides housing, inhalation exposure within vehicles has increased and become a more important risk for many people in built environments. On one hand, evaporation of fuel, engine oil, and lubricants as well as from fabrics may contribute to

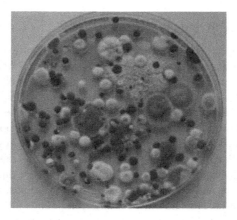

Figure 10.6 Indoor molds.
Picture obtained from World Health Organization (2012).

personal exposure in vehicles; and increased ventilation may reduce passengers' exposure to toxic vapors evaporated by the vehicle. On the other hand, the background air on roads, especially on congested highways, contains concentrated pollutants that are inevitably transported inside vehicles. The resulting exposure in a vehicle on congested highways may exceed the risk from cigarette smoking.

10.5 Techniques for simulating air toxins

10.5.1 Dispersion modeling

Simulation of air pollution at toxic levels has been typically conducted by models that include detailed emission, transport, and deposition processes, as well as corresponding chemical reactions. As many air toxins react slowly before being dispersed and falling to non-toxic levels, dispersion models with detailed representations of local topography and building structures but simplified representations of chemical transformation may be applied in hotspot areas, such as ports, train stations, highways, power plants, refineries, oil fields, landfills, and other factories with significant amounts of emissions.

10.5.2 Chemical transport modeling

For air toxins that do not have a safe level, such as mercury, three-dimensional chemical transport models have been applied to simulate regional and global distributions of mercury species and their responses to changes in anthropogenic emissions. Detailed simulations have also been conducted for radon, organic pesticides, and PAH compounds in regional and global atmospheres (e.g. Friedman and Selin, 2012).

The transformation of a subset of PAH compounds with two and three rings into secondary organic aerosols in Texas, USA was simulated by a regulatory air quality model, US EPA CMAQ, in a recent study (Zhang and Ying, 2012). The oxidation of nine PAH compounds by OH was explicitly represented, and all other PAH compounds were represented by a general model species, "PAH2". The "PM yield" method was used to calculate the rates of formation of three secondary aerosol species, namely non-volatile (NVp), low-volatility (LVg), and high-volatility (HVg) secondary organic aerosol compounds, from the OH oxidation of PAH (R10.5.1). The two semi-volatile secondary organic aerosol compounds (LV and HV) were partitioned between gas and aerosol phases following equilibrium reactions (R10.5.2) and (R10.5.3), using estimated molecular weights, partitioning coefficients, and enthalpies of vaporization, which determine the temperature dependence of partitioning coefficients or equilibrium constants.

$$PAH + OH \rightarrow Products + c_1 \cdot NVp + c_2 \cdot LVg + c_3 \cdot HVg \qquad (R10.5.1)$$

$$LVg \leftrightarrow LVp \qquad (R10.5.2)$$

$$HVg \leftrightarrow HVp \qquad (R10.5.3)$$

Oxidation by other oxidants, such as O_3, O and NO_3, and the partitioning of PAH compounds between gas and aerosol phases were not considered in the study due to lack of reaction product and coefficient data. When such data is available, three reactions in the form of (R10.5.1)–(R10.5.3) may be used to describe the oxidation reaction of a PAH compound if the "PM yield" method is used. Simulations over a 2-week period in summer suggested that wildfires resulted in the peak concentration (20 ng m^{-3}) of secondary organic aerosol from PAH compounds in Texas, while emissions from solvent evaporation, vehicle exhausts, and industrial production resulted in PAH-originated SOA concentrations up to 2, 1.4, and 1.6 ng m^{-3} respectively in various locations.

Summary

In this chapter, a number of chemical and biological toxins in ambient air and built environments have been discussed. Important levels for exposure assessment have been defined. The sources, risks, and safe levels of 11 toxic metals have been tabulated, with elaborations of Be, Cd, Pb, and Hg. Sixty-five toxic aromatic compounds are introduced, following the sequence of one-ring aromatics, poly-chlorinated biphenyls, and polycyclic aromatic compounds. An additional 103 inorganic and organic air toxins as well as important toxic mixtures have been tabulated. Sources of air pollutants inside houses and vehicles have been discussed. Techniques for simulating air toxins in hotspots and on regional and global scales have been outlined; a simplified representation of OH oxidation of PAH compounds has been elaborated.

Exercises

1. Survey the literature to find the major form of the 11 metals listed in Table 10.2. Write several major sources of a handful of the metals in your neighborhood, and analyze how their emissions could be reduced feasibly and cost-effectively.
2. Choose a PCB molecule and write a set of chemical reactions to describe its transformation into dicarboxylic acids. Based on this, estimate the reaction products and coefficients for reactions (R10.5.1)–(R10.5.3) for the PCB.
3. Build a box model to simulate the mixing ratio of HCHO in a house, using the following data. The outdoor air contains 10 ppbv HCHO and 0.1 pptv OH. The ventilation rate of the house is such that the indoor air of the house is completely replaced twice a day. The emission rate of HCHO from indoor sources is 1 ppbv a day, and the deposition rate of HCHO on indoor surfaces is 0.5 ppbv a day. Photolysis of HCHO indoors is negligible. Starting from the initial value of 10 ppbv HCHO at 7 a.m., calculate the hourly HCHO level until 7 p.m. using a computer program or analytical formula.
4. Find the formula of a Gaussian dispersion model in the fumigation scenario, and estimate the maximum emission rate of As allowing for a power plant situated in a valley with a volume of 1 km × 1 km × 200 m for the airshed during a 3-day high-pressure system. Use the safe level listed in Table 10.2. Assume that the initial value and deposition loss

are negligible. The stack height is 100 m and the stack diameter is 2 m, while the flow velocity is 5 m s^{-1}.

5. Repeat the calculations in problem 4 for the PCB in problem 2, by assuming a safe level of 10 ng m^{-3} for the PCB. Also assume that the power plant in the valley was replaced by a chemical plant. The stack height of the chemical plant is 10 m and the stack diameter is 0.5 m, while the flow velocity is 1 m s^{-1}. Estimate the SOA level from the PCB.

Part Three

Analysis

11 Corroborative analysis tools

As modeling techniques advance, a spectrum of chemical phenomena in the air has been simulated. During the process of evaluating model results against observations, various problems have emerged. To extract as much information from model results and observational data as possible, in order to understand the mechanisms of formation of chemical phenomena in the air and to simulate chemical responses to changes in human activities, a handful of tools have been applied to corroborate analyses. The details are elaborated below.

11.1 Indicator ratios for ozone

The preferred method of surface O_3 pollution control is to reduce the emission of limiting precursors, either NO_x or ROGs, whenever feasible and cost-effective.

Chemical Modeling for Air Resources. http://dx.doi.org/10.1016/B978-0-12-408135-2.00011-2

However, it is not intuitive to quantify which method of precursor control is most effective at a specific location. A handful of so-called indicator ratios, which may be measured at a station, have been developed to empirically assess the O_3 control strategy for a location; the replication of observed values of similar indicator ratios by atmospheric chemical models is a further challenge (e.g. Xiao et al., 2007). The potential application and ranges of selected indicator ratios have been evaluated using a three-dimensional air quality model, the Comprehensive Air Quality Model with Extensions (CAMx, Environ, California, USA), using a pre-existing base case for central California.

11.1.1 Theoretical basis

Assume an air parcel sampled at a station near an oil field originated from a relatively pristine environment with little background pollution 3 days ago. Before arrival at the station, the air parcel picked up some ship emissions and passed through a metropolitan area with emissions of NO_x and anthropogenic ROGs, and an agricultural region with biogenic emissions from crops and anthropogenic emissions from vehicles. Hourly averaged mixing ratios of O_3, NO_x, and non-methane hydrocarbons (NMHCs) have been routinely measured at the station; during episodic weather conditions, hourly averaged mixing ratios of NO_y species, H_2O_2, and organic peroxides have also been measured.

To reduce O_3 pollution at the station during episodic days, it is important to understand which precursor limits O_3 production in the air parcel. Traditionally, the EKMA diagram from box modeling depicts three regions for O_3 control: (a) the NO_x-limited region with a ratio of ROGs/NO_x greater than ~10; (b) the ROG-limited region with a ratio of ROGs/NO_x less than ~6; (c) the region where O_3 is sensitive to both NO_x and ROGs. However, as individual reactive organic gases produce O_3 at different efficiencies and the composition of ROG mixture changes with time and location, alternative indicator ratios are desirable.

The ratios of $[O_3]/[NO_2]$, $[O_3]/[NO_y]$, $[O_3]/[NO_x]$, [total peroxide]/[HNO_3], [HCHO]/[NO_y], and the extent of reaction have been investigated for their suitability to distinguish whether O_3 pollution is sensitive to NO_x, ROGs, or both. As an air parcel travels to a station, photochemical reactions that increase O_3 pollution may also increase the mixing ratios of intermediate and terminal products from ROG degradation and NO_x oxidation. Thus, concurrent measurements of the mixings ratios of O_3 and photochemical reaction products contain cumulative information for an air parcel during the journey, which may be used to distinguish the photochemical regimes for O_3 formation. For example, if the ratio of $[O_3]/[NO_2]$ is relatively large, it may indicate that the formation of O_3 pollution in the air parcel was NO_x limited, using the EKMA diagram as reference. As measurements of NO_y and NO_x are more convenient than that of NO_z, the temptation exists to use the ratios $[O_3]/[NO_y]$ and $[O_3]/[NO_x]$. The ratio of [total peroxide]/[HNO_3] in an air parcel with O_3 pollution reflects the relative importance of radical terminations in low- and high-NO_x conditions respectively. In low-NO_x or NO_x-limited conditions, the ratio is expected to be large; the opposite is true in high-NO_x or ROG-limited conditions. The ratio

[HCHO]/[NO$_y$] is similar to [total peroxide]/[HNO$_3$] but with other difficulties regarding measurements.

The extent of reaction (ER) was introduced for smog chamber O$_3$ data analysis by Johnson (1984), and refined for ambient O$_3$ data analyses by Hess et al. (1992a–c) and Blanchard et al. (1999). The ER is defined as:

$$ER = SP(t)/SP(max) \tag{F11.1.1}$$

$$SP(t) = ([O_3] - [NO])_t - ([O_3] - [NO])_0 + \sum_{\tau=0}^{t} E(NO)_\tau + \sum_{\tau=0}^{t} D(O_3)_\tau \tag{F11.1.2a}$$

$$SP(t) \approx \int_0^t (k_1[HO_2][NO] + k_2[RO_2][NO] - k_3[O_3][OH] - k_4[O_3][HO_2])d\tau \tag{F11.1.2b}$$

$$SP(t) \approx A\{[NO_x(0)] - [NO_x(t)]\}^B \tag{F11.1.2c}$$

$$SP(max) = A \cdot [NO_x(0)]^B \tag{F11.1.3}$$

where ER at time t in formula (F11.1.1) is defined as the ratio of cumulative O$_3$ production at time t to the maximum O$_3$ production by cumulative NO$_x$ in the system. Formula (F11.1.2) shows three forms of expressions for "smog production" (SP). The four reactions involved in formula (F11.1.2b) are:

$$NO + HO_2 \rightarrow NO_2 + OH \tag{R11.1.1}$$

$$NO + RO_2 \rightarrow NO_2 + RO \tag{R11.1.2}$$

$$O_3 + OH \rightarrow O_2 + HO_2 \tag{R11.1.3}$$

$$O_3 + HO_2 \rightarrow 2O_2 + OH \tag{R11.1.4}$$

The values of parameters A and B in formulae (F11.1.2c) and (F11.1.3) depend on the units of NO$_x$ as well as the photochemical environment; when the units of NO$_x$ are ppmv, $A \approx 2$ and $B \approx 2/3$. When ER is close to 1, O$_3$ production is NO$_x$ limited; when ER is close to 0, O$_3$ production is ROG limited (Blanchard and Fairley, 2001).

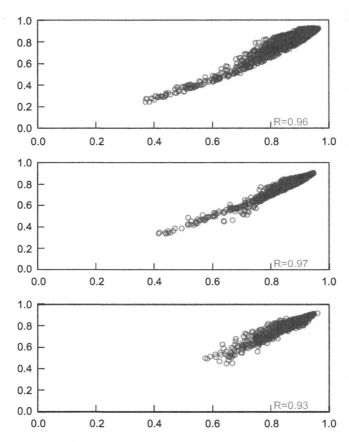

Figure 11.1 The extent of reaction vs. $(NO_z/NO_y)^B$ in grids with simulated ozone excess in the Central California Ozone Study domain at 1–5 pm (PDT) during July 31–August 2, 2000.

The ER values for transition from an ROG-limited to an NO_x-limited regime, and for regimes that show lack of benefit of NO_x to benefit of NO_x, may vary with time and location. For ambient data analysis, ER defined by formulae (11.1.2c) and (11.1.3) may be closely correlated with the observed $(NO_z/NO_y)^B$ at a station during episodic periods, as shown in Figure 11.1 for an O_3 episode observed in central California from July 31 to August 2, 2000. The episode is also used to analyze other indicators, as discussed below.

11.1.2 A numerical experiment

To identify useful ranges of indicator ratios in a region, detailed photochemical air quality modeling needs to be conducted. Liang et al. (2006) reported such an effort for

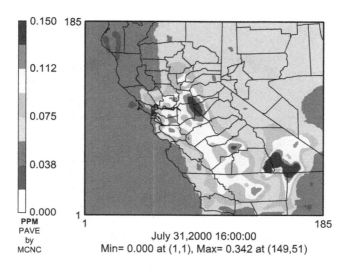

Figure 11.2 Simulated surface hourly average O_3 mixing ratio in central California, USA.

assessing O_3 pollution in central California during a 3-day O_3 episode observed in the summer of 2000.

The observed O_3 episode was simulated by a state-of-science air quality modeling system developed for the Central California O_3 Study program, which was part of the ~30 million CRPAQS program in California. The modeling system consisted of CAMx, an emission modeling ensemble, and a meteorological modeling ensemble. The simulated surface O_3 distribution in central California is shown in Figure 11.2 for the base case, which captured observed peak values and peak areas fairly closely.

To obtain ranges for indicator ratios in central California, two sensitivity runs were conducted with emissions of NO_x and ROG both reduced by 25%. Thus, each indicator ratio in the base case is coupled with simulated changes in O_3 mixing ratio due to NO_x and ROG reductions respectively. At most three ranges may be present for an indicator ratio: a range where O_3 decreases unequivocally due to reduced emission of a precursor, a range where O_3 increases unequivocally due to reduced emission of a precursor, and a range where O_3 change is statistically insignificant. These ranges may be calculated precisely from model results.

Figure 11.3 shows a scatterplot of the simulated change in 8-hour ozone concentration: (a) as a function of base-case 8-hour ozone concentration in grids with 8-hour average ozone exceeding 80 ppb, and (b) due to emission controls as a function of $[O_3]/[NO_z]$. It is seen that ROG control was always beneficial for 8-hour average O_3 in central California during the episode, while NO_x control resulted in benefit in some grids and lack of benefit in other grids. The indicator ratio $[O_3]/[NO_z]$ showed three ranges in the figure.

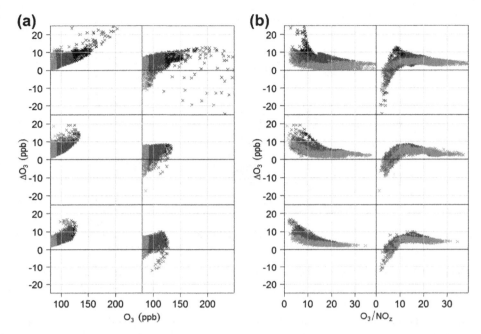

Figure 11.3 Illustration of an indicator ratio in central California. (a) The change in 8-hour ozone concentration as a function of base-case 8-hour ozone concentration in grids with 8-hour average ozone exceeding 80 ppb. (b) The change in 8-hour ozone concentration due to emission controls as a function of $[O_3]/[NO_z]$. The upper panels are for July 31, 2000 and the central panels for August 1, 2000. The lower panels are for August 2, 2000. The left panels are for 25% reduction in VOC emissions and the right panels are for the same reduction for NO_x. ΔO_3 is the difference in $[O_3]$ (ppb) between base and controlled cases, so that it is positive when precursor controls are beneficial and negative when they lack benefit. Red circles denote 8-hour ozone in the baseline simulation between 80 and 90 ppb, and blue circles larger than 90 but less than 120 ppb. Black circles denote 8-hour ozone concentrations larger than 120 ppb in the baseline simulation.
From Liang et al. (2006a).

11.1.3 Ranges of indicator ratios

For each indicator ratio, there are two ranges that may be useful: (a) the range that distinguishes benefit and lack of benefit of NO_x control; (b) the range that distinguishes between NO_x- or ROG-limited scenarios. Table 11.1 lists these two ranges in columns 3 and 4 for five indicator ratios in central California, as described above. Results from another study in the eastern USA are listed in the last column (Sillman, 2002).

   It is seen that indicator ratios $[O_3]/[NO_y]$, $[O_3]/[NO_z]$, and $[ROOH]/[HNO_3]$ are monotonic, and thus may be useful for determining whether O_3 pollution is NO_x or ROG limited and whether NO_x control is beneficial or not at a station. Indicator ratio $[HCHO]/[NO_y]$ and the extent of reaction (ER) are not monotonic; their usefulness

Table 11.1 Ranges of indicator ratios

Indicator ratio	Monotonic	Range for NO_x	Transition range	Reference range
O_3/NO_y	Yes	3.3–7.4	5.1–40	4–8
O_3/NO_z	Yes	8.8–15	12–48	5–10
$ROOH/HNO_3$	Yes	0.18–0.40	0.34–3.6	0.3–0.5
$HCHO/NO_y$	No	0.12–0.76	0.15–0.90	0.2–0.4
ER	No	0.34–0.81	0.43–0.98	0.6–0.8

must be scrutinized. In combination with local measurements and intermittent control measures, such as the spare-the-air program implemented in the California San Francisco Bay area of the USA, indicator ratios may help prevent O_3 excess in the short term.

11.2   Charge balance for PM

The ratio of cation to anion equivalent concentrations and the "excess NH_3" indicator, to be defined below, may be calculated directly from observations. These two parameters may be used to evaluate modeling results for practical applications.

11.2.1   Ionic information

Inhalable PM contains a number of soluble ions, the major ions presently being ammonium, sulfate, and nitrate over most land areas. In ideal conditions, cationic and anionic equivalences are the same. However, ambient measurements have rarely captured the ionic balance, as shown in Table 11.2.

Table 11.2 Selected ionic information of PM at global stations

NH_4^+	SO_4^{2-}	NO_3^-	Other (C–A)	C/A ratio	Location, year	Unit	Ref.
6.11	4.52		−0.81	1.13	China, 2006–9	ueq mg^{-1}, PM$_{10}$	a
0.11	0.13	0.02	0.00	0.74	USA, 2001–2	ueq m^{-3}, PM$_{2.5}$	b
0.09	0.07	0.04	0.00	0.84	Swiss, 1998–9	ueq m^{-3}, PM$_{2.5}$	c
0.12	0.12	0.00	−0.01	1.05	Italy, 2007	ueq m^{-3}, PM$_{2.5}$	d
0.12	0.13	0.02	0.03	1.14	Thailand, 2006	ueq m^{-3}, PM$_{2.5}$	e
112	137	17	0.69	0.77	Germany, 1985	neq m^{-3}, PM$_1$	f
0.90	1.03	0.25	0.09	1.08	Taiwan, 2004–5	relative, PM$_{2.5}$	g
0.72	3.54	0.26	0.52	1.21	Mexico, 1997–8	relative, PM$_{2.5}$	h
0.03	0.10	0.03	0.08	1.05	India, 2000–2	ueq m^{-3}, PM	i
0.30	0.21	0.10	0.00	0.96	Asia, 2001–4	ueq m^{-3}, PM$_{2.5}$	j

a. Huo et al. (2012); b. Rees et al. (2004); c. Hueglin et al. (2005); d. Perrone et al. (2011); e. Li et al. (2012); f. Mehlmann and Warneck (1995); g. Wen and Fang (2007); h. Vega et al. (2001); i. Rastogi and Sarin (2005); j. Kim et al. (2006).

The ratio of cationic to anionic equivalences ranged from 0.74 to 1.21 in measurements at global locations. As not all ions were measured at all locations all the time, the difference between cationic and anionic equivalences may simply reflect the presence of unmeasured ions. For example, inorganic PM ions reported in recent literature include K^+, Na^+, Mg^{2+}, Ca^{2+}, NH_4^+, H^+, OH^-, SO_4^{2-}, HCO_3^-, CO_3^{2-}, NO_3^-, Cl^-, and F^-, and organic PM ions reported in recent literature include $HC(O)O^-$, $CH_3C(O)O^-$, $C_2O_4^{2-}$, isoprene oxidation product (2-methylglycerate), α-pinene oxidation products (*cis*-pinonate, pinate, 3-methyl-1,2,3-butane tricarboxylate, 3-hydroxyglutarate, 3-hydroxy-4,4-dimethylglutarate, 2-hydroxy-4,4-dimethylglutarate, 3-acetyl hexanedioic acid, 2-hydroxy-4-ispropyladipate), and sesquiterpene oxidation product (β-caryophyllinate).

11.2.2 Precursor limitation

In California and some other areas where $PM_{2.5}$ pollution is mainly due to ammonium nitrate, it is useful to have an indicator to determine if the formation of $PM_{2.5}$ nitrate is ammonia or nitrate limited. Blanchard et al. (2000) defined the "excess NH_3" indicator as:

$$\text{Excess } NH_3 = [NH_3] + [NH_4^+] - 2[SO_4^{2-}] - [NO_3^-] - [HNO_3] - [HCl]$$
$$+ 2[Ca^{2+}] + 2[Mg^{2+}] + [Na^+] + [K^+] - [Cl^-]$$

(F11.2.1)

where all concentrations are in units of $\mu mol\ m^{-3}$. When the "excess NH_3" is positive, $PM_{2.5}$ ammonium nitrate formation is limited by nitrate, and the control of NO_x is expected to be more effective than controlling NH_3 for reducing $PM_{2.5}$ ammonium nitrate. When there is no excess NH_3, $PM_{2.5}$ ammonium nitrate formation is limited by NH_3. In southern and central California, $PM_{2.5}$ ammonium nitrate formation was NH_3 rich during episodes.

11.3 Factor analysis

Chemical models simulating the modern atmosphere are complex due to detailed representations of various processes afforded by advanced computing power. In a typical three-dimensional simulation, hundreds of gas-phase photochemical reactions and heterogeneous reactions in aerosol and cloud droplets, line-by-line radiative transfers, emissions and depositions of dozens of species, and three-dimensional atmospheric transport processes need to be solved thousands of times for millions of boxes in a model domain.

Programmers and modelers make human errors from time to time in typing, formulations, configurations, and interpretations. Thus, simulated results need to be evaluated as much as possible with observational constraints, for practical applications such as new source reviews and ambient air quality attainment demonstrations.

From a chemical perspective, there are three levels of evaluations with increasing degrees of difficulty: (a) comparing a single compound in a model and via observations; (b) comparing an indicator ratio in a model and via observations; (c) comparing a factor characterized by many variables and parameters in a model and via observations.

11.3.1 Methods

Factor analysis may be conveniently conducted using a statistical software package, such as the commercial S-plus software or the free R software (http://www.r-project.org).

Factor analysis refers to a group of statistical methods that extract a small number of important vectors to explain the major features of complex variations in a relatively large database. An input data matrix $A[V, S]$ may be approximated using factor analysis as the product of two smaller matrices $B[V, p]$ and $C[p, S]$, namely:

$$A[V, S] \approx B[V, p] \times C[p, S] \qquad (F11.3.1)$$

where p is much smaller than V and S. To solve for B and C from A, two major methods have been applied in chemical data analyses. The first method uses measurements to estimate B and then solve C; the second method solves various combinations of B and C, and provides the best estimate of a pair of B and C using certain criteria. Representative models for the first method include the Chemical Mass Balance method (Winchester and Nifong, 1971; Liang and Wang, 1990), and the non-negative matrix factorization with optimization (NMFROC) method (Liang and Fairley, 2006). Representative models of the second method include the empirical orthogonal function (EOF), the positive matrix factorization (PMF) method (Paatero, 1997; Paatero and Hopke, 2003), principal component analysis (PCA), and their derivatives (Thurston and Spengler, 1985; Henry et al., 1991; Thurston et al., 2005). Both methods may be programmed in R or S-plus using built-in functions for matrix operations such as singular value decomposition and solving for a matrix equation with optimized non-negative values.

When an initial B is estimated, C in the formula (F11.3.1) may be solved by using an optimization package in R with non-negative constraints. Then iterations may be performed using formula (F11.3.2) to reduce solution error:

$$C_{pS} = C_{pS} \frac{(B'A)_{pS}}{(B'BC)_{pS}} \qquad (F11.3.2a)$$

$$B_{Vp} = B_{Vp} \frac{(AC')_{Vp}}{(BCC')_{Vp}} \qquad (F11.3.2b)$$

where B' and C' are transposes of B and C respectively. In formula (F11.3.2), the division is applied to each element, and there is no matrix inversion. The above updating rule allows simultaneous update of one row of **C**, followed by one column of **B**. It is provable that this updating rule ensures the Euclidean distance between **A** and (**B** × **C**), defined as the square root of the sum of squared differences of the elements, to be non-increasing. Since the updating rule is multiplicative, the solution matrices are guaranteed to be non-negative during the iteration. It is conceivable that any element of solution matrices initialized with true zero will remain unchanged if numerical techniques are not used. A numerically very small value (e.g. 10^{-100}) may be used to replace true zero in solution matrices for environmental data (Liang and Fairley, 2006).

When there is no reliable data for **B** and **C**, the singular value decomposition method (the "svd" function in R), as shown in formula (F11.3.3), may be used to generate a pair of initial matrices:

$$\mathbf{A} = \mathbf{U} \times \mathbf{D} \times \mathbf{V}' \tag{F11.3.3}$$

where **D** is a diagonal matrix with decreasing eigenvalues, and both **U** and **V** are orthogonal, normalized matrices. A somewhat arbitrary choice has to be made to decide the number of eigenvalues, or the value of p in formula (F11.3.1). Then the corresponding number of columns of **U** and **V** are reduced to p. Thus, initial $\mathbf{B} = \mathbf{U}[V, p]$, and initial $\mathbf{C} = \mathbf{D}[p, p] \times \mathbf{V}'[p, S]$.

The principal component analysis (PCA) makes use of eigenvectors of the correlation matrix of input matrix **A**, to split normalized input matrix **Å** (F11.3.4a) into two matrices, namely an eigenvector matrix $\mathbf{E}[V, V]$ that is also termed PC coefficients matrix, and a PC score matrix ($\mathbf{E}' \times \mathbf{\mathring{A}}$). Eigenvectors with eigenvalues less than 1 are usually discarded, so that only p ($<V$) factors are retained. The absolute PCA method rotates the $\mathbf{E}[V, p]$ matrix with a scheme called "varimax" to reach a final coefficient matrix $\mathbf{E}^*$, and calibrates the corresponding PC score matrix ($\mathbf{S} = \mathbf{E}^{*\prime} \times \mathbf{\mathring{A}}$) to reach the absolute PC score matrix **X**, as shown in formula (F11.3.4b). For factor identification purposes, the correlation between variables and PCs in the samples is calculated to form a PC loading matrix. The absolute PC score matrix $\mathbf{X}[p, S]$ can be used in subsequent regression against variables of interest related to samples, which corresponds to **C**.

$$\mathbf{\mathring{A}}[iv, is] = \frac{\mathbf{A}[iv, is] - \overline{\mathbf{A}[iv]}}{\sigma[iv]}, \quad is = 1:S; \; iv = 1:V \tag{F11.3.4a}$$

$$\mathbf{X}[ip, is] = \mathbf{S}[ip, is] + \sum_{iv=1}^{v} \mathbf{E}^* \frac{\overline{\mathbf{A}[iv]}}{\sigma[iv]}, \quad is = 1:S; \; ip = 1:p \tag{F11.3.4b}$$

11.3.2 Air quality model diagnosis

Factor analysis methods may be applied to corroborate analyses of three-dimensional air quality modeling results. A few recent examples are discussed below, and interested readers are referred to the references therein.

11.3.2.1 Debugging an error

The absolute principal component analysis method and an efficient non-negative matrix factorization method have been used to improve simulations of PM pollution in central California (Liang et al., 2006).

The California Central Valley of the USA experienced severe PM pollution, with peak daily average $PM_{2.5}$ concentration reaching 160 µg m^{-3}, between December 25, 2000 and January 7, 2001. In order to mitigate PM pollution at such a level, a series of air quality modeling was conducted by scientists and engineers at a number of institutions with national and international specialists, as a part of the CRPAQS program. In the beginning simulations gave abnormally high, unrealistic concentrations of fine PM sulfate in the model domain, as shown in Figure 11.4(a).

In order to debug the error, the widely used absolute PCA method and the NMFROC method were employed. For both factor analysis methods, the input data matrix **A** consisted of hourly quantities of emissions, simulated concentrations of trace gases and fine PM components, and important meteorological parameters at three anchor stations of the CRPAQS program from December 25, 2000 to January 7, 2001. The deposition parameters used in the PM model were excluded from the input matrix, and emissions in corresponding grids were used as surrogates for all emissions to keep the input matrix at a manageable size. The resulting input matrix contained 76 variables (Table 11.3) and 1008 hourly samples.

For the absolute PCA method, the input matrix **A** was normalized to have the standard (0, 1) distribution for all variables. In the NMFROC method, steps were taken to convert a raw species from US EPA CMAQ inputs and outputs to a variable in the input matrix **A**: (a) each species was subtracted with its minimum value; (b) each species was then divided by its mean quantity. The conversion had little effect on PM sulfate, but enabled inclusions of meteorological variables, such as wind components and the inverse of the Monin–Obukov length (MOLI), which can be negative.

The absolute PCA method was applied to the input data matrix **A** using the standard normalization and singular value decomposition techniques. Nine principal components (PCs) with eigenvalues larger than 1 were retained and subjected to "varimax" rotation. The coefficients of the nine PCs, together with their eigenvalues and correlation coefficients with variables, also called loadings, in the input matrix, are listed in Table 11.4.

In Table 11.4, Factors 1 and 2 indicate clustering of emissions and primary species. Factor 3 reflects that high ozone was related to rural air with high PAN, NH_3, dust, and low primary anthropogenic PM emissions. Factor 4 reflects that secondary production of organic PM and nitrate was associated with high temperature and H_2O_2. Factor 5 reflects that PM sulfate was associated with PM water content, as well as ammonium, cloud fraction, low temperature and H_2O_2. Factor 6 reflects that ozone and HNO_3 were high in unstable conditions (low MOLI) when PBL and ground radiation was high. Factor 7 associates ALK1 and crustal PM in the air with emissions. Factor 8 associates cloud fraction and rainwater with northwesterly winds. Factor 9 reflects that high H_2O_2 was associated with northern, downward flows. The above factors are consistent with science related to ozone and PM in the model. When PC loading alone is considered,

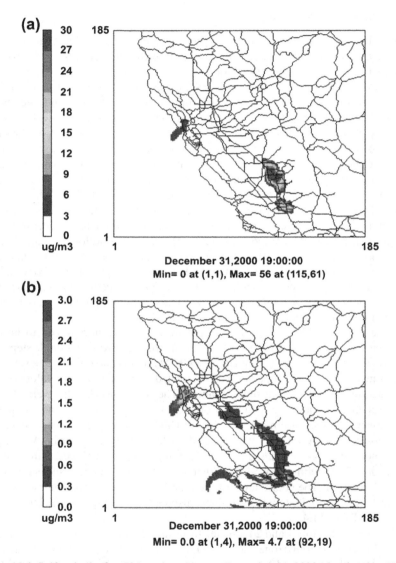

Figure 11.4 Sulfate in the fine PM mode at 11 a.m., December 31, 2000, simulated by US EPA CMAQ. (a) With MM5 rains/clouds. (b) Without rains but with PM liquid water content within 0.05 g m^{-3}. Liang et al. (2006b).

Factor 5 suggests that PM sulfate was highly associated with PM water only. Note that PCA tends to extract average information from samples, and the "varimax" rotation tends to push the results towards extreme rays, the direction of the NMFROC method (Thurston et al., 2005; Liang and Fairley, 2006). The apportionment results suggested that Factor 5 was responsible for the abnormally high PM sulfate.

Table 11.3 Variables used in input matrix **A** for factor analyses

Concentration: Gas	O_3, NO, NO_2, PAN, CO, HNO_3, SO_2, HCHO, HO_2H, ALK1, ALK2, ALK3, ALK4, ALK5, ARO1, ARO2, OLE1, OLE2, ETHENE, TRP1, NH_3
Concentration: $PM_{2.5}$	AH_2O, ANH_4, ANO_3, ASO_4, AEC, AORGPA, A25, AORGA, AORGB, NUMACC, SRFACC
Emission: Gas	NO, NO_2, HONO, SO_2, SULF, HCHO, CCHO, RCHO, ACET, MEK, CH_4, ETHENE, ISOPRENE, TRP1, MTBE, ETOH, ALK1, ALK2, ALK3, ALK4, ALK5, ARO1, ARO2, OLE1, OLE2, CO, NH_3
Emission: $PM_{2.5}$	PMFINE, PEC, POA, PNO_3, PSO_4, D_p, σ
Meteorological: Parameters	PBL, MOLI, RGRND, CFRAC, WWIND, TA, QR, QC, UWIND, VWIND

Note: Variable names follow nomenclatures in the US EPA CMAQ model, except for D_p and σ, which were added for the CRPAQS modeling study. D_p is the geometric mean volume diameter and σ the geometric standard deviation.

Table 11.4 Key coefficients and loadings from absolute PCA with "varimax" rotation

Factor	Eigenvalue	Variables with large values (PC coefficient, high loading)
1	35.0	**Emis.**: NO (0.22, **0.95**), NO_2 (0.22, **0.95**), HONO (0.22, **0.95**), SO_2 (**0.72**), SULF (**0.72**), HCHO (0.21, **0.98**), CCHO (0.18, **0.97**), RCHO (**0.85**), ACET (**0.76**), MEK (**0.91**), C_2H_4 (0.24, **0.99**), ISOP (**0.86**), TRP1 (**0.78**), MTBE (0.22, **0.95**), ALK2 (0.20, **0.97**), ALK3 (**0.92**), ALK4 (0.20, **0.96**), ALK5 (**0.92**), ARO1 (0.21, **0.98**), ARO2 (0.22, **0.98**), OLE1 (0.23, **0.98**), OLE2 (0.24, **0.96**), CO (0.22, **0.96**), PEC (0.21, **0.94**), POA (**0.79**)
2	13.1	**Chem.**: NO (0.28, **0.87**), NO_2 (0.20, **0.89**), CO (0.24, **0.97**), SO_2 (0.22, **0.89**), HCHO (0.27, **0.89**), ALK1 (**0.74**), ALK2 (0.24, **0.97**), ALK3 (0.22, **0.96**), ALK4 (0.23, **0.98**), ALK5 (0.21, **0.95**), ARO1 (0.25, **0.98**), ARO2 (0.24, **0.97**), OLE1 (0.24, **0.97**), OLE2 (0.26, **0.96**), C_2H_4 (0.24, **0.97**), TRP1 (**0.88**), EC (0.25, **0.98**), AORGPA (0.18, **0.91**), A25 (**0.81**), NUMACC (**0.76**), SRFACC (**0.86**)
3	5.42	**Chem.**: O_3 (0.20), PAN (0.19), NH_3 (0.28) **Emis.**: NH_3 (0.42, **0.90**), PNO_3 (−0.36, **−0.95**), PSO_4 (−0.27, **−0.88**), D_m (0.31, **0.87**), σ (0.28, **0.71**)
4	3.38	**Chem.**: H_2O_2 (0.26), ANH_4 (0.29), ANO_3 (0.44, **0.74**), AORGA (0.42, **0.86**), AORGB (0.43, **0.74**) **Met.**: TA (0.29)
5	3.23	**Chem.**: H_2O_2 (−0.21), ANH_4 (0.44, **−0.75**), AH_2O (0.45, **0.85**), ASO_4 (0.52, **0.88**) **Met.**: QC (0.31), TA (−0.24)
6	2.31	**Chem.**: O_3 (0.31), HNO_3 (0.23), NH_3 (−0.24) **Emis.**: σ (0.24) **Met.**: PBL (0.41, **0.79**), MOLI (−0.39, **−0.82**), RGRND (0.46, **0.89**)

(Continued)

Table 11.4 Key coefficients and loadings from absolute PCA with "varimax" rotation—*cont'd*

Factor	Eigenvalue	Variables with large values (PC coefficient, high loading)
7	1.85	**Chem.**: CO (**0.76**), ALK1 (0.25, **0.76**), ALK4 (**0.75**), ALK5 (**0.75**), ARO1 (**0.73**), ARO2 (**0.76**), OLE1 (**0.74**), OLE2 (**0.71**), C$_2$H$_4$ (**0.73**), A25 (0.17), NUMACC (0.19, **0.87**), SRFACC (**0.75**) **Emis.**: SO$_2$ (**0.83**), SULF (**0.83**), CCHO (**0.76**), RCHO (0.20, **0.93**), ACET (0.25, **0.95**), CH$_4$ (0.29, **0.96**), TRP1 (0.23, **0.92**), MTBE (**0.76**), ETOH (0.28, **0.94**), ALK1 (0.30, **0.93**), ALK2 (**0.70**), ALK3 (**0.76**), ALK4 (**0.80**), ARO2 (**0.70**), OLE1 (**0.72**), OLE2 (**0.72**), CO (**0.76**), PMFINE (0.25, **0.94**), POA (0.21, **0.92**)
8	1.43	**Chem.**: AH2O (−0.20) **Met.**: CFRAC (−0.62, **−0.83**), QR (−0.26), QC (−0.33, **0.61**), UWIND (0.40), VWIND (−0.29)
9	1.32	**Chem.**: H$_2$O$_2$ (0.22) **Met.**: WWIND (−0.74, **−0.87**), VWIND (−0.57, **−0.72**)

Note: |Coef.| $\geq 1.5/\sqrt{v}$ are shown, and $v = 76$ in this study. The 88% variance was explained by the first nine factors listed here. Correlation coefficients (R) between variables and PCs are listed in bold when $R \geq 0.7$.

The NMFROC method was also applied to input matrix **A**. The p value was set at 9, and Table 11.5 lists variables with coefficients larger than 2 in columns of **B**, and variables with factor loadings $\geq$0.7. Variables with values between 1 and 2 in **B** are listed in parentheses only when there are few variables with larger values.

In Table 11.5, some factors correspond to unique processes, such as rain (Factor 1), anthropogenic/diesel emissions (Factor 2), fog (Factor 3), and chemical concentrations (Factor 4), while other factors combine several processes. For example, Factor 7 features MTBE and OLE2 as well as other anthropogenic and isoprene emissions. Factor 5 represents conditions with relatively strong solar insolation, deep PBL height, large biogenic emissions, and large concentrations of nitric acid and ozone, as well as hydrogen peroxide. Factor 6 associates high concentration of sulfate in fine PM with very high aerosol water content, and with terpene and organic PM to a much lesser degree. Factor 8 indicates that high cloud fraction is associated with elevated PBL and surface concentrations of ozone and PAN. Factor 9 associates high ozone with a stable, rural condition with large emissions of ammonia and a coarser portion of fine PM as well as a number of other variables.

It is interesting to compare factors from the two factor analysis methods. Table 11.6 lists correlations between the factors identified by the NMFROC and APCA methods.

It is seen that these two methods identified some common factors, but significant differences existed in other factors, presumably due to the differences in methodology. Nevertheless, both factor analysis methods linked abnormally high PM sulfate to liquid water in their results. Furthermore, the NMFROC method was able to partition the PM sulfate among nine selected factors, as illustrated in Figure 11.5. It is seen that Factor 6 dominated the contribution to samples with high PM sulfate in the

Table 11.5 Major (coefficients, loadings) for **B** from the NMFROC method

Factor	Variables with high values in matrix B
1	QR (65.4, **1.0**) [CFRAC (1.6), PBL (1.5)]
2	**Emis.**: NO (**0.81**), NO$_2$ (**0.81**), HONO (**0.81**), NH$_3$ ($-$**0.89**), PNO$_3$ (3.8, **0.93**), PSO$_4$ (2.8, **0.73**), PEC (2.4), ALK2 (2.3), ALK3 (2.0), ALK5 (2.2), HCHO (2.8), NO$_x$ (3.4)
3	QC (42.7, **1.0**), CFRAC (7.7), AH$_2$O (12.3, **0.73**)
4	**Chem.**: VOC (2.3–3.7), NO (3.3, **0.83**), NO$_2$ (2.0, **0.83**), CO (2.8, **0.91**), SO$_2$ (3.4, **0.90**), HCHO (2.3, **0.80**), ALK1 (**0.78**), ALK2 (**0.94**), ALK3 (**0.94**), ALK4 (**0.95**), ALK5 (**0.87**), ARO1 (**0.91**), ARO2 (**0.90**), OLE1 (**0.92**), OLE2 (**0.91**), C$_2$H$_4$ (**0.91**), AEC (2.4, **0.92**), AORGPA (3.2, **0.94**), A25 (2.1, **0.86**), SRFACC (**0.80**)
5	**Met.**: RGRND (12.2, **0.83**), PBL (9.9, **0.72**)
	Emis.: ISOPRENE (5.7, **0.63**), ALK5 (2.0), (σ(1.7), PEC(1.7))
	Chem.: HNO$_3$ (16.2, **0.88**), O$_3$ (4.7), [HO$_2$H(1.7)]
6	**Chem.**: SRFACC (2.7), AORGB (2.3), AORGPA (2.0), AH$_2$O (21.8, **0.92**), ASO$_4$ (15.3, **0.85**), TRP1 (2.0)
7	**Emis.**: SO$_2$ (2.1, **0.85**), SULF (2.1, **0.85**), CCHO (**0.89**), RCHO (**0.93**), ACET (2.0, **0.92**), MEK (**0.79**), CH$_4$ (2.0, **0.87**), ETHENE (2.2, **0.88**), ISOPRENE (2.0, **0.74**), TRP1 (**0.81**), MTBE (2.4, **0.91**), ETOH (2.2, **0.92**), ALK1 (2.3, **0.87**), ALK2 (**0.75**), ALK3 (**0.75**), ALK4 (2.1, **0.90**), ARO1 (2.0, **0.82**), ARO2 (2.1, **0.86**), OLE1 (2.3, **0.90**), OLE2 (2.4, **0.91**), CO (2.3, **0.91**), PMFINE (2.1, **0.94**), POA (**0.81**)
8	**Chem.**: O$_3$ (3.9), PAN (2.4), PBL (5.0), CFRAC (41.5, **0.93**)
9	**Chem.**: O$_3$ (6.5), PAN (4.1), HO$_2$H (3.2), NH$_3$ (5.8), ANH$_4$ (3.1), ANO$_3$ (3.9), AORGA (2.4), AORGB (2.8), NUMACC ($-$**0.81**)
	Emis.: NO ($-$**0.77**), NO$_2$ ($-$**0.77**), HONO ($-$**0.77**), HCHO ($-$**0.80**), CCHO ($-$**0.76**), RCHO ($-$**0.74**), MEK ($-$**0.71**), TRP1 ($-$**0.86**), ALK2 ($-$**0.83**), ALK3 ($-$**0.87**), ALK4 ($-$**0.74**), ALK5 ($-$**0.71**), ARO1 ($-$**0.72**), NH$_3$ (5.3), POA ($-$**0.86**), PNO$_3$ ($-$**0.74**), PSO$_4$ ($-$**0.84**), D_p (7.9, **0.92**), σ(4.5)
	Met.: MOLI (3.62), WWIND (3.2), TA (2.9), UWIND (3.1), VWIND (3.1)

Note: variable names follow Table 11.3. Variables are listed when their values in **B** (in parentheses) are greater than 2, while 1 is the average value. A few variables with values less than but close to 2 are listed in brackets. Correlation (R) between variables and factors is listed in bold when $R \geq 0.7$.

Table 11.6 Correlations between factors from the NMFROC and APCA methods

NMFROC	1	2	3	4	5	6	7	8	9
PCA	8	3	5, 8	2	6	5	1, 7	8	3, 7
R	0.39	-0.85	-0.64, 0.61	0.95	0.84	-0.90	0.82, -0.86	0.68	0.85, 0.73

input matrix **A**, with a minor contribution from Factor 7. Thus, the PM liquid water content appeared to be the major factor for the anomaly in PM sulfate.

Following the information extracted from factor analyses, a simulation was conducted without rain and with aerosol water content limited to be within 0.1 g m^{-3}. In addition, catalytic pathways that produce sulfate in the aqueous phase with no dependence on other chemical reactions were ignored. The resulting sulfate in the fine PM mode became realistic, as shown in Figure 11.4(b).

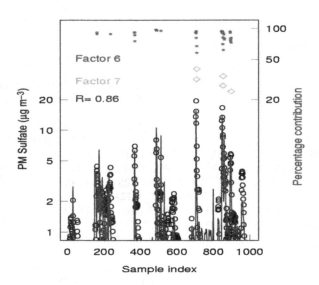

Figure 11.5 PM sulfate in the input (circles) and prediction (lines) matrices and major contributing factors in samples with PM sulfate larger than 4 μg m^{-3}.

11.3.2.2   Model versus observations

The positive matrix factorization method was applied to analyze the 2-year time series of observed and simulated concentrations of chemicals in the gas and aerosol phases at two urban and two rural stations in the southeastern USA in 2000–2001 (Marmur et al., 2009). Nine chemicals were included for PMF analysis, namely: sulfate, nitrate, and ammonium ions; EC and OC in particles; and CO, NO, HNO$_3$, and NO$_y$ gases. Four

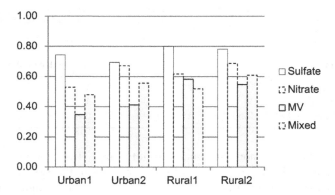

Figure 11.6 Correlation coefficients between PMF-resolved source contributions to observed and simulated concentrations of chemical species in the gas and aerosol phases at two urban and two rural stations in the southeastern USA.
Data were calculated from Marmur et al. (2009).

factors, identified as "secondary sulfate", "secondary nitrate", "fresh motor vehicle", and "mixed primary species", were able to explain most variations in the observed and simulated concentrations. Comparing time series of each factor at each station, it was found that factors characterized by secondary species showed better agreement between model and observations, while factors characterized by primary species showed significant differences. Figure 11.6 illustrates the 16 correlation coefficients between observations and simulations for each factor at the four stations. Thus, it appears that the co-variation of secondary chemicals in observations was better simulated by the model than the co-variation of primary chemicals. This is consistent with the fact that primary chemicals vary on a smaller scale than secondary chemicals; at the spatial resolution used for the modeling, the model represented secondary chemicals more realistically than primary chemicals.

11.4 Sensitivity analysis

Sensitivity analysis refers to multiple modeling exercises with changed inputs or model parameters, to identify areas for refinement, or to simulate realistic concentration fields, or to explore source–receptor relationships.

11.4.1 Brute-force method

The "brute-force" method simply repeats the base-case simulation with changed parameters and inputs. This method is accurate and is the only way to demonstrate air quality attainment with improved conditions in future years. For example, to generate an EKMA diagram for a region with O_3 pollution, hundreds of three-dimensional O_3 simulations need to be conducted with a range of NO_x–ROG emission changes. As each additional simulation in the brute-force method uses the same amount of computer time as the base-case simulation, when a large number of parameters and inputs need to be analyzed, such as for a new field, alternative methods are often useful.

11.4.2 Direct sensitivity modeling

Sensitivity of concentration fields to inputs and parameters may be directly modeled by solving corresponding sensitivity equations, which is termed the "direct method". The sensitivity of concentrations of n chemicals, C, in a three-dimensional field to fractional changes of m model parameters and input variables, P, may be described by:

$$\partial C/\partial t = -\nabla(uC) + \nabla(K\nabla C) + R[C, T, t] + E = F(C, P, t) \qquad \text{(F11.4.1a)}$$

$$\partial S_{ij}/\partial t = (\partial R_i/\partial C_k)S_{kj} + \{ -\nabla(uS_{ij}) + \nabla(K\nabla S_{ij}) + \partial R/\partial \varepsilon_j \\ + \partial E/\partial \varepsilon_j - [\nabla(uC) + \nabla(K\nabla C)]\delta_{ij}\} \qquad \text{(F11.4.1b)}$$

where formula (F11.4.1a) denotes the change of concentration field due to advection, diffusion, chemical reactions, and emission terms; deposition may be treated as a reaction. Formula (F11.4.1b) denotes the corresponding equation governing the sensitivity of the concentration of a chemical (C_i) to a parameter, $p_j(x, t)$; $S_{ij}(t) = \partial C_i(t)/\partial \varepsilon_j$, and $\varepsilon_j = p_j(x, t)/P_j(x, t)$, while $P_j(x, t)$ is the unperturbed field.

The left-hand side of the sensitivity formula (F11.4.1b) may be approximated by a first-order finite difference method, namely, $\partial S_{ij}/\partial t = [S_{ij}(t) - S_{ij}(t_0)]/(t - t_0)$, and solved separately from the concentration formula (F11.4.1a). A fast and stable, decoupled direct method (DDM-3D) was introduced by Yang et al. (1997) for O_3 air quality sensitivity analysis, to calculate many sensitivity coefficients in one three-dimensional simulation. For example, computing the sensitivity coefficients of all model chemicals to 20 parameters in an O_3 simulation over southern California consumed only 80% more computer time than a base-case simulation (Yang et al., 1997). Napelenok et al. (2006) extended the method to simulate sensitivity for PM air quality using US EPA CMAQ. The DDM-3D provided reasonable estimates on sensitivity parameters with fractional change of up to 25%.

An alternative method to solve the sensitivity equation is by using the kernel or Green function (Vuilleumier et al., 1997). The sensitivity equation may be abbreviated to:

$$\partial S_j/\partial t = JS_j + \partial F/\partial P_j, \quad \text{where } J = \partial F/\partial C \tag{F11.4.1c}$$

and the solution is given by

$$S_j(t) = \mathbf{K}(t, t_0)S_j(t_0) + \int_{t_0}^{t} \mathbf{K}(t, \tau)\partial F/\partial P_j(\tau)\mathrm{d}\tau$$

where

$$\mathbf{K}(t, \tau) = \exp[\mathbf{\Omega}(t, \tau)] = \mathbf{Z} \cdot \exp(\mathbf{D}\Delta t)\mathbf{Z}^{-1} \text{ and}$$

$$\Omega(t, \tau) \approx \int_{\tau}^{t} J(t_1)\mathrm{d}t_1 \tag{F11.4.1d}$$

where $\mathbf{K}[n \times n]$ is called the kernel or Green function of the problem. $\mathbf{Z}$ is an $n \times n$ matrix containing the right-hand eigenvectors of Ω as its columns, and $\mathbf{D}$ is a diagonal matrix containing the eigenvalues of $\Omega/\Delta t$. This method may increase the validity of the sensitivity up to a 50% change in parameters, according to comparison with brute-force sensitivity calculations.

11.4.3 Source–receptor relationship

The source–receptor relationship refers to the relationship of the chemical concentration at a location to emission sources, which is important in establishing a new source assessment and for mitigating air pollution problems. Besides the brute-force method, the direct method may provide quantitative estimates for the source–receptor relationship for linear chemicals. For example, US EPA CMAQ with DDM was applied to estimate the effects of anthropogenic emissions of SO_2 in China on downwind areas including South Korea and Japan; in certain weather conditions, China's SO_2 emissions could elevate daily sulfate concentration over an island in western Japan by 10 µg m^{-3} (Itahashi et al., 2012). In addition, a method called "concentration source apportionment" is often used to calculate the instantaneous contribution to a chemical concentration from various source regions and categories.

The concentration source apportionment method uses different model species to represent a chemical emitted from different source regions and categories. Using "EC" as an example, four model species (ECrc, ECRC, ECrC, and ECRc) may be first created to represent EC emitted from subregions "r" and "R" and categories "c" and "C" respectively. Then, the four model species are simulated according to their own concentration equations, such as formula (F11.4.1a). At the end of a simulation, the concentrations of the four model species at the locations of interest are summarized to provide simulated source apportionment for EC concentration. As both this method and the direct method use emission information at the base case only, their estimates on emission control benefits for nonlinear chemicals may be realistic only for a small change in base-case emissions, as illustrated in Figure 11.7. It is seen that, when the concentration response to emission reduction is nonlinear, such as for O_3 and PM nitrate, the brute-force method estimated a 25% reduction in concentration in response to a 50% reduction in emissions. For the same emission change, the decoupled direct method estimated only a 15% reduction in concentration, but the concentration source apportionment method estimated a 50% reduction in concentration. A discussion on this can be found in Koo et al. (2009).

11.4.4 Mechanism uncertainty analysis

Statistical sensitivity analysis is commonly conducted to analyze concentration ranges simulated from a photochemical mechanism. The concentration of a chemical in a mechanism may depend on the rate constants of a number of reactions and the concentrations of a number of other chemicals. As rate constant parameters and chemical concentrations are measured with uncertainty ranges, any chemical

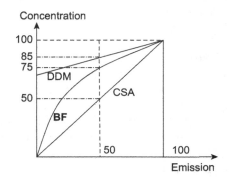

Figure 11.7 Illustration of major sensitivity analysis methods for source–receptor relationship. BF, brute force; CSA, concentration source apportionment; DDM, decoupled direct method.

concentration simulated from a photochemical mechanism also has a corresponding uncertainty range. Major contributors to the resulting uncertainty in simulated chemical concentrations may be analyzed by the sensitivity analysis methods described previously, as well as the "global sensitivity analysis" method based on screened Monte Carlo numerical experiments.

For a target chemical dependent on n parameters, the local sensitivity of its concentration to parameters may be first calculated by the brute-force method or the direct method, and used to screen the final list of parameters for global sensitivity analysis. To estimate the global sensitivity, $(n + 1)r$ parameter values are first selected from $(n + 1)M$ randomly generated values to efficiently represent the maximum variety of parameters in their uncertainty ranges, where M is much larger than r. Then, corresponding $(n + 1)r$ simulations may be conducted, and concentrations of the target chemical (C) are regressed against n parameters using:

$$C = C(0) + \sum_{i=1}^{n} F(P_i) + \sum_{1 \leq i < j \leq n} F(P_i, P_j) + \dots + F(P_1, P_2, \dots, P_n)$$

(F11.4.2)

The terms on the right-hand side of formula (F11.4.2) represent the average concentration, the first-order sensitivity components, the second-order sensitivity components, and so on, up to the nth order sensitivity component. The contribution of each term on the right-hand side to the total variance of the simulated concentration of the target chemical is defined as global sensitivity.

A statistical sensitivity analysis was recently conducted to analyze the origins of major uncertainty for OH and HO_2 in the Regional Atmospheric Chemical Mechanism updated with the most recent rate constants (Chen et al., 2012). It was found that about 90% of the global sensitivity of the OH concentration was due to the first-order sensitivity components. Of nearly 600 mechanism parameters, the simulated concentrations of OH and HO_2 were mostly sensitive to reactions of xylenes and isoprene with OH, NO_2 with OH, NO with HO_2, and internal alkenes with O_3. The simulated range of OH and HO_2 concentrations using the RACM mechanism and a box model overlapped observed ranges at a tower 60 m above the ground on the campus of the University of Houston, Texas, USA on a polluted late summer day in 2006.

11.5 Adjoint method

Simulated chemical fields need to be compared with observed values at a network of stations, which is similar to the use of observed meteorological parameters for calibrating weather forecasts. The adjoint method, widely used for meteorological studies, is also useful for atmospheric chemical analyses, as discussed below.

11.5.1 A backward problem

Surface air pollution is often featured with an excessive O_3 mixing ratio and PM concentration at a location for a short time. In order to mitigate for the excess, air quality simulations need to establish a base case that uses a state-of-science formulation, parameters, and input data to replicate observed features. To understand the delicate spatial and temporal factors that contributed to a simulated excess, a backward sensitivity analysis method that may apportion the excess of a pollutant at a location to all known factors before the event is desirable. The problem may be described by:

$$\Delta R = \int_{t=0}^{T} \oint h \cdot f \, dt \, dV \qquad (E11.5.1)$$

where ΔR denotes the change of a chemical, e.g. O_3, in a region from time 0 to time T, in general; h denotes the change of a chemical at a location at a time, and is the numerator of the local sensitivity S_{ij} governed by formula (F11.4.1b); f is a value dependent on time and location.

To analyze a simulated hourly O_3 excess in a box at time T, ΔR may be set for O_3 in the box at the hour only. In that case, $h \equiv \Delta O_3$; $f = 1/(\text{box volume} \times 1 \text{ hour})$ when V denotes the box location and $t = T$, and $f = 0$ in all other cases. ΔR needs to be expressed in terms of all known factors in a region, or model domain, for a (short) period of time, e.g. a few days, before an event.

11.5.2 Formulation

The governing equation for h in the three-dimensional domain during a time period is similar to formula (F11.4.1b) for S_{ij}:

$$\partial h_i / \partial t - \mathbf{J}_{ik} h_k + \nabla(u h_i) - \nabla(K \nabla h_i) = \Delta Q_i \qquad (E11.5.2)$$

where $h_i = \Delta O_3$ and h_k denotes all chemicals related to O_3. The chemical partial derivative matrix (Jacobian) $\mathbf{J}_{ik} = (\partial R_i / \partial C_k)$ and the "source term" $\Delta Q_i = \mathbf{F}_{il} \Delta P_l + \Delta E_i + \nabla(\Delta u C_i) + \nabla(\Delta K \cdot \nabla C_i)$, while the parameter partial derivative matrix $\mathbf{F}_{il} = (\partial R_i / \partial P_l)$. P_l denotes all parameters related to photochemical reactions including deposition.

In order to insert equation (E11.5.2) into equation (E11.5.1) using integration by parts, an adjoint sensitivity (function) ψ_i may be defined for h_i using the value of f:

$$\partial \psi_i / \partial \tau - \mathbf{J}_{ik} \psi_k + \nabla(\mu \psi_i) - \nabla(K \nabla \psi_i) = f_i \qquad (E11.5.3)$$

where τ starts from 0 at $t = T$ backward to T at $t = 0$, and μ denotes $-u$, while all operators updating the spatial field of h_i in equation (E11.5.2) are present. To make ψ_i meaningful physically, its initial value is set to zero in model domain at $t = T$, and boundary conditions may be set accordingly. For example, horizontal boundary conditions may be set as $\nabla \psi_i = 0$ for inflow and $\mu \psi_i = K \nabla \psi_i$ for outflow. The vertical boundary conditions may be set as $\nabla \psi_i = 0$ at the bottom and $\mu \psi_i = K \nabla \psi_i$ at the top. The four-dimensional field of ψ_i may be numerically integrated in reverse using the

same algorithms and results from three-dimensional O_3 simulations, after a base case
has been established.

ΔR in equation (E11.5.1) may be rewritten for sensitivity apportionment purposes,
by replacing f with the left-hand side of equation (E11.5.3), which results in:

$$\Delta R_i = \int_{t=0}^{T} \oint \psi_i \cdot \Delta Q_i \, dt \, dV + R_{ic} - R_{hb} + R_{vb} \qquad (E11.5.4a)$$

where the index i in R_i, ψ_i, and Q_i may denote O_3, PM, or other chemicals of interest
calculated in the base-case simulation. The first term on the right-hand side represents
contributions from advection, eddy diffusion, chemical reactions, and emission
processes throughout the time period from $t = 0$ to $t = T$; it is seen that the adjoint
sensitivity serves as a weighting factor in time and space. The contributions from the
initial condition (R_{ic}), horizontal boundary condition (R_{hb}), and vertical boundary
condition including deposition and emission (R_{vb}) may be calculated from ψ and
archived values from the base-case simulation for O_3, PM, or other chemicals using:

$$R_{ic} = \oint \psi_i(t = 0) \cdot \Delta C_i(t = 0) dV \qquad (E11.5.4b)$$

$$R_{hb,x} = \int_{t=0}^{T} \int_{z=0}^{Z} \int_{y=0}^{Y} \{\psi_i[\Delta U(C_i(b) - C_i + U\Delta C_i(b) + \Delta K_{xx}\nabla C_i]\}\Big|_0^X \, dt \, dz \, dy$$

$$(E11.5.4c)$$

$$R_{vb} = \int_{t=0}^{T} \int_{x=0}^{X} \int_{y=0}^{Y} [\psi_i(\Delta E_i + \Delta K_{zz}\nabla C_i - \Delta D_i C_i)]_{z=0} dt \, dx \, dy \qquad (E11.5.4d)$$

where $\Delta C_i(t = 0)$ denotes the change in initial concentration of chemical i, and $dV \equiv$
$(dx \, dy \, dz)$. $R_{hb,x}$ denotes the contribution from the horizontal boundary condition
along the west–east direction; the contribution from the south–north direction, $R_{hb,y}$,
can be obtained by swapping (x, X, U) with (y, Y, V). $C_i(b)$ denotes the boundary
concentration of chemical i. ΔD_i denotes the change in deposition velocity of
chemical i. Note that the units used for all variables in equation (E11.5.4) need to be
consistent. A detailed derivation of the formula can be found in Martien (2004).

11.5.3 Applications

The adjoint sensitivity analysis method described above is uniquely suitable for
analyzing contributions to peak O_3 or PM from hundreds or more input parameters to
a three-dimensional atmospheric chemical model, compared with other sensitivity

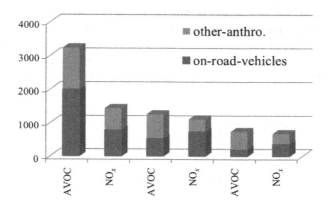

Figure 11.8 Anthropogenic emissions of VOCs and NO$_x$ from on-road vehicles and other sources in California South Coast Air Basin, USA, in 1987, 1997, and 2010 respectively (from left to right). The units are tons per day, and values are for typical summer weekdays. Data are from Martien and Harley (2006).

analysis methods. Martien and Harley (2006) reported its application at three stations in southern California, USA, and apportioned O$_3$ sensitivities to nearly 900 model inputs, which include surface emissions, reaction rate coefficients, dry deposition velocities, boundary conditions, and initial conditions, using emission inventories pertinent to summer weekdays of 1987, 1997, and 2010 respectively. Figure 11.8 shows anthropogenic emissions of VOCs and NO$_x$ from on-road vehicles and other sources in the California South Coast Air Basin, USA, in 1987, 1997, and 2010 respectively. It is seen that anthropogenic volatile organic compound (AVOC) emissions decreased dramatically while NO$_x$ emissions decreased only a little between 1987 and 1997; from 1997 to 2010, both AVOC and NO$_x$ decreased to some extent, about evenly. They found that emission reductions from 1987 to 1997 resulted in a large reduction in O$_3$ mixing ratio and changes in O$_3$ sensitivities at the three sites, while emission reductions from 1997 to 2010 resulted in little change in the O$_3$ mixing ratio at the three stations. The sensitivity of O$_3$ change to NO$_x$ emissions became larger when the sensitivity of O$_3$ mixing ratio to NO$_x$ emissions changed sign from NO$_x$ benefit in 1987 to lack of NO$_x$ benefit in 1997 and 2010 at Rubidoux, where high O$_3$ was often observed. They further plotted the evolution of the relative importance of emission source regions and boundary condition altitudes for 1-hour versus 8-hour peak O$_3$, which is not readily obtainable from forward and statistical sensitivity analysis methods.

The adjoint sensitivity analysis method has been implemented into US EPA CMAQ, to analyze sensitivities of population exposure and environmental stress as well as O$_3$ mixing ratio (Hakami et al., 2007). To analyze sensitivities of population exposure and environmental stress, equation (E11.5.1) may be generalized to:

$$F = \int_{t=0}^{T} \oint g(C, t, V) \mathrm{d}t \, \mathrm{d}V \qquad (E11.5.5)$$

where F is called the cost function. For O_3 sensitivity, equations (11.5.1)–(11.5.4) correspond to the case when $g = (h \cdot f)$, while $\psi_i = \partial F/\partial C_i$, and $f_i = \partial g/\partial C_i$. To analyze sensitivities of population exposure and environmental stress, different cost functions, as shown in equations (E11.5.6) and (E11.5.7), need to be used, while adjoint equation (E11.5.3) and sensitivity apportionment equation (E11.5.4) need to be updated carefully.

$$F(\text{pe}) = \sum_{x,y,t}[P(C_i - \gamma)]_{x,y,t} \tag{E11.5.6}$$

$$F(\text{es}) = \sum_{x,y,t}\{O_3/[1 + 4403\exp(-126 \cdot O_{3,\text{ppm}})]\}_{x,y,t} \tag{E11.5.7}$$

Sensitivities of population exposure and environmental stress to NO_x emissions and air temperature via photochemical reactions were demonstrated for the southern USA using US EPA CMAQ.

11.6 Current issues

Chemical modeling provides a powerful tool to integrate numerous scientific understandings and engineering data for exploring emerging technical issues related to the air. A number of technical debates have been elucidated via chemical modeling in the past few decades. For example, consensus has been built worldwide to amend the O_3 hole in the stratosphere, and to mitigate anthropogenic climate change via sustained collaborative research to control greenhouse gas emissions. There is little doubt that anthropogenic emissions dominate O_3 production in the global troposphere, especially in breathing zones, and that many toxic chemicals emitted into the atmosphere may be transported to another region or nation. Simulating the source–receptor relationship for air pollutants has become more and more credible.

A number of technical issues remain unexplored. As the number of trace chemicals in the air is numerous, only a small number of air toxins have been thoroughly simulated so far. Diseases transmitted in breathing zones have not been well investigated by state-of-science chemical models. The composition, toxicity, and evolution of particles suspended in the air are still at the semi-quantitative stage, compared with atmospheric oxidants that also have uncertainties as revealed by recent measurements. As a result, corroborative analysis tools may play an important role in corresponding research. A systematic integration of corroborative analysis tools with chemical models is desirable.

Summary

In this chapter, a number of analytical methods in support of simulating chemical fields in the air have been introduced. Firstly, indicator ratios for O_3 were elaborated for their theoretical bases and simulated ranges. Secondly, the charge balance for soluble PM was described using ambient data measured at global locations in a recent decade, and an example was given to illustrate its use for determining precursor limitations. Thirdly, factor analysis methods including PCA, NMFROC, and PMF were elaborated to interpret chemical transport modeling results from the perspectives of debugging and comparison with observations. Fourthly, forward and backward sensitivity analysis methods were elaborated to analyze the sensitivity of a target chemical to model inputs and parameters; the methods considered were the brute-force method, direct method, concentration source apportionment, global sensitivity analysis, and adjoint sensitivity analysis. Finally, some comments on current issues were given.

Exercises

1. Indicator ratios may provide convenient guidance for effective local emission mitigation efforts, in order to avoid an expected excess of pollutants according to air quality standards. However, from an atmospheric chemical perspective, the existence of an accurate indicator ratio may make model simulations more difficult. Discuss the pros and cons of an indicator, using an example in your air district.
2. Charge balance may never be perfect as a measure of PM composition due to instrumental uncertainties and human error. Survey the PM composition data in your air district, and calculate the ratios of cations to anions in each sample. List the fractions of PM mass speciated together with the ionic ratios, and analyze the possible reasons for imbalances.
3. Use a fraction or all variables listed in Table 11.3, and estimate their hourly values at a few stations in your air district over a period of a few weeks. Conduct a factor analysis to extract at least three factors, and analyze their environmental implications.
4. Suppose there is an interesting new persistent organic pollutant about to be marketed in a nation. Before approval for marketing, the environmental protection agency of the nation needs to conduct a thorough evaluation of the new chemical. Which sensitivity analysis tools are suitable as part of the assessment? Write a one-page proposal on how to evaluate the chemical using the sensitivity analysis tools.
5. Using the cost function in equation (E11.5.6) rather than equation (E11.5.1), derive governing and solution equations for adjoint sensitivity analysis corresponding to equations (E11.5.3) and (E11.5.4).
6. Extra credit: Repeat the above exercise using the cost function in equation (E11.5.7).

References

Books

California Air Resources Board, June 2012. 2012 Annual Monitoring Network Report for Small Districts in California. CARB, California, USA.

Crouch, E.A.C., Wilson, R., 1982. Risk/Benefit Analysis. Ballinger Publishing Company, Cambridge, MA.

European Commission, July 27, 2001. Ambient Air Pollution by Polycyclic Aromatic Hydrocarbons (PAH). Position Paper Annexes. European Commission, Luxembourg.

Gear, C.W., 1971. Numerical Initial Value Problems in Ordinary Differential Equations. Prentice-Hall, Englewood Cliffs, NJ, USA.

Holdren, J.P., 2011. National Acid Precipitation Assessment Program Report to Congress 2011: An Integrated Assessment. Washington, DC, USA.

Holland, H.D., 1978. The Chemistry of the Atmosphere and Oceans. John Wiley & Sons, New York, USA.

Horowitz, L.W., 1997. The Influence of Boundary Layer Chemistry on Global Tropospheric Ozone and Nitrogen Oxides (Harvard University Ph.D. Dissertation). UMI Dissertation Services, Ann Arbor, MI, USA.

Intergovernmental Panel on Climate Change (IPCC), 1990. Climate Change: The IPCC Scientific Assessment. In: Houghton, J.T., Jenkins, G.J., Ephraums, J.J. (Eds.). Cambridge University Press, Cambridge, UK. Subsequent reports available at: http://www.ipcc.ch.

Jacob, D.J., 1999. Introduction to Atmospheric Chemistry. Princeton University Press, Princeton, NJ, USA.

Jacobson, M.Z., 1999. Fundamentals of Atmospheric Modeling. Cambridge University Press, Cambridge, UK.

Leighton, P.A., 1961. Photochemistry of Air Pollution. Academic Press, New York, USA.

Liang, J., 1997. Tropospheric Ozone: Effects of Cloud Chemistry and Boundary Layer Processes (Harvard University Ph.D. Dissertation). UMI Dissertation Services, Ann Arbor, MI, USA.

Liou, K.N., 1992. Radiation and Cloud Processes in the Atmosphere: Theory, Observation, and Modeling. Oxford University Press, New York, USA.

Martien, P.T., 2004. Forward and Adjoint Sensitivity Analysis in Eulerian Photochemical Air Quality Models (Ph.D. dissertation). University of California at Berkeley, CA, USA.

Sanders, S.P., Abbatt, J., Barker, J.R., Burkholder, J.B., Friedl, R.R., Golden, D.M., Huie, R.E., Kolb, C.E., Kurylo, M.J., Moortgat, G.K., Orkin, V.L., Wine, P.H., 2011. Chemical Kinetics and Photochemical Data for Use in Atmospheric Studies. Evaluation No. 17, JPL Publication 10-6. Jet Propulsion Laboratory, Pasadena CA, USA.

Seinfeld, J.H., Pandis, S.N., 1998. Atmospheric Chemistry and Physics: From Air Pollution to Climate Change. John Wiley & Sons, New York, USA.

US EPA, 1999. Science Algorithms of the EPA Models-3 Community Multiscale Air Quality (CMAQ) Modeling System. EPA/600/R-99/030, USA.

Warneck, P., 1988. Chemistry of the Natural Atmosphere. Academic Press, San Diego, CA, USA.

Journal articles

Arndt, R.L., Carmichael, G.R., Streets, D.G., Bhatti, N., 1997. Sulfur dioxide emissions and sectorial contributions to sulfur deposition in Asia. Atmos. Environ. 31, 1553–1572.

Atkinson, R., Baulch, D.L., Cox, R.A., Crowley, J.N., Hampson, R.F., Hynes, R.G., Jenkin, M.E., Rossi, M.J., Troe, J., 2006. Evaluated kinetic and photochemical data for atmospheric chemistry: Volume II – gas phase reactions of organic species. Atmos. Chem. Phys. 6, 3625–4055.

Atkinson, R., Baulch, D.L., Cox, R.A., Crowley, J.N., Hampson, R.F., Hynes, R.G., Jenkin, M.E., Rossi, M.J., Troe, J., 2004. Evaluated kinetic and photochemical data for atmospheric chemistry: Volume I – gas phase reactions of O_x, HO_x, NO_x and SO_x species. Atmos. Chem. Phys. 4, 1461–1738.

Aumont, B., Szopa, S., Madronich, S., 2005. Modeling the evolution of organic carbon during its gas-phase tropospheric oxidation: Development of an explicit model based on a self-generating approach. Atmos. Chem. Phys. 5, 2497–2517.

Aymoz, G., Jaffrezo, J.L., Chapuis, D., Cozic, J., Maenhaut, W., 2007. Seasonal variation of PM_{10} main constituents in two valleys of the French Alps. I: EC/OC fractions. Atmos. Chem. Phys. 7, 661–675.

Bae, M.-S., Schwab, J.J., Hogrefe, O., Frank, B.P., Lala, G.G., Demerjian, K.L., 2010. Characteristics of size distributions at urban and rural locations in New York. Atmos. Chem. Phys. 10, 4521–4535.

Baek, S.O., Field, R.A., Goldstone, M.E., Kirk, P.W., Lester, J.N., Perry, R., 1991. A review of atmospheric polycyclic aromatic hydrocarbons: sources, fate and behavior. Water, Air, Soil Pollut. 60, 279–300.

Barletta, B., Meinardi, S., Rowland, F.S., Chan, C., Wang, X., Zou, S., Chan, L., Blake, D.R., 2005. Volatile organic compounds in 43 Chinese cities. Atmos. Environ. 39, 5979–5990.

Blanchard, C.L., Fairley, D., 2001. Spatial mapping of VOC and NO_x-limitation of ozone formation in central California. Atmos. Environ. 35, 3861–3873.

Blanchard, C.L., Roth, P.M., Tanenbaum, S.J., Ziman, S.D., Seinfeld, J.H., 2000. The use of ambient measurements to identify which precursor species limit aerosol nitrate formation. J. Air Waste Manage. Assoc. 50, 2073–2084.

Blanchard, C.L., Lurmann, F.W., Roth, P.M., Jeffries, H.E., Korc, M., 1999. The use of ambient data to corroborate analyses of ozone strategies. Atmos. Environ. 33, 369–381.

Bukowiecki, N., Dommen, J., Prevot, A.S.H., Weingartner, E., Baltensperger, U., 2003. Fine and ultrafine particles in the Zurich (Switzerland) area measured with a mobile laboratory: an assessment of the seasonal and regional variation throughout a year. Atmos. Chem. Phys. 3, 1477–1494.

Cai, H., Xie, S., 2007. Estimation of vehicular emission inventories in China from 1980 to 2005. Atmos. Environ. 41, 8963–8979.

Cao, G., Zhang, X., Zheng, F., 2006. Inventory of black carbon and organic carbon emissions from China. Atmos. Environ. 40, 6516–6527.

Cao, J.J., Wu, F., Chow, J.C., Lee, S.C., Li, Y., Chen, S.W., An, Z.S., Fung, K.K., Watson, J.G., Zhu, C.S., Liu, S.X., 2005. Characterization and source apportionment of atmospheric organic and elemental carbon during fall and Winter of 2003 in Xi'an, China. Atmos. Chem. Phys. 5, 3127–3137.

Carmichael, G.R., Streets, D.G., Calori, G., Amann, M., Jacobson, M.Z., Hansen, J., Ueda, H., 2002a. Changing trends in sulfur emissions in Asia: Implications for acid deposition, air pollution, and climate. Environ. Sci. Technol. 36, 4707–4713.

Carmichael, G.R., et al., 2002b. The MICS-Asia study: model intercomparison of long-range transport and sulfur deposition in East Asia. Atmos. Environ. 36, 175–199.

Carslaw, N., Mota, T., Jenkin, M.E., Barley, M.H., McFiggans, G., 2012. A significant role for nitrate and peroxide groups on indoor secondary organic aerosol. Environ. Sci. Technol. 46, 9290–9298.

Cartalis, C., Varotsos, C., 1994. Surface ozone in Athens, Greece, at the beginning and at the end of the twentieth century. Atmos. Environ. 28, 3–8.

Carter, W.P.L., 2000. Documentation of the SAPRC-99 chemical mechanism for VOC reactivity assessment. A report to the California Air Resources Board, Contracts 92–329 and 95–308. CA, USA.

Carter, W.P.L., Heo, G., Cocker, D.R., Nakao, S., 2012. SOA Formation: Chamber study and model development. Draft final report to the California Air Resources Board Contract No. 08–326. CA, USA.

Cess, R.D., et al., 1991. Interpretation of snow-climate feedback as produced by 17 general circulation models. Science 253, 888–891.

Cess, R.D., et al., 1990. Intercomparison and interpretation of climate feedback processes in 19 atmospheric general circulation models. J. Geophys. Res. 95, 16601–16615.

Cess, R.D., et al., 1989. Interpretation of cloud-climate feedback as produced by 14 atmospheric general circulation models. Science 245, 513–516.

Chan, C.K., Yao, X., 2008. Air pollution in mega cities in China. Atmos. Environ. 42, 1–42.

Chan, C.Y., Xu, X.D., Li, Y.S., Wong, K.H., Ding, G.A., Chan, L.Y., Cheng, X.H., 2005. Characteristics of vertical profiles and sources of $PM_{2.5}$, PM_{10} and carbonaceous species in Beijing. Atmos. Environ. 39, 5113–5124.

Chan, L.Y., Liu, H.Y., Lam, K.S., Wang, T., Oltmans, S.J., Harris, J.M., 1998. Analysis of the seasonal behavior of tropospheric ozone at Hong Kong. Atmos. Environ. 32, 159–168.

Chang, J., et al., 2012. An inventory of biogenic volatile organic compounds for a subtropical urban–rural complex. Atmos. Environ. 56, 115–123.

Chang, J.S., Brost, R.A., Isaksen, I.S.A., Madronich, S., Middleton, P., Stockwell, W.R., Walcek, C.J., 1987. A three-dimensional Eulerian acid deposition model: Physical concepts and Formulation. J. Geophys. Res. 92, 14681–14700.

Chen, S., Brune, W.H., Oluwole, O., Kolb, C.E., Bacon, F., Li, G., Rabitz, H.A., 2012. Global sensitivity analysis of the regional atmospheric chemical mechanism: an application of random sampling-high dimensional model representation to urban oxidation chemistry. Environ. Sci. Technol. 46 (20), 11162–11170.

Chen, T., Li, X., Yan, J., Lu, S., Cen, K., 2011. Distribution of polychlorinated dibenzo-p-dioxins and Dibenzofurans in ambient air of different regions in China. Atmos. Environ. 45, 6567–6575.

Cheng, H.R., Guo, H., Saunders, S.M., Lam, S.H.M., Jiang, F., Wang, X.M., Simpson, I.J., Blake, D.R., Louie, P.K.K., Wang, T.J., 2010. Assessing photochemical ozone formation in the Pearl River Delta with a photochemical trajectory model. Atmos. Environ. 44, 4199–4208.

Cheng, I., Zhang, L., Blanchard, P., Graydon, J.A., Louis, V.L.S., 2012. Source–receptor relationships for speciated atmospheric mercury at the remote Experimental Lakes Area, northwestern Ontario, Canada. Atmos. Chem. Phys. 12, 1903–1922.

Cheng, Z.L., Lam, K.S., Chan, L.Y., Wang, T., Cheng, K.K., 2000. Chemical characteristics of aerosols at coastal station in Hong Kong. I. Seasonal variation of major ions, halogens and mineral dusts between 1995 and 1996. Atmos. Environ. 34, 2771–2783.

Cheung, V.T.F., Wang, T., 2001. Observational study of ozone pollution at a rural site in the Yangtze Delta of China. Atmos. Environ. 35, 4947–4958.

China MEP, 2012. Ambient air quality standards (GB 3095-2012). P.R. China (in Chinese).

Cicerone, R.J., Elliott, S., Turco, R.P., 1991. Reduced Antarctic ozone depletions in a model with hydrocarbon injections. Science 254, 1191–1194.

Clean Air Status and Trends Network (CASTNET), April 2012. 2010 Annual Report. EPA Contract No. EP-W-09–028, USA.

Cohen, D.D., Crawford, J., Stelcer, E., Bac, V.T., 2010. Long range transport of fine particle windblown soils and coal fired power station emissions into Hanoi between 2001 to 2008. Atmos. Environ. 44, 3761–3769.

Collins, W.D., et al., 2006. Radiative forcing by well-mixed greenhouse gases: Estimates from climate models in the Intergovernmental Panel on Climate Change (IPCC) Fourth Assessment Report (AR4). J. Geophys. Res. 111, D14317.

Cooper, O.R., Stohl, A., Eckhardt, S., Parrish, D.D., Oltmans, S.J., Johnson, B.J., Nedelec., P., Schmidlin, F.J., Newchurch, M.J., Kondo, Y., Kita, K., A., 2005. Spring comparison of tropospheric ozone and transport pathways on the east and west coasts of the United States. J. Geophys. Res. 110 D05S90.

Deandreis, C., Balkanski, Y., Dufresne, J.L., Cozic, A., 2012. Radiative forcing estimates of sulfate aerosol in coupled climate-chemistry models with emphasis on the role of the temporal variability. Atmos. Chem. Phys. 12, 5583–5602.

Debaje, S.B., Kakade, A.D., Jeyakumar, S.J., 2010. Air pollution effect of O_3 on crop yield in rural India. J. Hazard Mater. 183, 773–779.

DeBauer, M.L., Hernández-Tejeda, T., 2007. A review of ozone-induced effects on the forests of central Mexico. Environ. Pollut. 147, 446–453.

Deng, J., Dua, K., Wang, K., Yuan, C.S., Zhao, J., 2012. Long-term atmospheric visibility trend in Southeast China, 1973–2010. Atmos. Environ. 59, 11–21.

Ding, W., Cai, Z., Wang, D., 2004. Preliminary budget of methane emissions from natural wetlands in China. Atmos. Environ. 38, 751–759.

Ding, X., Wang, X.M., Zheng, M., 2011. The influence of temperature and aerosol acidity on biogenic secondary organic aerosol tracers: Observations at a rural site in the central Pearl River Delta region, South China. Atmos. Environ. 45, 1303–1311.

Driscoll, C.T., Lawrence, G.B., Bulger, A.J., Butler, T.J., Cronan, C.S., Eagar, C., Lambert, K.F., Likens, G.E., Stoddard, J.L., Weathers, K.C., 2001. Acid rain revisited: Advances in scientific understanding since the passage of the 1970 and 1990 Clean Air Act Amendments, Hubbard Brook Research Foundation, Science Links™ Publication 1(1), NH, USA.

Egorova, T., Rozanov, E., Zubov, V., Manzini, E., Schmutz, W., Peter, T., 2005. Chemistry–climate model SOCOL: a validation of the present-day climatology. Atmos. Chem. Phys. 5, 1557–1576.

Eichler, H., et al., 2008. Hygroscopic properties and extinction of aerosol particles at ambient relative humidity in South-Eastern China. Atmos. Environ. 42, 6321–6334.

Fagnano, M., Maggio, A., Fumagalli, I., 2009. Crops' responses to ozone in Mediterranean environments. Environ. Pollut. 157, 1438–1444.

Fahey, K., Liang, J., Allen, P.D., Gürer, K., Kaduwela, A., October 16–18, 2006. Long-term one-atmosphere CMAQ modeling in central California: Model performance evaluation and predicted trends. 5th Annual CMAS Models-3 User's Conference, UNC-Chapel Hill, NC, USA.

Falkovich, A.H., Graber, E.R., Schkolnik, G., Rudich, Y., Maenhaut, W., Artaxo, P., 2005. Low molecular weight organic acids in aerosol particles from Rondonia, Brazil, during the biomass-burning, transition and wet periods. Atmos. Chem. Phys. 5, 781–797.

Fang, S., Mu, Y., 2009. NOx fluxes from several typical agricultural fields during Summer-autumn in the Yangtze Delta, China. Atmos. Environ. 43, 2665–2671.

Fang, S., Mu, Y., 2006. Air/surface exchange of nitric oxide between two typical vegetable lands and the atmosphere in the Yangtze Delta, China. Atmos. Environ. 40, 6329–6337.

Faoro, F., Iriti, M., 2009. Plant cell death and cellular alterations induced by ozone: key studies in Mediterranean conditions. Environ. Pollut. 157, 1470–1477.

Farman, J.C., Gardiner, B.G., Shanklin, J.D., 1985. Large losses of total ozone in Antarctica reveal seasonal ClOx/NOx interaction. Nature 315, 207–210.

Feichter, J., Kjellstrom, E., Rodhe, H., Dentener, F., Lelieveld, J., Roelofs, G., 1996. Simulation of the tropospheric sulfur cycle in a global climate model. Atmos. Environ. 30, 1693–1707.

Fiore, A.M., et al., 2009. Multimodel estimates of intercontinental source–receptor relationships for ozone pollution. J. Geophys. Res. 114 D04301.

Fischer, A.M., et al., 2008. Interannual-to-decadal variability of the stratosphere during the 20th century: ensemble simulations with a chemistry–climate model. Atmos. Chem. Phys. 8, 7755–7777.

Fisseha, R., Dommen, J., Gutzwiller, L., E.WeingartnerGysel, M., Emmenegger, C., Kalberer, M., Baltensperger, U., 2006. Seasonal and diurnal characteristics of water soluble inorganic compounds in the gas and aerosol phase in the Zurich area. Atmos. Chem. Phys. 6, 1895–1904.

Friedman, C.L., Selin, N.E., 2012. Long-range atmospheric transport of polycyclic aromatic hydrocarbons: a global 3-d model analysis including evaluation of arctic sources. Environ. Sci. Technol. 46, 9501–9510.

Friedrich, R., 2009. Natural and biogenic emissions of environmentally relevant atmospheric trace constituents in Europe. Atmos. Environ. 43, 1377–1379.

Fu, Q., Zhuang, G., Wang, J., Xu, C., Huang, K., Li, J., Hou, B., Lu, T., Streets, D.G., 2008. Mechanism of formation of the heaviest pollution episode ever recorded in the Yangtze River Delta, China. Atmos. Environ. 42, 2023–2036.

Fu, Y., Liao, H., 2012. Simulation of the interannual variations of biogenic emissions of volatile organic compounds in China: Impacts on tropospheric ozone and secondary organic aerosol. Atmos. Environ. 59, 170–185.

Gao, G.W.W., Tang, Y.H., Tam, C.M., Gao, X.F., 2008. Anti-radon coating for mitigating indoor radon concentration. Atmos. Environ. 42, 8634–8639.

Gerasopoulos, E., Koulouri, E., Kalivitis, N., Kouvarakis, G., Saarikoski, S., Makela, T., Hillamo, R., Mihalopoulos, N., 2007. Size-segregated mass distributions of aerosols over Eastern Mediterranean: seasonal variability and comparison with AERONET columnar size-distributions. Atmos. Chem. Phys. 7, 2551–2561.

Gong, P., Wang, X., Sheng, J., Yao, T., 2010. Variations of organochlorine pesticides and polychlorinated biphenyls in atmosphere of the Tibetan Plateau: Role of the monsoon system. Atmos. Environ. 44, 2518–2523.

Grell, G., Dudhia, J., Stauffer, D., 1995. A Description of the Fifth-Generation Penn State/ NCAR Mesoscale Model (MM5). NCAR/TN-398+STR, USA.

Gu, B., Ge, Y., Ren, Y., Xu, B., Luo, W., Jiang, H., Gu, B., Chang, J., 2012. Atmospheric reactive nitrogen in China: Sources, recent trends, and damage costs. Environ. Sci. Technol. 46, 9420–9427.

Guo, H., et al., 2010. Carbonyl sulfide, dimethyl sulfide and carbon disulfide in the Pearl River Delta of southern China: Impact of anthropogenic and biogenic sources. Atmos. Environ. 44, 3805–3813.

Guo, H., Zhang, Q., Shi, Y., Wang, D., 2007. On-road remote sensing measurements and fuel-based motor vehicle emission inventory in Hangzhou, China. Atmos. Environ. 41, 3095–3107.

Guo, H., Wang, T., Simpson, I.J., Blake, D.R., Yu, X.M., Kwok, Y.H., Li, Y.S., 2004. Source contributions to ambient VOCs and CO at a rural site in eastern China. Atmos. Environ. 38, 4551–4560.

Guo, S., Hu, M., Guo, Q., Zhang, X., Zheng, M., Zheng, J., Chang, C.C., Schauer, J.J., Zhang, R., 2012. Primary sources and secondary formation of organic aerosols in Beijing, China. Environ. Sci. Technol. 46, 9846–9853.

Haagen-Smit, A.J., Darley, E.F., Zaitlin, M., Hull, H., Nobel, W., 1951. Investigation on injury to plants from air pollution in the Los Angeles area. Plant Physiol. 27, 18.

Hagler, G.S.W., Bergin, M.H., Salmon, L.G., Yu, J.Z., Wan, E.C.H., Zheng, M., Zeng, L.M., Kiang, C.S., Zhang, Y.H., Schauer, J.J., 2007. Local and regional anthropogenic influence on $PM_{2.5}$ elements in Hong Kong. Atmos. Environ. 41, 5994–6004.

Hakami, A., Henze, D., Seinfeld, J.H., Singh, K., Sandu, A., Kim, S., Byun, D., Li, Q., 2007. The adjoint of CMAQ. Environ. Sci. Technol. 41, 7807–7817.

Hamilton, J.F., Webb, P.J., Lewis, A.C., Hopkins, J.R., Smith, S., Davy, P., 2004. Partially oxidised organic components in urban aerosol using GCxGC-TOF/MS. Atmos. Chem. Phys. 4, 1279–1290.

He, K., Yang, F., Ma, Y., Zhang, Q., Yao, X., Chan, C.K., Cadle, S., Chan, T., Mulawa, P., 2001. The characteristics of $PM_{2.5}$ in Beijing, China. Atmos. Environ. 35, 4959–4970.

Henry, R.C., Wang, Y.J., Gebhart, K.A., 1991. The relationship between empirical orthogonal functions and sources of air pollution. Atmos. Environ. 25A, 503–509.

Henze, D.K., Shindell, D.T., Akhtar, F., Spurr, R.J.D., Pinder, R.W., Loughlin, D., Kopacz, M., Singh, K., Shim, C., 2012. Spatially refined aerosol direct radiative forcing efficiencies. Environ. Sci. Technol. 46, 9511–9518.

Hess, G.D., Carnovale, F., Cope, M.E., Johnson, G.M., 1992a. The evaluation of some photochemical smog reaction mechanisms – I. Temperature and initial composition effects. Atmos. Environ. 26A, 625–641.

Hess, G.D., Carnovale, F., Cope, M.E., Johnson, G.M., 1992b. The evaluation of some photochemical smog reaction mechanisms – II. Initial addition of alkanes and alkenes. Atmos. Environ. 26A, 642–650.

Hess, G.D., Carnovale, F., Cope, M.E., Johnson, G.M., 1992c. The evaluation of some photochemical smog reaction mechanisms – III. Dilution and emissions effects. Atmos. Environ. 26A, 651–658.

Ho, K.F., Lee, S.C., Chan, C.K., Yu, J.C., Chow, J.C., Yao, X.H., 2003. Characterization of chemical species in $PM_{2.5}$ and PM_{10} aerosols in Hong Kong. Atmos. Environ. 37, 31–39.

Hofzumahaus, A., Rohrer, F., Lu, K.D., Bohn, B., Brauers, T., Chang, C.C., Fuchs, H., Holland, F., Kita, K., Kondo, Y., Li, X., Lou, S.R., Shao, M., Zeng, L.M., Wahner, A., Zhang, Y.H., 2009. Amplified trace gas removal in the Troposphere. Science 324, 1702–1704.

Holloway, T., et al., 2008. MICS-Asia II: Impact of global emissions on regional air quality in Asia. Atmos. Environ. 42, 3543–3561.

Horowitz, L.W., 2006. Past, present, and future concentrations of tropospheric ozone and aerosols: Methodology, ozone evaluation, and sensitivity to aerosol wet removal. J. Geophys. Res. 111, D22211.

Hu, D., Zhang, W., Chen, L., Chen, C., Ou, L., Tong, Y., Wei, W., Long, W., Wang, X., 2012. Mercury emissions from waste combustion in China from 2004 to 2010. Atmos. Environ. 62, 359–366.

Hu, M., Peng, J., Sun, K., Yue, D., Guo, S., Wiedensohler, A., Wu, Z., 2012. Estimation of size-resolved ambient particle density based on the measurement of aerosol number, mass, and chemical size distributions in the Winter in Beijing. Environ. Sci. Technol. 46, 9941–9947.

Hu, M., He, L.Y., Zhang, Y.H., Wang, M., Kim, Y.P., Moon, K.C., 2002. Seasonal variation of ionic species in fine particles at Qingdao, China. Atmos. Environ. 36, 5853–5859.

Hueglin, C., Gehrig, R., Baltensperger, U., Gysel, M., Monn, C., Vonmont, H., 2005. Chemical characterisation of $PM_{2.5}$, PM_{10} and coarse particles at urban, near-city and rural sites in Switzerland. Atmos. Environ. 39, 637–651.

Huo, M., Sun, Q., Bai, Y., Li, J., Xie, P., Liu, Z., Wang, X., 2012. Influence of airborne particles on the acidity of rainwater during wash-out process. Atmos. Environ. 59, 192–201.

Ion, A.C., Vermeylen, R., Kourtchev, I., Cafmeyer, J., Chi, X., Gelencser, A., Maenhaut, W., Claeys, M., 2005. Polar organic compounds in rural $PM_{2.5}$ aerosols from K-puszta, Hungary, during a 2003 Summer field campaign: Sources and diel variations. Atmos. Chem. Phys. 5, 1805–1814.

Itahashi, S., Uno, I., Kim, S., 2012. Source contributions of sulfate aerosol over East Asia estimated by CMAQ-DDM. Environ. Sci. Technol. 46, 6733–6741.

Ito, K., Mathes, R., Ross, Z., Nádas, A., Thurston, G., Matte, T., 2011. Fine Particulate Matter constituents associated with cardiovascular hospitalizations and mortality in New York City. Environ. Health Perspect. 119, 467–473.

Jackson, B., Chau, D., Gurer, K., Kaduwela, A., 2006. Comparison of ozone simulations using MM5 and CALMET/MM5 hybrid meteorological fields for the July/August, 2000 CCOS Episode. Atmos. Environ. 40, 2812–2822.

Jacob, D.J., 1986. Chemistry of OH in remote clouds and its role in the production of formic acid and peroxymonosulfate. J. Geophys. Res. 91, 9807–9826.

Jacob, D.J., Gilliland, A.B., 2005. Modeling the impact of air pollution on global climate change. Environmental Manager, October, 24–27.

Jacob, D.J., et al., 1993. Simulation of summertime ozone over North America. J. Geophys. Res. 98, 14,797–14,816.

Jacobson, J.L., Jacobson, S.W., 1996. Intellectual impairment in children exposed to polychlorinated biphenyls in utero. New Engl. J. Med. 335 (11), 783–789.

Jacobson, M.Z., 1998a. Improvement of SMVGEAR II on vector and scalar machines through absolute error tolerance control. Atmos. Environ. 32, 791–796.

Jacobson, M.Z., 1998b. Studying the effects of aerosols on vertical photolysis rate coefficient and temperature profiles over an urban airshed. J. Geophys. Res. 103, 10,593–10,604.

Jacobson, M.Z., Ginnebaugh, D.L., 2010. Global-through-urban nested three-dimensional simulation of air pollution with a 13,600-reaction photochemical mechanism. J. Geophys. Res. 115, D14304.

Jacobson, M.Z., Turco, R.P., 1995. Simulating condensational growth, evaporation, and coagulation of aerosols using a combined moving and stationary size grid. Aerosol Sci. Technol. 22, 73–92.

Jacobson, M.Z., Turco, R.P., 1994. SMVGEAR: A sparse-matrix, vectorized Gear code for atmospheric models. Atmos. Environ. 28 (A), 273–284.

Jiang, M., et al., 2005. Vehicle fleet emissions of black carbon, polycyclic aromatic hydro-carbons, and other pollutants measured by a mobile laboratory in Mexico City. Atmos. Chem. Phys. 5, 3377–3387.

Johnson, D., Utembe, S.R., Jenkin, M.E., 2006. Simulating the detailed chemical composition of secondary organic aerosol formed on a regional scale during the TORCH 2003 campaign in the southern UK. Atmos. Chem. Phys. 6, 419–431.

Johnson, G.M., May 2, 1984. A simple model for predicting the ozone concentration of ambient air. Proceedings of the 8th International Clean Air Conference, 715–731. Melbourne, Australia.

Kacenelenbogen, M., Leon, J.-F., Chiapello, I., Tanr, D., 2006. Characterization of aerosol pollution events in France using ground-based and POLDER-2 satellite data. Atmos. Chem. Phys. 6, 4843–4849.

Karl, M., Gross, A., Leck, C., Pirjola, L., 2007. Intercomparison of dimethylsulfide oxidation mechanisms for the marine boundary layer: Gaseous and particulate sulfur constituents. J. Geophys. Res. 112, D15304.

Karnosky, D.F., Skelly, J.M., Percy, K.E., Chappelka, A.H., 2007. Perspectives regarding 50 years of research on effects of tropospheric ozone air pollution on US forests. Environ. Pollut. 147 (3), 489–506.

Ketzel, M., Wahlin, P., Kristensson, A., Swietlicki, E., Berkowicz, R., Nielsen, O.J., Palmgren, F., 2004. Particle size distribution and particle mass measurements at urban, near-city and rural level in the Copenhagen area and Southern Sweden. Atmos. Chem. Phys. 4, 281–292.

Keywood, M.D., Ayers, G.P., Gras, J.L., Boers, R., Leong, C.P., 2003. Haze in the Klang Valley of Malaysia. Atmos. Chem. Phys. 3, 591–605.

Kiehl, J.T., Trenberth, K.E., 1997. Earth's annual global mean energy budget. Bull. Am. Meteorol. Soc. 78, 197–208.

Kim Oanh, N.T., et al., 2006. Particulate air pollution in six Asian cities: Spatial and temporal distributions, and associated sources. Atmos. Environ. 40, 3367–3380.

Konovalov, I.B., Beekmann, M., Vautard, R., Burrows, J.P., Richter, A., Nu, H., Elansky, N., 2005. Comparison and evaluation of modelled and GOME measurement derived tropospheric NO_2 columns over Western and Eastern Europe. Atmos. Chem. Phys. 5, 169–190.

Koo, B., Wilson, G.M., Morris, R.E., Dunker, A.M., Yarwood, G., 2009. Comparison of source apportionment and sensitivity analysis in a particulate matter air quality model. Environ. Sci. Technol. 43, 6669–6675.

Kouyoumdjian, H., Saliba, N.A., 2006. Mass concentration and ion composition of coarse and fine particles in an urban area in Beirut: Effect of calcium carbonate on the absorption of nitric and sulfuric acids and the depletion of chloride. Atmos. Chem. Phys. 6, 1865–1877.

Kulikov, M.Y., Feigin, A.M., Sonnemann, G.R., 2009. Retrieval of water vapor profile in the mesosphere from satellite ozone and hydroxyl measurements by the basic dynamic model of mesospheric photochemical system. Atmos. Chem. Phys. 9, 8199–8210.

Laakso, L., et al., 2008. Basic characteristics of atmospheric particles, trace gases and mete-orology in a relatively clean Southern African Savannah environment. Atmos. Chem. Phys. 8, 4823–4839.

Lam, K.S., Ding, A., Chan, L.Y., Wang, T., Wang, T.J., 2002. Ground-based measurements of total ozone and UV radiation by the Brewer spectrophotometer #115 at Hong Kong. Atmos. Environ. 36, 2003–2012.

Larsen, L., September 2006. Air quality in California, A presentation for the Chinese Envi-ronmental Delegation. California Air Resources Board, Sacramento, CA, USA.

Lelieveld, J., Crutzen, P.J., 1990. Influence of cloud photochemical processes on tropospheric ozone. Nature 343, 227–233.

Li, C., Tsay, S.C., Hsu, N.C., Kim, J.Y., Howell, S.G., Huebert, B.J., Ji, Q., Jeong, M.J., Wang, S.H., Hansell, R.A., Bell, S.W., 2012. Characteristics and composition of atmospheric aerosols in Phimai, central Thailand during BASE-ASIA. Atmos. Environ. 60, 1–12.

Li, C., Yu, Y., Zhang, D., Feng, J., Wu, M., Sheng, G., Fu, J., 2010. Polychlorinated biphenyls in indoor and outdoor dust of Shanghai and exposure assessment of them to human. China Environ. Sci. 30 (4), 433–441.

Li, Q., Cheng, H., Zhou, T., Lin, C., Guo, S., 2012. The estimated atmospheric lead emissions in China, 1990–2009. Atmos. Environ. 60, 1–8.

Li, S.M., 2006. A deeper understanding of the ambient aerosols in the Lower Fraser Valley and its policy implications: The 2nd special issue on Pacific 2001 Air Quality Study. Atmos. Environ. 40, 2635–2636.

Li, S., Matthews, J., Sinha, A., 2008. Atmospheric hydroxyl radical production from electronically excited NO_2 and H_2O. Science 319, 1657–1660.

Liang, C., Pankow, J.F., Odum, J.R., Seinfeld, J.H., 1997. Gas/particle partitioning of semi-volatile organic compounds to model inorganic, organic, and ambient smog aerosols. Environ. Sci. Technol. 31, 3086–3092.

Liang, J., Fairley, D., 2006. Validation of an efficient non-negative matrix factorization method and its preliminary application in central California. Atmos. Environ. 40, 1991–2001.

Liang, J., Jacob, D.J., 1997. Effect of aqueous-phase cloud chemistry on tropospheric ozone. J. Geophys. Res. 102, 5993–6001.

Liang, J., Jacobson, M.Z., 2011. CVPS: An operator solving complex chemical and vertical processes simultaneously with sparse-matrix techniques. Atmos. Environ. 45, 6820–6827.

Liang, J., Jacobson, M.Z., 2000a. Effects of subgrid segregation on ozone production in a chemical model. Atmos. Environ. 34, 2975–2982.

Liang, J., Jacobson, M.Z., 2000b. Comparison of a 4000-reaction chemical mechanism with the Carbon Bond IV and an adjusted Carbon Bond IV-EX mechanism using SMVGEAR II. Atmos. Environ. 34, 3015–3026.

Liang, J., Jacobson, M.Z., 1999. A study of sulfur dioxide oxidation pathways over a range of liquid water contents, pH values, and temperatures. J. Geophys. Res. 104, 13749–13769.

Liang, J., Wang, W., 1990. A regional source resolution model. Acta Scient. Circumstant. 10, 17–25.

Liang, J., Jackson, B., Kaduwela, A., 2006a. A critical evaluation of the ability of observation based methods to diagnose ozone precursor limitations in central California. Atmos. Environ. 40, 5156–5166.

Liang, J., Kaduwela, A., Jackson, B., Gurer, K., Allen, P., 2006b. Off-line diagnostic analysis of a three-dimensional PM model using two matrix factorization methods. Atmos. Environ. 40, 5759–5767.

Lin, C.Y., Chang, C.C., Chan, C.Y., Kuo, C.H., Chen, W.C., Chu, D.A., Liu, S.C., 2010. Characteristics of Spring profiles and sources of ozone in the low troposphere over northern Taiwan. Atmos. Environ. 44, 182–193.

Lin, C.Y.C., Jacob, D.J., Fiore, A.M., 2001. Trends in exceedances of the ozone air quality standard in the continental United States, 1980–1998. Atmos. Environ. 35, 3217–3228.

Lisac, I., Grubišić, V., 1991. An analysis of surface ozone data measured at the end of the 19th century in Zagreb, Yugoslavia. Atmos. Environ. 25 (A), 481–486.

Liu, Z., Wang, Y., Gu, D., Zhao, C., Huey, L.G., Stickel, R., Liao, J., Shao, M., Zhu, T., Zeng, L., Amoroso, A., Costabile, F., Chang, C.-C., Liu, S.-C., 2012. Summertime photochemistry during CAREBeijing-2007: ROx budgets and O_3 formation. Atmos. Chem. Phys. 12, 7737–7752.

Livingstone, P.L., Gurer, K., Motallebi, N., Luo, D., Propper, R., Dec. 15, 2010. Simulating emission control impacts on summertime O_3 and PM in California. AGU Fall Meeting San Francisco, CA, USA.

Livingstone, P.L., et al., 2009. Simulating PM concentration during a winter episode in a subtropical valley: Sensitivity simulations and evaluation methods. Atmos. Environ. 43, 5971–5977.

Logan, J.A., 1985. Tropospheric ozone: Seasonal behavior, trends, and anthropogenic influence. J. Geophys. Res. 90, 10,463–10,482.

Longinelli, A., Lenaz, R., Ori, C., Langone, L., Selmo, E., Giglio, F., 2010. Decadal changes in atmospheric CO_2 concentration and $d^{13}C$ over two seas and two oceans: Italy to New Zealand. Atmos. Environ. 44, 4303–4311.

Lu, H., Zhu, L., Zhu, N., 2009. Polycyclic aromatic hydrocarbon emission from straw burning and the influence of combustion parameters. Atmos. Environ. 43, 978–983.

Lund, M.T., Eyring, V., Fuglestvedt, J., Hendricks, J., Lauer, A., Lee, D., Righi, M., 2012. Environ. Sci. Technol. 46, 8868–8877.

Lurmann, F.W., Wexler, A.S., Pandis, S.N., Musarra, S., Kumar, N., Hering, S.V., Seinfeld, J.H., 1997. Development of an acid deposition model for the South Coast Air Basin. Final Report to California Air Resources Board and the California Environmental Protection Agency, California, USA.

Madronich, S., Calvert, J.G., 1990. Permutation reactions of organic peroxy radicals in the troposphere. J. Geophys. Res. 95, 5697–5715.

Mahmud, A., Hixson, M., Kleeman, M.J., 2012. Quantifying population exposure to airborne particulate matter during extreme events in California due to climate change. Atmos. Chem. Phys. 12, 7453–7463.

Majestic, B.J., Schauer, J.J., Shafer, M.M., 2007. Application of synchrotron radiation for measurement of iron redox speciation in atmospherically processed aerosols. Atmos. Chem. Phys. 7, 2475–2487.

Marcy, T.P., et al., 2007. Measurements of trace gases in the tropical tropopause layer. Atmos. Environ. 41, 7253–7261.

Marley, N.A., Gaffney, J.S., Ramos-Villegas, R., Gonzalez, B.C., 2007. Comparison of measurements of peroxyacyl nitrates and primary carbonaceous aerosol concentrations in Mexico City determined in 1997 and 2003. Atmos. Chem. Phys. 7, 2277–2285.

Marmur, A., Liu, W., Wang, Y., Russell, A.G., Edgerton, E.S., 2009. Evaluation of model simulated atmospheric constituents with observations in the factor projected space: CMAQ simulations of SEARCH measurements. Atmos. Environ. 43, 1839–1849.

Martien, P.T., Harley, R.A., 2006. Adjoint sensitivity analysis for a three-dimensional photochemical model: application to southern California. Environ. Sci. Technol. 40, 4200–4210.

Martinez, M., Harder, H., Ren, X., Lesher, R.L., Brune, W.H., 2004. Measuring atmospheric naphthalene with laser-induced fluorescence. Atmos. Chem. Phys. 4, 563–569.

Mauzerall, D.L., October 2011. Methane mitigation – Benefits for air quality, health, crop yields and climate. IGAC Newsletter, 17–18.

McCabe, L.C. (Ed.), 1952. Air Pollution. McGraw-Hill, New York (Proc. U.S. Tech. Conf. Air Pollution, 1950), USA.

McElroy, M., 1971. The composition of planetary atmospheres. J. Quant. Spec. Radiative Transfer 11, 813–825.

Mehlmann, A., Warneck, P., 1995. Atmospheric gaseous HNO_3, particulate nitrate, and aerosol size distributions of major ionic species at a rural site in western Germany. Atmos. Environ. 29, 2359–2373.

Mickley, L.J., Leibensperger, E.M., Jacob, D.J., Rind, D., 2012. Regional warming from aerosol removal over the United States: Results from a transient 2010–2050 climate simulation. Atmos. Environ. 46, 545–553.

Miller-Schulze, J.P., Shafer, M.M., Schauer, J.J., Solomon, P.A., Lantz, J., Artamonova, M., Chen, B., Imashev, S., Sverdlik, L., Carmichael, G.R., Deminter, J.T., 2011. Characteristics of fine particle carbonaceous aerosol at two remote sites in Central Asia. Atmos. Environ. 45, 6955–6964.

Monks, P.S., and many coauthors, 2009. Atmospheric composition change – Global and regional air quality. Atmos. Environ. 43, 5268–5350.

Moreno, T., et al., 2011. Variations in time and space of trace metal aerosol concentrations in urban areas and their surroundings. Atmos. Chem. Phys. 11, 9415–9430.

Napelenok, S.L., Cohan, D.S., Hu, Y., Russell, A.G., 2006. Decoupled direct 3D sensitivity analysis for particulate matter (DDM-3D/PM). Atmos. Environ. 40, 6112–6121.

Nguyen, D.L., Kim, J.Y., Shim, S.G., Zhang, X.S., 2011. Ground and shipboard measurements of atmospheric gaseous elemental mercury over the Yellow Sea region during 2007–2008. Atmos. Environ. 45, 253–260.

Niemi, J.V., Tervahattu, H., Vehkamaki, H., Martikainen, J., Laakso, L., Kulmala, M., Aarnio, P., Koskentalo, T., Sillanpaa, M., Makkonen, U., 2005. Characterization of aerosol particle episodes in Finland caused by wildfires in Eastern Europe. Atmos. Chem. Phys. 5, 2299–2310.

Oanh, N.T.K., et al., 2006. Particulate air pollution in six Asian cities: Spatial and temporal distributions, and associated sources. Atmos. Environ. 40, 3367–3380.

Ohura, T., Amagai, T., Shen, X., Li, S., Zhang, P., Zhu, L., 2009. Comparative study on indoor air quality in Japan and China: Characteristics of residential indoor and outdoor VOCs. Atmos. Environ. 43, 6352–6359.

Paatero, P., 1997. Least squares Formulation of robust, nonnegative factor analysis. Chemometrics Intelligent Lab. Syst. 37, 23–35.

Paatero, P., Hopke, P.K., 2003. Discarding or downweighing high-noise variables in factor analysis models. Analyt. Chim. Acta 490, 277–289.

Pandis, S.N., Harley, R.A., Cass, G.R., Seinfeld, J.H., 1992. Secondary organic aerosol formation and transport. Atmos. Environ. 26A, 2269–2282.

Pang, X., Mu, Y., Lee, X., Fang, S., Yuan, J., Huang, D., 2009. Nitric oxides and nitrous oxide fluxes from typical vegetables cropland in China: Effects of canopy, soil properties and field management. Atmos. Environ. 43, 2571–2578.

Park, R.J., Jacob, D.J., Logan, J.A., 2007. Fire and biofuel contributions to annual mean aerosol mass concentrations in the United States. Atmos. Environ. 41, 7389–7400.

Peltier, R.E., Sullivan, A.P., Weber, R.J., Brock, C.A., Wollny, A.G., Holloway, J.S., Gouw, J.A.D., Warneke, C., 2007. Fine aerosol bulk composition measured on WP-3D research aircraft in vicinity of the Northeastern United States – results from NEAQS. Atmos. Chem. Phys. 7, 3231–3247.

Penner, J.E., Eddleman, H., Novakov, T., 1993. Towards the development of a global inventory for black carbon emissions. Atmos. Environ. 27 (A), 1277–1295.

Perrone, M.R., Piazzalunga, A., Prato, M., Carofalo, I., 2011. Composition of fine and coarse particles in a coastal site of the central Mediterranean: Carbonaceous species contributions. Atmos. Environ. 45, 7470–7477.

Porter, E., Johnson, S., 2007. Translating science into policy: Using ecosystem thresholds to protect resources in Rocky Mountain National Park. Environ. Pollut. 149, 268–280.

Qin, Y., Xie, S.D., 2011. Estimation of county-level black carbon emissions and its spatial distribution in China in 2000. Atmos. Environ. 45, 6995–7004.

Qiu, C., Khalizov, A.F., Zhang, R., 2012. Soot aging from OH-initiated oxidation of toluene. Environ. Sci. Technol. 46, 9464–9472.

Querol, X., et al., 2008. PM speciation and sources in Mexico during the MILAGRO-2006 Campaign. Atmos. Chem. Phys. 8, 111–128.

R Development Core Team, 2012. R: A language and environment for statistical computing, R Foundation for Statistical Computing. Austria, Vienna. http://www.R-project.org ISBN: 3-900051-00-3.

Rastogi, N., Sarin, M.M., 2005. Long-term characterization of ionic species in aerosols from urban and high-altitude sites in western India: Role of mineral dust and anthropogenic sources. Atmos. Environ. 39, 5541–5554.

Rees, S.L., Robinson, A.L., Khlystov, A., Stanier, C.O., Pandis, S.N., 2004. Mass balance closure and the Federal Reference Method for $PM_{2.5}$ in Pittsburgh, Pennsylvania. Atmos. Environ. 38, 3305–3318.

Reynolds, S.D., Roth, P.M., Seinfeld, J.H., 1973. Mathematical modeling of photochemical air pollution – I: Formulation of the model. Atmos. Environ. 7, 1033–1061.

Rodriguez, S., et al., 2007. A study on the relationship between mass concentrations, chemistry and number size distribution of urban fine aerosols in Milan, Barcelona and London. Atmos. Chem. Phys. 7, 2217–2232.

Saxena, P., Hildemann, L.M., 1997. Water absorption by organics: Survey of laboratory evidence and evaluation of UNIFAC for estimating water activity. Atmos. Environ. 31, 3318–3324.

Schaap, M., Loon, M.V., Brink, H.M.T., Dentener, F.J., Builtjes, P.J.H., 2004. Secondary inorganic aerosol simulations for Europe with special attention to nitrate. Atmos. Chem. Phys. 4, 857–874.

Shi, C., Roth, M., Zhang, H., Li, Z., 2008. Impacts of urbanization on long-term fog variation in Anhui Province, China. Atmos. Environ. 42, 8484–8492.

Shin, H.M., McKone, T.E., Bennett, D.H., 2012. Intake fraction for the indoor environment: a tool for prioritizing indoor chemical sources. Environ. Sci. Technol. 46, 10063–10072.

Shindell, D.T., et al., 2012. Radiative forcing in the ACCMIP historical and future climate simulations. Atmos. Chem. Phys. Discuss. 12, 21105–21210.

Siddaway, J.M., Petelina, S.V., Karoly, D., Klekociuk, A.R., Dargaville, R.J., 2012. Future Antarctic ozone recovery rates in September–December predicted by CCMVal-2 model simulations. Atmos. Chem. Phys. Discuss. 12, 18959–18991.

Sillman, S., June 2002. Evaluation of observation-based methods for analyzing ozone production and ozone–NOx–VOC sensitivity. A Report to U.S. EPA.

Song, W.W., He, K.B., Lei, Y., 2012. Black carbon emissions from on-road vehicles in China, 1990–2030. Atmos. Environ. 51, 320–328.

Sorooshian, A., Csavina, J., Shingler, T., Dey, S., Brechtel, F.J., Sáez, A.E., Betterton, E.A., 2012. Hygroscopic and chemical properties of aerosols collected near a copper smelter: Implications for public and environmental health. Environ. Sci. Technol. 46, 9473–9480.

Staehelin, J., Thudium, J., Buehler, R., Volz-Thomas, A., Graber, W., 1994. Trends in surface ozone concentrations at Arosa (Switzerland). Atmos. Environ. 28, 75–87.

Streets, D.G., et al., 2007. Air quality during the 2008 Beijing Olympic Games. Atmos. Environ. 41, 480–492.

Streets, D.G., Hao, J., Wu, Y., Jiang, J., Chan, M., Tian, H., Feng, X., 2005. Anthropogenic mercury emissions in China. Atmos. Environ. 39, 7789–7806.

Streets, D.G., Gupta, S., Waldhoff, S.T., Wang, M.Q., Bond, T.C., Bo, Y., 2001. Black carbon emissions in China. Atmos. Environ. 35, 4281–4296.

Sun, G., Yao, L., Jiao, L., Shi, Y., Zhang, Q., Tao, M., Shan, G., He, Y., 2013. Characterizing $PM_{2.5}$ pollution of a subtropical metropolitan area in China. Atmos. Climate Sci. 3, 100–110.

Sun, Y., Zhuang, G., Wang, Y., Han, L., Guo, J., Dan, M., Zhang, W., Wang, Z., Hao, Z., 2004. The air-borne particulate pollution in Beijing concentration, composition, distribution and sources. Atmos. Environ. 38, 5991–6004.

Szopa, S., Aumont, B., Madronich, S., 2005. Assessment of the reduction methods used to develop chemical schemes: Building of a new chemical scheme for VOC oxidation suited to three-dimensional multiscale HOx–NOx–VOC chemistry simulations. Atmos. Chem. Phys. 5, 2519–2538.

Thurston, G.D., Spengler, J.D., 1985. A quantitative assessment of source contributions to inhalable particulate matter pollution in metropolitan Boston. Atmos. Environ. 19, 9–25.

Thurston, G.D., et al., 2005. Workgroup report: Workshop on source apportionment of particulate matter health effects–intercomparison of results and implications. Environ. Health Perspect. 113 (12), 1768–1774.

Tian, H., Gao, J., Lu, L., Zhao, D., Cheng, K., Qiu, P., 2012a. Temporal trends and spatial variation characteristics of hazardous air pollutant emission inventory from municipal solid waste incineration in China. Environ. Sci. Technol. 46, 10364–10371.

Tian, H., Cheng, K., Wang, Y., Zhao, D., Lu, L., Jia, W., Hao, J., 2012b. Temporal and spatial variation characteristics of atmospheric emissions of Cd, Cr, and Pb from coal in China. Atmos. Environ. 50, 157–163.

Tkacik, D.S., Presto, A.A., Donahue, N.M., Robinson, A.L., 2012. Secondary organic aerosol formation from intermediate-volatility organic compounds: cyclic, linear, and branched alkanes. Environ. Sci. Technol. 46, 8773–8781.

Tørseth, K., Aas, W., Breivik, K., Fjæraa, A.M., Fiebig, M., Hjellbrekke, A.G., Myhre, C.L., Solberg, S., Yttri, K.E., 2012. Introduction to the European Monitoring and Evaluation Programme (EMEP) and observed atmospheric composition change during 1972–2009. Atmos. Chem. Phys. 12, 5447–5481.

US EPA, 2009. Technical assistance document for the reporting of daily air quality – the Air Quality Index (AQI). Report EPA-454/B-09-001 by D. Mintz, Environmental Protection Agency, USA.

US EPA, 1999. Science algorithms of the EPA MODELS-3 Community Multiscale Air Quality (CMAQ) Modeling System. In: Byun, D.W., Ching, J.K.S. (Eds.), Report EPA/600/R-99/030, USA.

Varotsos, C., Kondratyev, K.Y., Efstathiou, M., 2001. On the seasonal variation of the surface ozone in Athens, Greece. Atmos. Environ. 35, 315–320.

Vega, E., Mugica, V., Reyes, E., Sanchez, G., Chow, J.C., Watson, J.G., 2001. Chemical composition of fugitive dust emitters in Mexico City. Atmos. Environ. 35, 4033–4039.

Venier, M., Ma, Y., Hites, R.A., 2012. Bromobenzene flame retardants in the Great Lakes atmosphere. Environ. Sci. Technol. 46, 8653–8660.

Vingarzan, R., Li, S.M., 2006. The Pacific 2001 Air Quality Study – synthesis of findings and policy implications. Atmos. Environ. 40, 2637–2649.

Vuilleumier, L., Harley, R.A., Brown, N.J., 1997. First- and second-order sensitivity analysis of a photochemically reactive system (a Green's function approach). Environ. Sci. Technol. 31, 1206–1217.

Vutukuru, S., Dabdub, D., 2008. Modeling the effects of ship emissions on coastal air quality: a case study of southern California. Atmos. Environ. 42, 3751–3764.

Wang, C., Corbett, J.J., Firestone, J., 2008. Improving spatial representation of global ship emissions inventories. Environ. Sci. Technol. 42, 193–199.

Wang, G., Xie, M., Hu, S., Gao, S., Tachibana, E., Kawamura, K., 2010. Dicarboxylic acids, metals and isotopic compositions of C and N in atmospheric aerosols from inland China: implications for dust and coal burning emission and secondary aerosol formation. Atmos. Chem. Phys. 10, 6087–6096.

Wang, J., Zhu, L., 2005. Preliminary exploration of fate of polycyclic aromatic hydrocarbons (PAHs) in air of dry and wet deposition. China Environ. Sci. 25 (4), 471–474.

Wang, J., Chen, S., Tian, M., Zheng, X., Gonzales, L., Ohura, T., Mai, B., Simonich, S.L.M., 2012. Inhalation cancer risk associated with exposure to complex polycyclic aromatic hydrocarbon mixtures in an electronic waste and urban area in South China. Environ. Sci. Technol. 46, 9745–9752.

Wang, L., et al., 2010. Assessment of air quality benefits from national air pollution control policies in China – Part II: Evaluation of air quality predictions and air quality benefits assessment. Atmos. Environ. 44, 3449–3457.

Wang, T., Wei, X.L., Ding, A.J., Poon, C.N., Lam, K.S., Li, Y.S., Chan, L.Y., Anson, M., 2009. Increasing surface ozone concentrations in the background atmosphere of Southern China, 1994–2007. Atmos. Chem. Phys. 9, 6217–6227.

Wang, T., Cheung, V.T.F., Lam, K.S., Kok, G.L., Harris, J.M., 2001a. The characteristics of ozone and related compounds in the boundary layer of the South China coast: temporal and vertical variations during autumn season. Atmos. Environ. 35, 2735–2746.

Wang, T., Wu, Y.Y., Cheung, T.F., Lam, K.S., 2001b. A study of surface ozone and the relation to complex wind flow in Hong Kong. Atmos. Environ. 35, 3203–3215.

Wang, T.J., Jin, L.S., Li, Z.K., Lam, K.S., 2000. A modeling study on acid rain and recommended emission control strategies in China. Atmos. Environ. 34, 4467–4477.

Wang, X., Mauzerall, D.L., 2004. Characterizing distributions of surface ozone and its impact on grain production in China, Japan and South Korea: 1990 and 2020. Atmos. Environ. 38, 4383–4402.

Wang, X., Mauzerall, D.L., Hu, Y., Russell, A.G., Larson, E.D., Wood, J.H., Streets, D.G., Guenther, A., 2005. A high-resolution emission inventory for eastern China in 2000 and three scenarios for 2020. Atmos. Environ. 39, 5917–5933.

Wang, Y., Jacob, D.J., Logan, J.A., 1998. Global simulation of tropospheric O_3-NOx-hydrocarbon chemistry, 1. Model formulation. J. Geophys. Res. 103, 10713–10726.

Wei, W., Wang, S., Hao, J., Cheng, S., 2011. Projection of anthropogenic volatile organic compounds (VOCs) emissions in China for the period 2010–2020. Atmos. Environ. 45, 6863–6871.

Wei, W., Wang, S., Chatani, S., Klimont, Z., Cofala, J., Hao, J., 2008. Emission and speciation of non-methane volatile organic compounds from anthropogenic sources in China. Atmos. Environ. 42, 4976–4988.

Wen, C.C., Fang, G.C., 2007. Characterization of size-differentiated particle composition of ionic species between Taichung Harbor (TH) and WuChi Traffic (WT) near Taiwan Strait during 2004–2005. Atmos. Environ. 41, 3853–3861.

Wennberg, P.O., Dabdub, D., 2008. Rethinking ozone production. Science 319, 1624–1625.

Wexler, A.S., Lurmann, F.W., Seinfeld, J.H., 1994. Modeling urban and regional aerosols – I. Model development. Atmos. Environ. 28, 531–546.

WHO guidelines for indoor air quality: dampness and mold, 2009.

Winchester, J.W., Nifong, G.D., 1971. Water pollution in Lake Michigan by trace elements from pollution aerosol fallout, Water. Air, Soil Pollution 1, 50–64.

Winchester, J.W., Lü, W.X., Ren, L.X., Wang, M.X., Maenhaut, W., 1981. Fine and coarse aerosol composition from a rural area in north China. Atmos. Environ. 15, 933–937.

Wong, H.L.A., Wang, T., Ding, A., Blake, D.R., Nam, J.C., 2007. Impact of Asian continental outflow on the concentrations of O_3, CO, NMHCs and halocarbons on Jeju Island, South Korea during March 2005. Atmos. Environ. 41, 2933–2944.

Xiao, Y., Jacob, D.J., Turquety, S., 2007. Atmospheric acetylene and its relationship with CO as an indicator of air mass age. J. Geophys. Res. 112, D12305.

Xie, X., Shao, M., Liu, Y., Lu, S., Chang, C., Chen, Z., 2008. Estimate of initial isoprene contribution to ozone formation potential in Beijing, China. Atmos. Environ. 42, 6000–6010.

Xiu, G., Cai, J., Zhang, W., Zhang, D., Bueler, A., Lee, S., Shen, Y., Xu, L., Huang, X., Zhang, P., 2009. Speciated mercury in size-fractionated particles in Shanghai ambient air. Atmos. Environ. 43, 3145–3154.

Yan, P., Zhang, R., Huan, N., Zhou, X., Zhang, Y., Zhou, H., Zhang, L., 2012. Characteristics of aerosols and mass closure study at two WMO GAW regional background stations in eastern China. Atmos. Environ. 60, 121–131.

Yan, X., Ohara, T., Akimoto, H., 2006. Bottom-up estimate of biomass burning in mainland China. Atmos. Environ. 40, 5262–5273.

Yang, F., Tan, J., Zhao, Q., Du, Z., He, K., Ma, Y., Duan, F., Chen, G., Zhao, Q., 2011. Characteristics of $PM_{2.5}$ speciation in representative megacities and across China. Atmos. Chem. Phys. 11, 5207–5219.

Yang, Y.J., Wilkinson, J.G., Russell, A.G., 1997. Fast, direct sensitivity analysis of multidimensional photochemical models. Environ. Sci. Technol. 31, 2859–2868.

Yao, X., Chan, C.K., Fang, M., Cadle, S., Chan, T., Mulawa, P., He, K., Ye, B., 2002. The water-soluble ionic composition of $PM_{2.5}$ in Shanghai and Beijing, China. Atmos. Environ. 36, 4223–4234.

Yttri, K.E., et al., 2007. Elemental and organic carbon in PM_{10}: A one year measurement campaign within the European Monitoring and Evaluation Programme EMEP. Atmos. Chem. Phys. 7, 5711–5725.

Yuan, Z.B., Yu, J.Z., Lau, A.K.H., Louie, P.K.K., Fung, J.C.H., 2006. Application of positive matrix factorization in estimating aerosol secondary organic carbon in Hong Kong and its relationship with secondary sulfate. Atmos. Chem. Phys. 6, 25–34.

Zakey, A.S., Solmon, F., Giorgi, F., 2006. Implementation and testing of a desert dust module in a regional climate model. Atmos. Chem. Phys. 6, 4687–4704.

Zhang, H., Ying, Q., 2012. Secondary organic aerosol from polycyclic aromatic hydrocarbons in Southeast Texas. Atmos. Environ. 55, 279–287.

Zhang, J.M., et al., 2009. Continuous measurement of peroxyacetyl nitrate (PAN) in suburban and remote areas of western China. Atmos. Environ. 43, 228–237.

Zhang, Q., Xu, J., Wang, G., Tian, W., Jiang, H., 2008a. Vehicle emission inventories projection based on dynamic emission factors: A case study of Hangzhou, China. Atmos. Environ. 42, 4989–5002.

Zhang, Q., Wei, Y., Tian, W., Yang, K., 2008b. GIS-based emission inventories of urban scale: A case study of Hangzhou, China. Atmos. Environ. 42, 5150–5165.

Zhang, Q., R.Worsnop, D., Canagaratna, M.R., Jimenez, J.L., 2005. Hydrocarbon-like and oxygenated organic aerosols in Pittsburgh: insights into sources and processes of organic aerosols. Atmos. Chem. Phys. 5, 3289–3311.

Zhang, X., Liu, Z., Hecobian, A., Zheng, M., Frank, N.H., Edgerton, E.S., Weber, R.J., 2012. Spatial and seasonal variations of fine particle water-soluble organic carbon (WSOC) over the southeastern United States: implications for secondary organic aerosol formation. Atmos. Chem. Phys. 12, 6593–6607.

Zhang, Y., Tao, S., 2009. Global atmospheric emission inventory of polycyclic aromatic hydrocarbons (PAHs) for 2004. Atmos. Environ. 43, 812–819.

Zhang, Y., Song, L., Liu, X.J., Li, W.Q., Lü, S.H., Zheng, L.X., Bai, Z.C., Cai, G.Y., Zhang, F.S., 2012. Atmospheric organic nitrogen deposition in China. Atmos. Environ. 46, 195–204.

Zhang, Y., Tao, S., Ma, J., Simonich, S., 2011. Transpacific transport of benzo[a]pyrene emitted from Asia. Atmos. Chem. Phys. 11, 11993–12006.

Zhang, Y., Zheng, L., Liu, X., Jickells, T., Cape, J.N., Goulding, K., Fangmeier, A., Zhang, F., 2008a. Evidence for organic N deposition and its anthropogenic sources in China. Atmos. Environ. 42, 1035–1041.

Zhang, Y.H., Hu, M., Zhong, L.J., Wiedensohler, A., Liu, S.C., Andreae, M.O., Wang, W., Fan, S.J., 2008b. Regional integrated experiments on air quality over Pearl River Delta 2004 (PRIDE-PRD2004): Overview. Atmos. Environ. 42, 6157–6173.

Zhang, Y.H., et al., 2008c. Regional ozone pollution and observation-based approach for analyzing ozone-precursor relationship during the PRIDE-PRD2004 campaign. Atmos. Environ. 42, 6203–6218.

Zhao, Y., Wang, S., Duan, L., Lei, Y., Cao, P., Hao, J., 2008. Primary air pollutant emissions of coal-fired power plants in China: Current status and future prediction. Atmos. Environ. 42, 8442–8452.

Zheng, J., Tan, M., Shibata, Y., Tanaka, A., Li, Y., Zhang, G., Zhang, Y., Shan, Z., 2004. Characteristics of lead isotope ratios and elemental concentrations in PM_{10} fraction of airborne particulate matter in Shanghai after the phase-out of leaded gasoline. Atmos. Environ. 38, 1191–1200.

Index